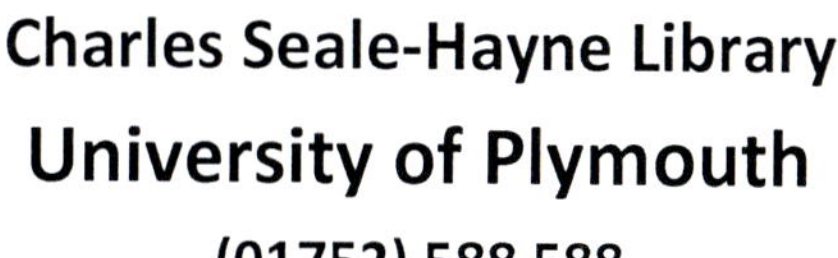

WITHDRAWN
FROM
UNIVERSITY OF PLYMOUTH
LIBRARY SERVICES

# PROJECT WORK IN THE GEOGRAPHY CURRICULUM: AN ADVANCED LEVEL PRIMER

# Project Work in the Geography Curriculum

AN ADVANCED LEVEL PRIMER

JOHN R. BEAUMONT and STEPHEN W. WILLIAMS

CROOM HELM
London & Canberra

Croom Helm Ltd, Provident House, Burrell Row,
Beckenham, Kent BR3 1AT

Croom Helm Australia, PO Box 391, Manuka,
ACT 2603, Australia

British Library Cataloguing in Publication Data

Beaumont, John R.
Project work in the geography curriculum.
1. Geography – Study and teaching (secondary)
– Great Britain
2. Project method in teaching
I. Title II. Williams, S.W.
910'71'241 G74

ISBN 0-7099-3211-1



Printed and bound in Great Britain
by Billing & Sons Limited, Worcester.

CONTENTS

CONTENTS

CONTENTS

CONTENTS

CONTENTS

# Preface

In the 1960's and 1970's, the discipline of geography experienced a so-called 'conceptual revolution', which involved fundamental alterations in both its philosophy and methodology. These changes have percolated into school curricula from the universities, resulting in rapid changes in both course content and teaching methods. One alteration has been the increasing emphasis placed on student project work, and it is this component that is examined in this book. Whilst this book is aimed primarily at teachers and students in schools, it is hoped that the ideas will stimulate some undergraduate projects.

It can be argued that a major benefit of a geographic education is derived from the opportunity for students to focus on real issues. In many respects, this orientation is especially true for project work, particularly if the study is undertaken in a student's local area. From a general educational perspective, project work helps students to develop and enhance their organisational and study skills. Moreover, it is one of the few examinable pieces of work that offer students scope for individual initiatives, and it also greatly assists them to learn how to present analytical and synthetical arguments in a coherent written form. It would be incorrect, however, to see project work as a separate, additional part of a geography course; throughout this book, particular attention is given to demonstrate how project work is an integral component of a whole course. In fact, in addition to highlighting the potential of projects in which real problems are addressed, small projects would be helpful for students to comprehend apparently abstract, geographic concepts. Obviously, the generality of specific concepts could be tested by a detailed case study. For completeness, methodological issues are discussed in the Appendices, although, at this stage, it must be stressed that techniques must be employed only when they provide additional insights into a problem and add force to an argument.

As a disclamer, it must be appreciated that the purpose

of this book is not to provide a set of well-defined, self-contained and complete projects that can be transferred directly to a local investigation. A variety of topics are discussed in sufficient detail to indicate a vast range of potential, manageable student projects. References are given to enable teachers and students to develop the themes further.

The encouragement of a number of people must be acknowledged. This book stems from a short-course run for local school-teachers at Keele University by the authors. We are very grateful to Philip Boden and John Bale for their administrative assistance in organising this course; they also provided helpful advice on the draft of the manuscript. In the final production of this book, we owe a great debt to Sue Davies, who deciphered and typed various instalments of a difficult manuscript. Muriel Patrick and Don Morris provided excellent cartographic and photographic assistance. As usual, we alone are responsible for any final shortcomings.

Finally, special thanks to our families, particularly our wives, Margaret and Fiona, for their continual support.

# ACKNOWLEDGEMENTS

The authors and publishers gratefully acknowledge permission received to make use of copyright material in diagrams and tables. Full citation to books, journals and authors may be found in the bibliography; sources of materials are given in the captions.

Chapter One

# PROJECT WORK WITHIN A GEOGRAPHICAL CURRICULUM

> 'Project: 'a significant, practical unit of activity having educational value and aimed at one or more definite goals of understanding; involves investigation and solution of problems... planned and carried to completion by the pupil and teacher in a natural "real-life" manner'
>
> (Good, 1973).

> 'A man cannot by teaching cause knowledge in another man'
>
> (Thomas Aquinas, 1225-74).

## 1.1 Introduction

During the last twenty years, the subject of geography has undergone a number of radical changes in both structure and content. Initially, this redirection took place in institutions of higher education. The 1970's, however, witnessed the diffusion of the so-called 'new geography' into school syllabuses, and by 1980 '... no examination board retained the syllabuses at O- and A- level which it had used in 1970' (Walford, 1981).

Undoubtedly, the changes which occurred in school syllabuses reflected primarily the innovative work undertaken earlier in university departments (which involved the adoption of the 'scientific method' and a greater emphasis on systematic, rather than regional, courses). However, it is necessary also to acknowledge the importance of the prevailing educational environment, which, in general, provided a context that was sympathetic to the needs and requirements of the 'new geography'. Two features can be noted. First, the reconsideration of educational objectives had cast doubt on the primacy of information recall and the acquisition of facts as important skills. In contrast, greater emphasis was to be placed on higher-order skills such as analysis, synthesis and evaluation. Second, the introduction of comprehensive education and the raising of the school leaving age to sixteen increased the demand for teachers, thus providing opportunities for the development of new courses and methods of teaching (see, for example, the 14-18 Project based at Bristol and Geography for the Young School Leavers based at Avery Hill).

It is within this broader context, therefore, that the introduction and development of the project approach, as a method of inquiry and evaluation, must be placed. Students who follow a syllabus whose primary objectives are the acquisition and comprehension of knowledge can be assessed by a variety of objective or non-objective tests and examinations. However, a syllabus which is organised around concepts, principles and theories (and the 'new geography' was, essentially, a 'conceptual revolution') requires, in addition, methods of assessment which enable students to be assessed in terms of a range of higher-order skills and abilities. It is argued that the project approach contributes substantially to the attainment of this educational objective.

## 1.2 The educational value of project work

At the outset, it must be stated that the use of projects is not intended to replace other forms of student evaluation and assessment. Indeed, an adequate assessment of a student's ability must involve a variety of methods. What is being advocated, however, is the view that, for the assessment of many skills and abilities (which also have relevance in the world outside the classroom), the project approach is the only satisfactory method available. Projects complement, rather than replace, the more traditional components of courses.

The educational project is essentially an American invention dating from the early years of this century, and it is associated closely with the philosopher and educationalist John Dewey. In contrast to the nineteenth century emphasis on the ability to absorb and reproduce material by rote, Dewey advocated what he called the 'problem method' of teaching. This included a sequence of operations including problem definition, formulation of hypotheses, and the selection and application of the optimum solution. During the 1920's, the project method attained widespread popularity in the United States, filtering through to the United Kingdom during the late 1920's and early 1930's. Essentially, the project method is a form of inquiry or discovery learning, and is associated with a problem-solving approach. This is in contrast to expository methods whereby concepts, facts and relationships are described by a teacher or presented in a textbook.

The project method, however, makes a number of assumptions concerning the mental development of the student. In particular it is assumed that the student has reached the stage of propositional thinking. Up to about the age of fourteen years, it is necessary to move from the particular, known and concrete to the general, unknown and abstract. This approach could be varied by testing out generalisations by the examination of particular cases which would involve the introduction of more advanced thinking skills. The full development of a problem-solving approach (including model-building and hypothesis-testing), however, requires that the student has reached the stage of what educational psychologists refer to as formal or propositional operations. The main features of this stage are listed in Table 1.1. The crucial feature

- assumptions can be accepted for the sake of argument;
- a succession of hypotheses can be made which can be expressed propositionally and tested against reality;
- general propositions can be sought which allow exhaustive definitions and the statement of general laws to be made;
- to go beyond the tangible, finite and familiar in spatial concepts, to conceive the infinitely small and to invent imaginery systems;
- to become conscious of one's own thinking, relecting on it to provide logical justification for judgements made;
- to deal with a variety of complex relations such as proportionality or correlation

TABLE 1.1 CHARACTERISTIC FEATURES OF FORMAL OR PROPOSITIONAL OPERATIONS

Source: Beard (1969)

to note is that the precise age at which formal operations is reached is difficult to determine. The Swiss psychologist Jean Piaget, suggests that this stage is reached between about eleven and twelve years; Rhys, however, in a study of student responses to problems in spatially remote areas concluded that true formal operations (involving comprehensive judgements based on hypothetico - reasoning) are not achieved until the age of about fifteen years. However, in certain instances, the full development of formal operations in an individual does not occur until much later, and in some cases certain aspects of propositional thought are not developed at all. These comments clearly have important implications with regard to the applicability of project methods to certain groups of students.

## 1.3 The Characteristics and aims of Project Work

Simply stated, most projects display five distinctive characteristics. First, they involve the solution of a problem often, though not necessarily, set by the student. Second, they involve initiative by the student or group of students, and necessitate a variety of educational activities. Third, they usually result in an end product (dissertation, report, oral report, for example). Fourth, work can continue for a considerable length of time (in a school context, this may range from a week to a couple of months). Fifth, teaching staff are involved in an advisory, rather than authoritarian, role at any or all of the stages - initiation, conduct and conclusion.

In traditional forms of practical work in geography the level of structuring is high, and the outcome of a particular exercise can be predicted by a teacher. These exercises often tend to emphasise the abstraction of information and the application of quantitative techniques under standardised conditions (see Table 1.2). In contrast, project work is relatively unstructured and stresses the student's problem-solving and interpretive abilities. The major purposes of

| LEVEL OF STRUCTURING BY THE TEACHER | DESCRIPTION | EXAMPLES |
|---|---|---|
| High | Simple reapplication | In-class practical topic, e.g. Standard deviation, correlation coefficient. |
| Medium | Structured reapplication | In-class or take-away exercises. |
| Low | Unstructured reapplication | Project work. |

TABLE 1.2 A SIMPLE DESCRIPTION OF PROJECT AND PRACTICAL WORK

Source: (Silk and Bowlby, 1981)

project work, as stated by a number of educationalists, are summarized in Table 1.3. In the remainder of this section, a more detailed examination of some of the specific aims associated with project work is undertaken.

- To encourage students to make their own choice of a subject of study and thus encourage a sense of commitment and personal responsibility for the task.
- To give students practice in learning to learn by undertaking a piece of personal research involving activities such as: planning the work, hunting out sources, collecting material, selecting from it and deciding on presentation.
- To enable students to experience the satisfaction of working on a complex task over a period of time with the possibility of producing a result of permanent value and interest to themselves and others.
- To provide scope (where possible) for a degree of cooperation among students in an atmosphere of emulation.
- To provide opportunities for the practice of communication skills in a framework where language is used in a number of ways for real communication: seeking information, oral and written reporting, discussing, synthesizing, revising and editing etc.

TABLE 1.3 THE PURPOSES OF PROJECT WORK

Source: Adderly et.al. (1975)

One of the major aims of the project method is to involve the student in a piece of work which has been chosen to a large extent by the student him (or her) self. This necessitates the selection of areas within which work can be realistically pursued and the acceptance of responsibility for making decisions which must be justified later. Hopefully, this involvement will help to engender a degree of commitment and enthusiasm for the work being undertaken. Although project work may require more sustained effort than other forms of assessment, experience has shown that students find such an approach satisfying and enjoyable.

In terms of its position within a geographical curriculum the project method contributes to the attainment of three, commonly identified, general educational objectives (*viz.*, knowledge, skills and values). During the early stages of a project the student will have to acquire information which will enable him or her to pursue the topic which has been selected for study. This will involve knowing where to find and how to use the relevant literature; thus, the student acquires factual knowledge relating to their chosen subject and, therefore, will become familiar with a range of facts, views or situations. In addition, the student will have to become familiar with the methodology associated with the project topic, and, if required, with certain devices (such as the computer) or methodological tools (such as statistical analysis). Certain practical knowledge will also be acquired; for example, an awareness of the costs of (or constraints on) solving problems in terms of time, materials, labour and overheads. In a sense project work is complementary to traditional forms of teaching and assessment. Thus, for example, it is probably appropriate to have a set of practicals to introduce quantitative techniques and to test students' comprehension. A project, however, must be more than the employment of techniques; it should be concerned with a well-defined problem which can be studied using various components of a geography course.

The recognition, definition and solution of a problem or task requires the use (and development) of skills which are different in character to those employed in traditional forms of practical and course work. Whereas practical work helps to develop skills of basic importance, project work emphasises the ability to apply principles and knowledge already acquired to the solution of new problems. Individual initiative and resourcefulness, which tend to be obscured by traditional methods of assessment, are encouraged. As a form of discovery/learning, project work is characterised by activity rather than passivity; as an ancient Chinese proverb says:

I hear and I forget
I see and I remember
I do and I understand

Arguably, if students have to learn to think for themselves, they must be given the opportunity to do so. In particular the following skills would appear to be of importance:

1. to be able to plan and organise his or her work effectively: problem specification, data collection, analysis, presentation and interpretation.

2. to hunt out sources of information, and to collect data or materials in a systematic manner.

3. to select relevant material and reject the remainder, and

to integrate information from a number of sources.

4. to generate material or data by making investigations and analysing the results.

5. to synthesize findings and formulate conclusions.

6. to examine critically his or her own work and the work of others.

7. to use commonsense when, for example, seeking further advice or knowing when to stop.

8. to present the findings of the project in an appropriate form (numerically, graphically and via the written word).

In addition, by being involved with the world outside the classroom, the student has a greater opportunity to develop an understanding of 'real problems'.

The skills noted above refer, essentially, to individual skills, indeed, most geographical syllabuses refer only to individual projects. However, there is no reason why group projects should not be permitted (for example, on field courses), and from the perspective of vocational training the development of group skills would be extremely valuable. Specific skills which would be developed in a group context include the ability to cooperate with others; to manage people and direct the operations they perform; to chair a discussion group; to participate in discussion and to share in decision making processes; and the exercise of tact and diplomacy.

Finally, project work provides an appropriate context for the extension of values education, which it is being argued increasingly should be an integral part of a geographic curriculum. (Huckle, 1981). Specific areas might include values analysis (for example, to encourage students to use logical thinking and scientific investigation to decide value issues and questions); moral development (to help students develop more complex moral reasoning patterns based on a higher set of values); values clarification (to help students become aware of and identify their own values and those of others, and to communicate with others about their values); action learning (to provide students with opportunities for personal and social action based on their values).

In addition, project work allows students to develop certain personal disciplines such as resourcefulness, self-confidence, clear thinking and the ability to work with others. Indeed, a realistic assessment of one's own capabilities is an essential qualification for successful work in almost any profession.

## 1.4 The geographical value of project work

Although the educational value of project work is considerable, it is necessary to relate these broad objectives to the specific needs and requirements of geography students.

As noted previously, new geography syllabuses generally reflect developments in the discipline in the universities, particularly the adoption of the scientific method and the emphasis on systematic studies. Perhaps the most obvious manifestation of the scientific method at the school level has been the introduction of a variety of quantitative techniques. However, the scientific approach involves far more than the application of an assorted collection of techniques. In particular, it provides a specific methodology (and philosophy) for the organisation and pursuit of research. The hallmark of this approach is that it is active rather than passive; the researcher is responsible for the design and execution of each stage in the study. In the context of the discipline, therefore, project work should be seen as a way of introducing the student to the principal method of inquiry which provides the basis for most research carried out in geography.

In general, the scientific method consists of a number of well-defined stages:

1. The clear statement of the problem to be examined. This is derived from our perceptual experiences and image of the structure of the real world.

2. The conceptualisation of the image, possibly using a model to formally represent the image (for example, Burgess' and Hoyt's models of urban structure). Explicit attention should be given to the underlying assumptions.

3. The generation of testable hypotheses (for example, the zone and sectoral patterns manifested by particular land-use or social characteristics).

4. The definition, classification and measurement of principal concepts and ideas.

5. The collection of relevant data (from land use maps, population censuses, field observations, and so on).

6. The utilization of appropriate techniques to confirm or reject the initial hypothesis or hypotheses.

7. If the hypothesis is verified or confirmed then an attempt at explanation should be made in terms of the model or theory from which the hypothesis was derived. Equally, if the hypothesis is rejected an attempt should be made

to explain why this occurred. Arguably, in the context of the project method, greater relative emphasis is placed on identifying, analysing and correctly defining the problem than finding a 'correct' solution. Project work like science is ultimately the 'art of the soluble' (Medawar, 1967), however, the discovery of what is not (or perhaps not yet) soluble is an important part of learning.

## 1.5 The Supervision of Projects

Although project work is desirable in both educational and geographical terms, there are a number of practical problems associated with the teaching and supervision of projects.

One of the major problems likely to be encountered involves the initial selection of a project topic. However, it is not possible to establish rigid rules for project selection; nevertheless, some general quidelines can be noted which might be helpful in the process of selecting a specific project. First, the choice of project is likely to involve both the student and the teacher. Ideally, the student should make the final selection insofar as a problem is not a problem for a student if he or she does not perceive it as such. Second, project topics should be suited to the ability and maturity of the student concerned. Although this consideration may directly influence the selection of a project topic (insofar as there is a natural tendency to be over ambitious; if the topic is too complex the student may become discouraged), it will invariably be an important factor in determining how a student tackles a specified problem and utilizes appropriate materials. Third, whenever possible a project topic should involve the choice of a course of action from among two or more possible solutions, thereby calling decision-making techniques into play. Fourth, the selection of specific projects will have to be balanced by the availability of suitable resources and materials (books, relevant data, access to computers, reprographic facilities, for example). In this context, the teacher's role as supervisor is crucial in ensuring that information is available so that students do not waste time looking for inaccessible or non-existent material. Fifth, in many instances a project topic may be multi-disciplinary in character (for example, the examination of certain environmental problems in a city). However, whereas the student should be encouraged to utilize viewpoints from other disciplines, the distinctively geographical perspective of the study should be foremost. Finally, whenever possible projects should be small-scale and locally based. This serves this purpose: first, a small scale project should be more manageable from a student's point of view than a large scale project; and, second, by focusing on the local area the student's understanding and appreciation of 'real world' problems would be enhanced.

This latter point is of particular importance given the increasing emphasis on geography as ' a useful and relevant discipline' (Coppock, 1974) and the growing concern of many geographers with the formation of public policy.

The teacher's role is not only crucial in the initial formulation of the problem but also during subsequent stages of the project's development. However, this role is principally consultative and collaborative rather than authoritative in character. This will involve being alert to the problems which face students: guiding students indirectly by asking questions which stimulate thoughtfulness; and providing assistance only when a student reaches the point of frustration in the approach to the problem. In addition, opportunities should be made available for individual (and group) reports. It might also be interesting and useful for students to discuss each others work as their individual projects progress, thereby encouraging students to examine ideas critically, and improve or set new directions in their work. Obviously, very often the quality of a student's project can reflect the amount of supervision and guidance provided by the teacher. Thus, during a consultation session teachers should attempt to be equally enthusiastic about all projects (even when this proves difficult!)

Although there is some flexibility with the timing of projects, it is suggested that the most suitable time for initial preparation would be during the summer term of the first year of the syllabus (for Advanced Level students). Sufficient material should have been covered already to provide the necessary background for students, and the summer vacation may offer an appropriate opportunity for some data collection. The student, however, should be dissuaded from getting over-involved in the project at the expense of other sections of the course (and other courses). In this context, the teacher's role is particularly important in guiding the student in the selection of the task; being selective in terms of content, and in organising his or her time economically. From the teacher's perspective, therefore, because of the need for individual guidance and supervision, project work can be both time-consuming and demanding.

## 1.6 Assessment of Projects

In terms of formal external examinations at school level, project work represents one component of the syllabus, and commonly accounts for approximately twenty per cent of the overall mark at Advanced Level. Very often, however, the project is assessed internally by the teacher involved; it is usually at this stage that a number of problems emerge. In fact, the Joint Matriculation Board recognised the general problem when they published their revised *Instructions and Guidance for Teachers* (1978). The following comments are clearly not intended to replace the examination board's own instructions, but to provide additional guidance on the assessment of projects and should be considered in conjunction with sections 1.3 and 1.4.

1.6.1 *Comprehension and Statement of the Problem*
This is a crucial stage which may be overlooked in an assessment where attention focuses primarily on the collection, organisation and interpretation of data. First, in formulating the problem the student should be fully aware of the theoretical background, geographical implications and practical difficulties associated with the selected topic. Second, the student should be able to state unambiguously the aims of the project, together with an indication of the procedure which he or she intends to use. An essential feature of this stage, therefore, involves the student thinking about both how principles and knowledge already acquired can be applied to new problems and the possibility of generating testable hypotheses. It is suggested that the student, before proceeding to the collection and analysis of data, should submit a fairly detailed outline of the project which should also be taken into account in the final assessment. The final title of the project should embody a clear statement of the problem to be tackled.

1.6.2 *Application*
At this stage, the focus of assessment is the manner in which the student actually applies principles and their knowledge to a particular problem. The following features would appear to be of importance:

a) The demonstration of certain skills.
In terms of problem-solving capabilities, has the problem been expressed in terms of a general hypothesis, generalisation or model. Where appropriate, aspects of the general problem should be examined by the formulation of specific hypotheses. The student should then analyse each hypothesis to determine the type of information required to test each one. Skills which are more specific in character may be summarized as numeracy, graphicacy and literacy. Numeracy involves the observation, extraction, measurement and recording of data. Attention should also be paid to the relevance of the data to the problem and the correct use of statistical methods. In terms of graphical skills attention should focus on the intelligent and correct use of maps and graphs and their integration with the text. Literacy is a rather vague criterion of assessment, but grammatical errors should not be ignored; a project is not simply a technical exercise but should reflect a variety of skills, including the ability to write clearly

and grammatically.

b) Geographical content.
Students should emphasise the geographical content of the study. This stricture, however, should not be interpreted too narrowly, particularly in studies where other factors (historical, economic, for instance) are clearly of significance. Nevertheless, projects which, for example, consist purely of historical description or use statistical techniques for their own sake should be avoided.

### 1.6.3 *Evaluation*

As well as being able to apply knowledge and principles to new problems, the student should be able to evaluate his or her results in terms of the initial hypothesis or hypotheses. At the simplest level, this would refer to the interpretation of the data, the quality of the conclusions and an estimation of the success of of the project. At a higher level, assessment would focus specifically on the student's problem-solving skills and abilities. For example, the ability to resolve a problem into its component parts (analysis), complemented by the ability to bring together ideas and conclusions into a connected whole (synthesis). In general, the hypothesis-testing approach to geographical inquiry, rather than purely descriptive accounts is to be preferred.

### 1.6.4 *Presentation*

The final assessment should also reflect the student's ability to present a substantial piece of work in a clear and pleasing manner. Although the actual standard attained will depend on the individual student, each project would benefit from the inclusion of the following features: a contents page; a list of tables and a list of maps and diagrams; acknowledgments indicating the names and positions of any persons consulted and an indication of the help and information received; a list of references that have been useful in compiling the project, which will include both books and secondary source material.

It is suggested that assessments should not be based simply on one final assessment, but that there should be a continuous monitoring of the student's performance, from initial statement of the problem through to application and the evaluation of results. Moreover, insofar as projects are usually internally assessed a brief oral examination (although time-consuming) would be a beneficial and valuable supplement. This form of assessment, for instance, may be of benefit to students who have difficulty in expressing themselves in writing but who are reasonably fluent, verbally. An oral examination allows the examiner to probe more deeply into how the student conceived and executed the project. Clearly, the structure of an oral examination will depend on what objectives are being tested. Graves (1982) suggests that in the context of project work the following abilities should be considered: justification of the procedures used to collect data; understanding the analysis to which the data were subjected; the interpretation of the results in different ways. Greater weight should be given to the latter element. Graves also notes that an examiner may find it difficult to be totally detached during the oral examination itself. A possible solution would involve recording the interview and to assess the candidates performance later.

Finally, the following allocation of marks for the overall assessment is suggested:

| | | |
|---|---|---|
| Presentation | (neatness, clarity, appropriate illustrations) | 10 |
| Research | (references, bibliography, personal investigations) | 25 |
| Content | (coherence, relevance, continuity and significance) | 25 |
| Conclusion | (interpretation of evidence presented) | 25 |
| Oral examination | | 15 |

Chapter Two

# AN INTRODUCTORY OVERVIEW: PROJECTS IN ADVANCED LEVEL GEOGRAPHY

## 2.1 Introduction

The enormous changes in geographic research interests and undergraduate degree courses in the last two decades has recently resulted in major alterations of school syllabuses. Whilst it is clearly essential that geography taught in schools must take account of developments in universities, there are obvious dangers of such a filtering down process. The subject matter, even in a watered-down form, may be inappropriate for teaching in schools. Moveover, and in many ways associated with modifications of what is actually taught, care must also be taken with the transfer of teaching methods. For instance, an undergraduate dissertation is an integral and compulsory component of many degree courses; as demonstrated in the last chapter, the introduction of project work into school syllabuses raises important issues regarding potential topics, integration within the entire course, assessment and so on.

It is argued that geographic projects present a challenge to both school teachers and students, and, whilst the number of potential topics is enormous, a number of fundamental unifying characteristics can be recognised (which are partly associated with recent developments in the discipline). In general, student projects would be applied, rather than theoretical, and, moreover, they would be frequently related to 'relevant' policy issues or problems (see, for example, Kohn (1982) for a discussion of problem-solving approaches). In addition, given an obvious need to integrate project work within a current curriculum, it is suggested that specific project topics would probably be relatively broadly based using material from different parts of the course. (Simply stated, many projects would likely involve an examination of some particular aspect of man-environment interrelationships including the man-made, as well as the natural environment).

Indeed, given the inherent nature of geographic problems, a multidisciplinary approach, incorporating aspects of students' other courses, would be apposite. The teacher, however, must ensure the balance remains appropriate; ultimately, it is to be assessed as a project for Geography. As stated in the Preface, a major attraction and value of a geographic education stems from the attention given to real issues. Moreover, individual projects provide an opportunity for students to undertake a detailed, personal study on a topic of their choice. Whilst there is no conclusive evidence available, it can be argued that the individual, problem-solving approach of project work intrinsically motivates students and, consequently, as a teaching and assessment method, it is especially useful (although the teacher must know each student personally and be able to ensure they work to their full capabilities).

Clearly, the teacher will play a leading role at least in the initial stages, in the selection and development of feasible project topics (although a number of Examining Boards also provide lists to assist teachers in this direction). Whilst teachers should ensure a project is suitable for a specific student's particular background and potential, it would be desirable if students were able to have an active involvement in the selection from the outset. The original definition of the project area must exhibit sufficient flexibility to present a student with scope for individual analysis. Clearly, the availability of resources in particular schools, such as a computer, a well stocked library, measuring instruments, reprographic facilities and so on, constrains the range of feasible projects. In this book, well-defined, easily transferrable, complete projects have not been presented; the range of ideas and examples provide a firm foundation for project specification but they must be altered and adapted carefully to suit a specific student's needs. In terms of actual supervision, the requirements are well-known by qualified teachers. It is sufficient to state that students should be assisted, rather then directed, and it is part of the learning process to actually make mistakes! From experience, to help students schedule their projects, short interim, oral reports to the teacher (and perhaps even to their fellow students) are beneficial.

In terms of balance, although it has been argued that the many facets of man-environment interrelationships offer suitable topics for student projects, the bulk of this book is concerned with projects in human geography. This reflects closely the past experience of the relative incidence of human and physical geography projects; with obvious constraints of time and equipment, many aspects of research in physical geography are untenable and impractical. However, two from a total of seven chapters are concerned directly with projects in physical geography; five chapters are concerned directly with projects in human geography, and the chapter on urban-based projects is the longest because this is the most frequently undertaken type of study. The ideas are presented in such a way that a student's particular project could combine elements from different chapters. The range of topics described are used for illustrative purposes to help generate additional ideas for the teachers and students.

In this chapter, a discussion of Berry's geographic data provides a useful background for the remainder of the book. Given that it forces analysts to consider interrelationships explicitly, brief reference to this framework during the initial stages of project formulation should force students to plan their stages of data collection, data presentation and data analysis coherently. Whilst such aspects are considered briefly in relation to specific projects in later chapters and are examined in detailed, methodological discussions that are presented in the appendices, some general comments are made in this chapter.

## 2.2 Berry's Data Matrix

Extending this important idea of integration, a brief discussion of Berry's (1964) so-called data matrix is presented because it explicitly highlights the interrelatedness of geographic phenomena. This formal ordering of information also provides a useful and efficient operational structure for data collection, description and analysis. Given geographers' concern with 'areal differentiation', 'spatial organisation' or 'locational analysis', two dimensions or basic features which are problem-specific can be recognised: location and characteristics. In the definition of locations, the spatial scale of analysis must be defined; are we interested in countries, regions, cities, ....? Similarly, the required level of detail for the (agricultural, climatic, demographic, geological, industrial, political, ....) characteristics is also dependent on the subject of interest. If you flick through a 'traditional' regional geography textbook, you would find a systematic treatment of columns. Thus, this matrix consists of a set of rows (or locations) and a set of columns (or characteristics).

In fact, Berry proposed the employment of a 3-dimensional data matrix. The rows of this matrix represent geographical locations, the columns represent characteristics, and the third dimension represents this information in a series of time slices. Many projects would be improved by examining general changes over time, if only to provide the background for a detailed analysis of the contemporary situation. For example, a study of industrial linkages in North-east England would benefit from a brief description of the past importance

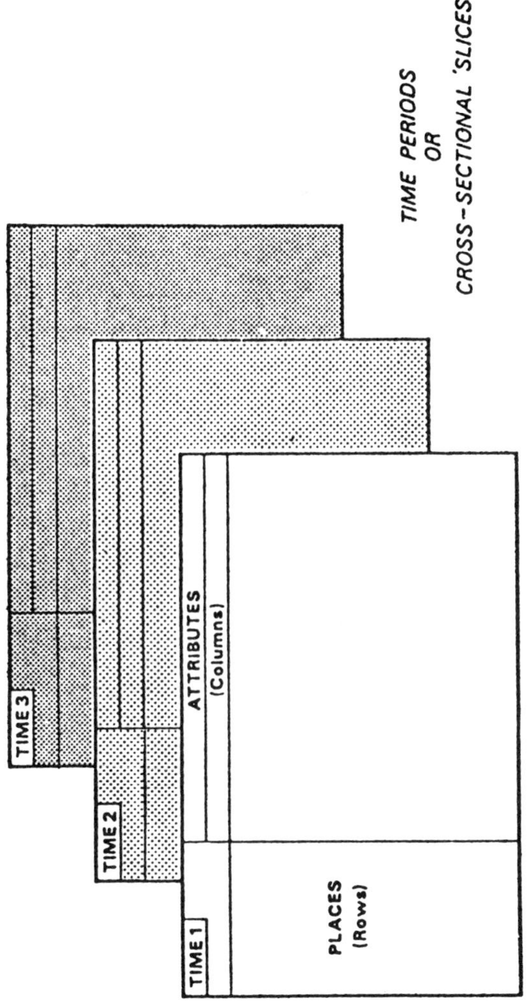


Figure 2.1 BERRY'S (1964) GEOGRAPHIC MATRIX

of coal-mining and ship building and of the impacts of the general structural changes in the economy in recent dacades.

It is possible to show that ten different types of traditional geographic study can be recognised by considering alternative operations on the matrix. An examination of the arrangement of cells within a specific column is concerned with the spatial distribution of characteristics, such as climatic regions, whereas an investigation of the arrangement of cells within a specific row is concerned with the association of different characteristics at a particular location, a location's inventory. Comparisons between a pair (or series) of columns or rows involves a study of the spatial covariance of characteristics or areal differentiation, respectively. These four kinds of geographies can be combined to produce a fifth, which highlights both functional and spatial interconnections, and the incorporation of a time dimension to each type of study produces five additional forms of analysis. By examining various slices of time, a so-called 'comparative static' approach has been adopted.

## 2.3 Data Collection

Although a geographic data matrix provides a sensible form for organising data, it must be the chosen topic that determines the actual information that is used. There is an obvious danger that too much or inappropriate data would be included; however, if a project is unambiguously defined, this problem should not arise. It should be appreciated that the data used directly determines the success of a project, and, consequently very careful planning is required at this stage; a particular data set presents opportunities and limitations. As the selective and correct use of appropriate statistical techniques can offer additional descriptive and inferential insights into a problem, it is important that this planning of data collection accommodates their needs; specifically, as discussed below, many of the techniques are restricted to the particular types of data.

Simply stated, there are two basic sources of data, primary and secondary. At the outset, attention is drawn to two simple, but important, and often neglected points. Firstly, when secondary sources are used, formal and full referencing of the fact is essential. Secondly, shortcomings of data sets should be highlighted, and students should be encouraged to briefly indicate what would be required for a more comprehensive study if there was sufficient time available.

### 2.3.1 *Secondary Data*

As the *Reviews of United Kingdom Statistical Sources* indicate, secondary data are obtainable from a variety of different sources. It should, however, be appreciated that two problems may arise. Firstly, as the data are unlikely to have been collected from the particular problem that a students studying has obvious shortcomings. For example, students are frequently undertaking local studies, but the relevant information has only been collected at a regional or national scale; however, this data can often be used to provide background material against which their own information can be compared. Secondly, and associated with this idea of spatial scale, students must be aware of the problems of the so-called 'ecological fallacy'. Given that aggregate level data (based on countries, regions or cities) are more readily available than data on individuals (which is often suppressed to maintain confidentiality), there are obvious desires to infer about individual behaviour from grouped data. However, Robinson (1956) in a much cited paper demonstrated possible invalidity. Specifically, he demonstrated that an aggregate level correlation coefficient need not equal the corresponding individual level coefficient. (It should be noted that similar discrepancies can exist between different spatial scales). The geography of elections is one topic which the ecological fallacy is important, because information about individuals' voting behaviour is not released.

Anyway, a variety of useful data sources exists for the students (see the table below) and these can often be obtained from local libraries and Local Authorities.

POSSIBLE SECONDARY DATA SOURCES

Census (see Dewdney, 1981) (decennial since 1801) (library)
Commercial directories (library)
Domesday Book ('... probably the most remarkable statistical document in the history of Europe' (Darby, 1973, p. 38))
Electoral Register (library/Local Authority)
Family Expenditure Survey (library)
General Household Survey (library)
Local Housing Statistics (quarterly publication for each Local Authority)
Rateable Value Records (Local Authority)
Structure Plans (Local Authority)
Timetables (local public transport organisations)
Yellow Pages (telephone directory)
Review of United Kingdom Statistical Sources (Pergamon Press)

Finally, the large range of different maps, covering

different phenomena and at different spatial scales, are a very important source of information for many student projects - (urban or agricultural) land use patterns, transport networks, rural settlement patterns and so on.

### 2.3.2 *Primary Data*

Alternatively, to obtain information about specific, often local, characteristics, it would often be necessary for students to undertake small personal surveys, which often provide useful and enjoyable experiences. Whilst these can be either personal interviews or mailed questionnaires, the latter may be impractical because of financial constraints (note - a stamped addressed envelope would have to be included in all the posted questionnaires); indeed, the typing and duplication of a number of questionnaires may stretch a school's secretarial resources. As no survey can be better than a questionnaire, it is obviously important to ensure the questionnaire is satisfactory. Although the design of questionnaires does not really benefit from formal, classroom teaching, because the process is essentially one of common sense and (geographical) experience, the teachers should carefully supervise the construction of a questionnaire to avoid potential pitfalls (see Appendix III). First, however, it is necessary to ensure that students adopt a suitable sampling procedure.

#### 2.3.2.1 *Sampling*

Unbiased and representative samples should be taken from the population of interest. A variety of sampling approaches are possible, and often a specific project requires a particular one. Two basic approaches - random or systematic - can be recognised, and a simple example concerned with consumer shopping behaviour highlights the distinction. Firstly, a simple, random sampling procedure involves interviewing shoppers at random, independent of those already questioned. Thus, every consumer has an equal probability of selection, and this is often an important assumption when applying inferential statistic assumptions. Secondly, a systematic sample involves, as the term suggests, sampling at regular intervals, say every twentieth customer walking out of a particular shop. Whilst it is possible to combine elements of these approaches, two additions are often useful for geographic projects. In a study of shopping behaviour, two types of stratification would be useful (unrelated to whether a random or systematic approach had been adopted). Different types of consumers (teenagers, pensioners, family shoppers and so on) have different patterns of behaviour, and, to ensure that a sample included all these types, it would be appropriate to stratify on the basis of customer type and have a number of sub-samples. Secondly, in geographic projects it is often useful to stratify on the basis of different locations; for instance, people who have travelled a long way to shop probably would have a distinctive pattern of behaviour in comparison to those who have just 'popped in' for a few things.

The second addition would be to sample (randomly or systematically) over time and space. With regard to consumer behaviour, for example, sampling on different days, particularly weekdays in comparison to on Saturdays, would produce different patterns of behaviour. Similarly, even if concerned exclusively with one store, it would be important to sample from different locations; the entrance near the car park, for example, probably would be biased towards motorists, whereas the entrance near the bus station probably would be biased towards those reliant on public transport.

#### 2.3.2.2 *Questionnaire Structure*

Broadly speaking, a good survey should flow like a natural conversation although the questions must be controlled to maintain a standard form. An explicit statement of the particular problem of interest must be given at the outset. Vague topics, such as problems of Newcastle's transport system, are not especially useful. Are we really interested in the potential success of a specific traffic management scheme, such as pedestrianisation, or the possible implications of a switch from private to public transport with regard to congestion problems? The project must be well-defined in terms of hypotheses, and a good idea of the types of analyses, particularly any cross-tabulations, that would be undertaken is essential. It is then possible to develop a logical structure which only includes pertinent questions. (There is a natural tendency to ask too many questions, and this makes the questionnaire unnecessarily long, a feature that puts off prospective respondents). Conclusive results from a short questionnaire are much better than inconclusive results from a long questionnaire! Some brief general background questions, such as age group and place of residence, should be asked at the beginning, and groups of related questions should be placed in sections; filtering should be introduced if specific questions are only applicable to particular respondents (such as, car owners).

Finally, and whilst appreciating the enormous time constraints on students, a pilot study should be undertaken because it would highlight loose ends and contradictory, ambiguous, leading and vague questions. In addition to actually trying out the questionnaire, it is perhaps worth asking people what they believe particular questions mean. Questions must have a standardised interpretation to permit comparative analysis.

#### 2.3.2.3 *Some Specific Characteristics of Questions*

The actual questions asked must be thought to be both important and interesting by the prospective respondents. Furthermore, simple short questions are usually less likely to be ambiguous. Two basic types of question exist, factual and opinion, and it is especially important to be careful about the latter. When collecting facts, it is often useful to provide pre-coded answers if the range of possible answers is limited or well-established (although too many categories are not useful). If possible, it is desirable to employ standardised classifications, such as the Standard Industrial Classification, the Register General's list of occupations. Open-ended questions are usually difficult to classify, but they are important because they add 'life' to a descriptive text and also permit individual characteristics/opinions to be highlighted.

In practical terms, students should be aware about difficulties associated with the temporal dimension: the accuracy of a person's memory about events in the past must be questioned, and, when asking about future behaviour, it should be appreciated that respondents' expectations may be modified by a degree of 'wishful thinking' and what they believe others think is proper. Moreover, in relation to opinion questions, it should be recognised in a student's interpretation that the issue is usually multi-sided. It is, therefore, essential to determine explicitly the main criteria on which respondents based their opinion. For example the siting of a nuclear power station is an emotive issue that can be considered from a number of perspectives: medical, legal, political, social and economic.

### 2.4 Data Description and Presentation

Given a suitably obtained data set, which, in practice, often takes longer than was envisaged initially, its actual description and presentation in the project is very important. Simply stated, variety overcomes potential monotony for the reader. In this section, two forms of description and presentation are referred to, but more detailed comments are made in relation to particular examples introduced in later chapters and in Appendices I and II. Visual presentations provide a clear and straight forward portrayal of information, whereas simple statistics are often useful to provide a numerical summary. No discussion of statistical techniques is undertaken because there are numerous introductory textbooks on this topic. However, attention is drawn to the fact that the type of data, specifically its level of measurement, directly determines which statistics are appropriate.

Graphs and pie charts can be used to summarise information; flow charts for movement. At this stage, it is of paramount importance to emphasise that maps and diagrams should be integrated within the overall text, and the standard of cartography should be high. Similarly, the use of statistical techniques must be discriminatory, enhancing a student's interpretation. Any application of techniques should involve a critical assessment of the justification behind their adoption and whether underlying assumptions have been satisfied. Students should not attempt to demonstrate a high degree of proficiency in the operation of quantitative methods by neglecting their geographical problems; the ultimate usefulness of a technique is demonstrated by a student's interpretation of the geography.

In terms of the particular statistical techniques that students are generally required to know, three kinds can be recognised: measures of central tendency; measures of dispersion, and measures of association. As Table 2.1 illustrates, different statistics are related to the type of data available.

Four levels of measurement to classify data can be recognised: nominal, ordinal, interval and ratio. The lowest level of measurement, the nominal scale, involves a simple categorisation of information. In contrast, in the next level of measurement, the ordinal scale, different categories are ranked in order of preference (although no indication of the numerical magnitude of the difference between categories is given). Many behavioural studies involve questionnaires designed to indicate various preferences. Nominal and ordinal data are low order data, and it is this kind that geographers possess most frequently. It should be noted, however, that the majority of simple statistics relate to high order, either interval or ratio, data. If the actual numerical differences between individuals are known, the data are termed interval. For example, using a standard measurement unit, such as degrees centigrade, it is possible to say how much hotter in March it is in Salisbury than in Stoke. In addition, if a non-arbitrary, fixed zero level exists, the measurement scale is called ratio. Zero distance (kilometres) is meaningful; it is the same location!

From this brief systematic description of different measurement levels it is clear that it is possible to convert ordinal level data into nominal, interval level into either ordinal or nominal, and so on. Such conversions involve an associated information loss, but they are often necessary to ensure the compatability of data and statistical technique. It is important to appreciate that the level of measurement of the data determines directly the specific statistic that is appropriate.

| TYPE OF STATISTIC | LEVEL OF MEASUREMENT: NOMINAL | ORDINAL | INTERVAL/RATIO |
|---|---|---|---|
| CENTRAL TENDENCY | MODE | MEDIAN | MEAN |
| DISPERSION | VARIATION RATIO | RANGE<br>QUARTILES | STANDARD DEVIATION<br>VARIANCE |
| ASSOCIATION | CONTINGENCY COEFICCIENT<br>PH1 COEFFICIENT | SPEARMAN'S RANK CORRELATION COEFFICIENT | PEARSON'S PRODUCT MOMENT COEFFICIENT |

Table 2.1 STATISTICAL TECHNIQUES FOR DIFFERENT LEVELS OF DATA

## 2.5 Some Concluding Comments

To impose any rigidity on the structure of projects would undermine their very essence. However, it is possible and useful, particularly in assisting the scheduling of work, to indicate a number of general characteristics of every project that should be considered. The check-list presented in this section, however, is not an exhaustive coverage, and it should not be thought to be a series of sequential stages. By the nature of a successful project, they must be interlinked, and, from initiation to completion, it is to be expected that the balance of a project would be altered a number of times.

At the outset, the teacher and student must ensure a practical project is specified explicitly. Whilst modifications and extensions obviously would occur over time, the topic area must be unambiguously defined as narrowly as possible to enable a student to direct his attention to specific issues. It is all too easy for a project to be a set of loosely-related facts and ideas, which fail to provide the coherence and the development of a logical argument that is required. It is important to appreciate that this exercise takes time to complete satisfactorily, and reference to Berry's geographic data matrix may be appropriate.

A major determinant of the ultimate nature of a project is the availability of suitable data for the study and time constraints on data collection by the individual student. Although it takes additional time, a sample survey is invaluable in finalising a questionnaire and it should be undertaken. Associated with any data collection, it is of fundamental importance that attention is given to the alternative ways of describing and representing information and, if necessary to the selection of appropriate techniques for analysis. (Background reading on the assumptions and shortcomings of specific procedures is essential). As stated earlier, the type of data available ultimately determines the kinds of presentations and analyses that are feasible. Clearly, it would be shortsighted to constrain these later stages through inadequate planning.

In the description and interpretation of the results, the argument is much clearer if it can be related to some loose structure provided by Berry's geographic data matrix or by appropriate geographical concepts. Organisational problems and difficulties in expressing an argument in written form are encountered frequently in student project work, and teachers should try to guide students away from obvious pitfalls. It may be necessary to comment on rough drafts of sections of a project.

Finally, in any assessment of a student's project, it is necessary to take account of the nature and scope of the topic in relation to the student's specific knowledge. However, particularly in relation to projects in which problems are addressed, brief suggestions for more detailed and associated investigation (including policy proposals) should be included, because students can indicate their breadth of comprehension of the topic in this way.

Chapter Three

# STUDIES IN URBAN GEOGRAPHY

## 3.1 Introduction

In this chapter, attention is focused on projects which have their basis in patterns and processes identified at the urban scale of analysis. This chapter is the longest in the book. The reasons for this emphasis are essentially twofold. First, in many human geography syllabi the urban element is usually the largest single component. This together with the fact that a majority of students live in an urban environment tends to result, quite naturally, in a preponderance of urban based projects in internally-assessed practical work. Second, many writers consider urban geography to be the leading sector in the development of human geographical thought and hence it mirrors the directions of the discipline as a whole. Within this context, the opportunity is taken in this chapter to introduce a number of projects which reflect the range of approaches characteristic of human geography in general and urban geography in particular.

Broadly speaking, it is possible to identify three major areas where urban geographers have concentrated their efforts in recent years.

(1) The first area consists of two groups of studies characteristic of the spatial science approach found in human geography in general (see Chapter One). At the inter-urban level, the study of urban places and urban systems has involved analysis of the distribution of city sizes and their distribution (for example, the rank-size rule), the location of cities of different sizes (for example, the central place theories of Lösch (1954) and Christaller (1966)) and urban functions and classification. Whereas studies of urban systems treat towns and cities as point features in the landscape, at the intra-urban scale, the point of attention is the internal structure of towns and cities. At this level two principal themes may be identified: first, the analysis of the locations and characteristics of non-residential

land uses, and second, the characteristics of residential areas.

(2) The behavioural approach provides the rationale for the second major area. Thus, urban geographers have examined the attitudes and behaviour of individuals and groups within the city and their perception of the urban environment (for example, consumer behaviour and preferences, the study of urban images and time-geography).

(3) The final approach emanates from a broad concern within human geography associated with welfare issues, human well-being and questions of social relevance. The overall perspective may be summarised as an interest in 'who gets what where and how' (Smith, 1977). The emphasis is on spatial patterns of inequality and how living standards vary according to place of residence, together with policy suggestions relating to the amelioration of such inequalities.

The themes and projects discussed in the following sections attempt to provide a balance between the approaches noted above. The majority of approaches focus on the intra-urban context. On the one hand this emphasis reflects recent developments in urban geography, and on the other, enables the student to construct a manageable project based on his or her local area. In the following section, the functional provision occurring within a central place system is examined, and a simple index is described which adequately measures the concept of centrality. This index could also be used at the intra-urban level. In the third section, some basic ideas of modelling are introduced via a consideration of a number of models which attempt to account for the internal structure of towns and cities. While the impact of the behavioural approach provides the context for section four (geographical space perception) and section five (time-geography). Both topics provide a rather novel range of possible projects for students. Finally, a brief outline of the welfare approach is followed by a straightforward method of describing urban deprivation.

## 3.2 Describing the Functional Provision of a Central Place: Index of Centrality

In association with central place theory, many empirical studies have been undertaken which consider the provision of different functions by centres in a particular area. Specifically, attention has focused on whether a hierarchy of centres exists, and, to summarise each centre's range of services, simple centrality indexes have been proposed. The index introduced in this section is only an extension of the location quotient which is modified to be consistent with central place theory.

Both Christaller and Lösch interpreted differences in centre size on the basis of functional characteristics, and it is important to appreciate that it was not explicitly assumed that functional complexity was mirrored by a centre's population. Indeed, Christaller (1966, p.17) stated that population did not accurately represent the importance of a centre:

> 'Neither (market) area nor population very precisely expresses the meaning of the importance of the town.'

Statistically significant correlations between a centre's population and its range of functions (or its number of establishments) have been found - larger centres usually provide a greater range of services - but there is no conceptual foundation to these studies. A similar comment can be made about the applications of the so-called rank-size rule to describe city size distributions.

Auerbach (1913) first commented on the regular relationship between the rank and size of centres, although its development can be primarily associated with the work of Zipf (1941). Simply stated, an inverse relationship exists between centre size and centre rank. In equation form, the rank-size rule is given by

$$Pr = \frac{P_1}{r^q}$$

Where Pr is the population of the $r^{th}$ centre in a series of all centres in an area which is ranked in decreasing order of population size; by definition, $P_1$ is the population of the largest ('primate') city and q is a constant (or parameter). This equation is commonly written in terms of logarithms to produce a linear relationship.

$$\log Pr = \log P_1 - q \log r$$

and this is clearly of the form

$$Y = a + bx$$

In a graph of log Pr against log r (population against rank), log $P_1$ is the value of the intercept and q is the gradient of the downward sloping line (see Figure 3.1). Figure 3.2 describes how the rank-size relationship has changed over time.

In fact, Christaller developed a centrality index in his application of central place theory to examine the distribution of central places in Southern Germany. Christaller's index was based on the number of telephone connections, which permitted centres' spheres of influences to be classified. Using Christaller's notation, the formula for the centrality of a centre (Zz) is

$$Zz = Tz - Ez \frac{Tg}{Eg}$$

and a non-positive value indicates that the centre does not function as a central place. Tz is the number of telephones at the central place; Ez is the number of inhabitants of the central place; Tg is the total number of telephones in the study area; and Eg is the total number of inhabitants of the study area. This provided an operational formulation of the centrality of a centre for empirical investigators that was consistent with the theoretical foundations.

It is important to appreciate that Christaller discriminated clearly between 'absolute importance' and 'relative importance'; centrality is measured by relative importance (whereas nodality is measured by absolute importance). This distinction relates to a division of the total demand for a centre's functions into two components: the internally generated demand (from the centre itself) and the externally generated demand. In an analysis of centrality, attention is concerned with the externally generated demand. Unfortunately, in many studies, this conceptual basis has been missing.

Indeed, Davies' functional index (for a discussion see Carter, 1981), which has proved particularly attractive to a number of authors, does not directly measure the demand generated from outside the centre. It is analogous to the industrial location coefficient, but it can be easily modified to provide a conceptually satisfactory centrality index. Davies defined the centrality index of any function as

$$C_f = \frac{t_f}{T_f} \cdot 100$$

where $C_f$ is the location coefficient of function f, $t_f$ is the number of outlets of function f in the particular centre, and $T_f$ is the total number of outlets of function f in the study area. Aggregation of these totals for all functions gives the centrality index for each centre, C, that is, the centre's so-called functional index

$$C = \sum_f C_f$$

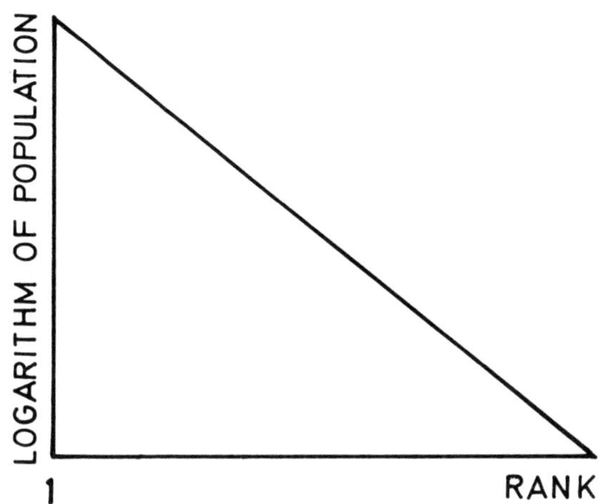


Figure 3.1 RANK-SIZE RULE

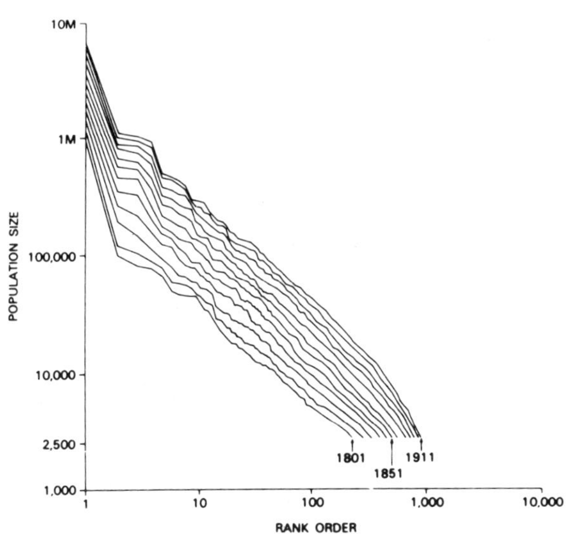


Figure 3.2 RANK-SIZE RELATIONSHIP OVER TIME
(Source: Carter, 1978)

An alternative index, $CI_f$, can be expressed, using the above notation,

$$CI_f = t_f - T_f \frac{p}{P}$$

where p is the centre's population and P is the total population of the study area. Summation of these totals over the function, once again, produces the centrality index for each centre.

It is important to appreciate that certain assumptions underlie this simple index. First, a centre's indigenous population uses the functions provided by the centre. This is consistent with the traditional hypothesis that consumers behave rationally and wish to minimise total travel costs, and, consequently, they patronise the nearest centre providing the required function.

Second, it is assumed that, irrespective of place of residence, each person's consumption of a particular function is the same. Although it would be a straightforward exercise, if the data were available, to take account of variations in consumption patterns with age and income, it should be appreciated that it is often argued that as the distance to a centre increases, an individual consumer's level of demand decreases because of the additional transport costs as in the case of Löschian demand cone analysis (see Figure 3.3)

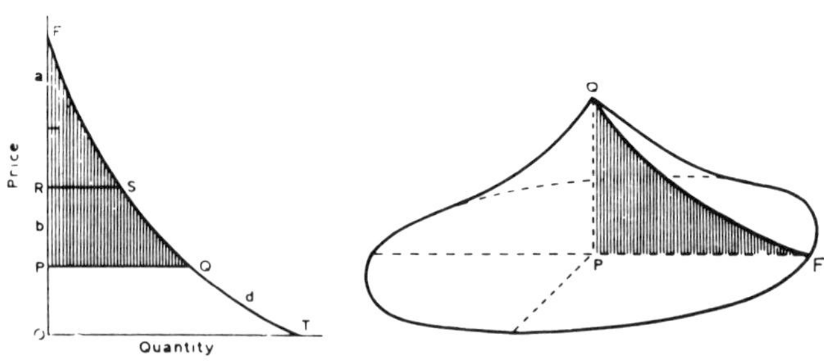


Figure 3.3 LOSCHIAN DEMAND CONE ANALYSIS (Source: Carter, 1981)

Third, a centre has a complementary region and its population obtain functions from the centre.

Fourth, there is the realistic assumption that all functions are not of equal significance. A relatively scarce activity contributes more to a centre's overall centrality index than the presence of a more ubiquitous function. For example, if there are 200 bread shops and 4 supermarkets in the study area, a centre which contains one of each of these establishments would have a greater centrality than one that only had two bread shops. This implicit weighting is an advantage in comparative analysis where similar functions possess different levels of significance. For instance, in a study of the changing centrality of centres over time, it accommodates the fact that different functions were important at different times - the blacksmiths, coopers and so on of the nineteenth century.

Fifth, it is assumed that each establishment providing a particular function is the same size. Whilst this is usually approximately true, the indexes could be easily weighted by using the occupied floor space or the number of employed persons.

Thus, although this index is simplistic, it is both intuitively and conceptually attractive. The individual treatment of each function takes account of variations in the centres' functional complexities. The student would define the particular set of functions of interest, trying to cover a range of activities. In contemporary work, assuming most facilities are on the telephone, much up-to-date data can be easily obtained from the 'yellow pages'. Alternatively, and particularly for studies at previous times, local commercial directories are usually available from the local public library. Clearly, such sources of information contain unavoidable biases, such as the specific way functions are classified, but these secondary sources could be complemented by some individual fieldwork.

## *Project Outline*

### Objective

To examine the functional provision occurring within a central place system for a particular year. Alternatively an historical study may be attempted to indicate the changing relative significance of centres of different sizes.

### Data

Kelly's Directory: 1845-1946 available for most counties and major provincial towns and cities; 1946-1976 available for main towns and cities only (published annually).
Yellow Pages: some establishments do not, however, enter their name.

Whilst this data are very useful, one is restricted to the definitions adopted.

Method
Develop index of centrality discussed above to provide an integrated theoretical and empirical investigation.

Discussion
If the centrality values for each centre are ranked and plotted on the vertical axis against the rank (in a similar way to the rank-size rule), is a stepped, hierarchical pattern produced? It may be possible to fit separate regression lines associated with each level.

## 3.3 Testing Models of the Internal Structure of Cities

In the foregoing section cities and towns were discussed in terms of point features in the landscape. In this section, attention is centred on the internal structure of towns and cities.

### 3.3.1 *Models of Urban Land Use*

Although there are a number of studies which fall under the heading 'urban land use' it is possible to distinguish two broad approaches:

The ecological

The economic: gradient studies.

The ecological models
These models are perhaps the more well known and include the formulations of Burgess, Hoyt and Ullman and Harris (for discussion of these models, see Carter (1981), Daniel and Hopkinson (1979) and Bradford and Kent (1977)).

In examining the growth and development of a number of American cities ( and Chicago in particular), Burgess adapted a number of concepts used by plant ecologists in their studies of plant associations (competition, invasion, dominance, succession). Thus, people compete for limited space. The more desirable locations for homes and businesses are acquired by those people best able to pay. This, Burgess suggested, leads to functional zoning and residential segregation. Within different areas of the city different functions form the dominant element. The zones are arranged concentrically around the city centre each being distinctive in age and character.

Burgess's zonal model was modified in the late 1930's by Homer Hoyt. He based his ideas on the same assumptions as Burgess with the exception of the transport factor. Thus the chief determinant of housing patterns in cities was still the ability of the wealthy to pay high rents; however, the influence of transport networks meant that the dominant feature was sectoral in nature, rather than distance from the city centre. (Hoyt acknowledged however that zoning may occur in particular sectors).

In 1945 Harris and Ullman proposed a model of urban structure which contrasted sharply with existing ideas. Whereas Burgess and Hoyt both envisaged zones developing outwards from a single centre, Harris and Ullman suggested new distinctive land use arrangements which develop around a number of other nuclei in addition to the C.B.D.. The number of nuclei depends on the size of the city, while the nuclei themselves may range from suburban shopping centres to small villages. Harris and Ullman suggested four factors which contribute to the development of functional areas:

(1) Specialized requirements of certain activities

(2) Tendencies for like activities to group together

(3) The repulsion of some activities by others

(4) Differences in the ability of various activities to pay rent and rates.

Harris and Ullman argued that these influences would condition the emergence of separate areas. These areas, they argued, would not be zonal or sectoral but cellular in character in which distinctive forms of land use would develop around certain nuclei.

The foregoing models were all developed in a North American context. In 1965 Mann produced a model, based on observations from Nottingham, Sheffield and Huddersfield, which he felt was applicable to a medium-sized British city which incorporated elements of both Burgess's and Hoyt's models. The main features of Mann's model are:

(1) A C.B.D.,

(2) surrounded by a zone in transition which is most apparent on the sides of cities leading to the more middle-class residential areas.

(3) From the C.B.D. extend sectors of different social status where industry is associated with the 'lower working-class sector'.

(4) The prevailing wind is from the South West, and, therefore, the most expensive residential sectors are in the west and industrial sectors in the east.

(5) The sectors of working-class houses are likely to be more extensive or numerous than those of middle-class houses as the working-class population outnumbers that of the middle-class by 3 : 1.

The concentric element in Mann's model is principally a reflection of houses of a particular age rather than a particular type. Thus houses become preogressively newer the farther they are from the C.B.D. Both Council and private houses will probably occur in Zone 4 but their distribution is likely to reinforce the class sector which already existed

## *Project Outline: The Ecological Models*

Objective
To assess the validity of the various ecological models to describe the distribution of land-use within a particular city/town.

Data
The relevant data for this project is in map form. Three major sources exist:

(1) Land use maps of Britain (1: 25000)

(2) The O.S. 1:50000 land use overlay used in conjunction with the relevant topographic map of the area.

(3) District Planning Authority maps (1: 10560).
By law Local Planning Authorities produce plans and maps for future development. These maps are always available for inspection and provide information on industrial, commercial, retail and residential functions.

Method
For each grid space on the map estimate the amount of land (in percentages) devoted to specific functions. Suggested functions are:

industrial, residential, retail, other.

For each land use category plot the relevant percentage at the centre of the grid square. An isoline map can now be drawn showing the spatial variation in land use.

Discussion
Do the distributions approximate the patterns suggested by any of the ecological models? Are combinations of features from different models found? If not, what factors may account for the deviation between the actual and predicted patterns?

Clearly such models were developed many years ago and, therefore, it may be appropriate to complement direct analyses of them by looking at the influence of more recent changes such as the establishment of light industrial estates and the growing influence of hypermarkets.

## Gradient Studies

These studies are firmly based on micro-economic theory and are essentially deductive in character. The fundamental idea concerns locational rent or bid-rent. Stated simply this means that as remoter locations are brought into use, central sites, which are more accessible, increase in value because transport costs are minimized. Locational rent therefore refers to the price potential users are prepared to pay for accessibility. Patterns of land use within cities are arrived at through the competition of functions for favourable locations. The use that can extract the greater return from a given site will be the successful bidder. Figure 3.4 indicates the pattern which emerges as a result of these processes, giving rise to an orderly pattern of land use arranged concentrically. Thus, the central locations are occupied by retail users and to a lesser extent by offices. In contrast industrial and residential uses place less value on centrality.

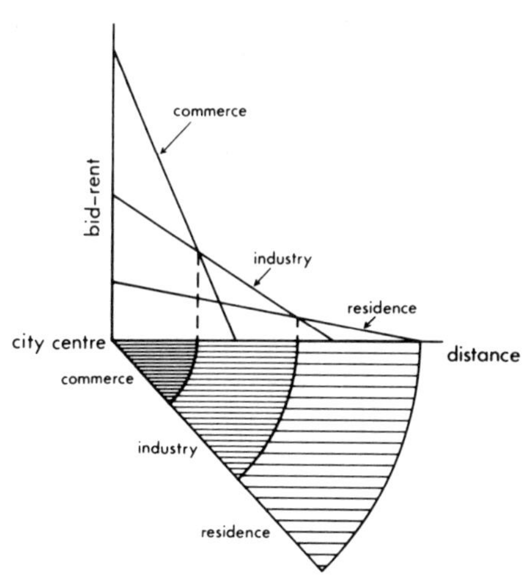


Figure 3.4 BID RENT ANALYSIS
(Source: Bradford and Kent, 1977)

A second, related, group of studies focuses on urban density gradients. Colin Clark, in a study of 36 cities throughout the world, found that population densities decline according to their distance from the city centre. The form of this decline is shown in Figure 3.5(a) and reveals a negative exponential relationship, that is, population densities decrease at a slower rate the farther the area is from the centre. Figure 3.5(b) reveals that if the natural logarithms of population densities are taken the relationship can be drawn as a straight line. The major reason for this relationship revolves around the idea that poorer households have steeper bid-rent curves than their richer counterparts. This means that the wealthy city dwellers are able to afford high commuting costs and live at lower residential densities on the urban fringe. In contrast the poorer households cannot afford the transport costs associated with suburban living and live at higher densities in inner city areas. It has been observed that, as cities grow they experience a decline in population density gradients (see Figure 3.6).

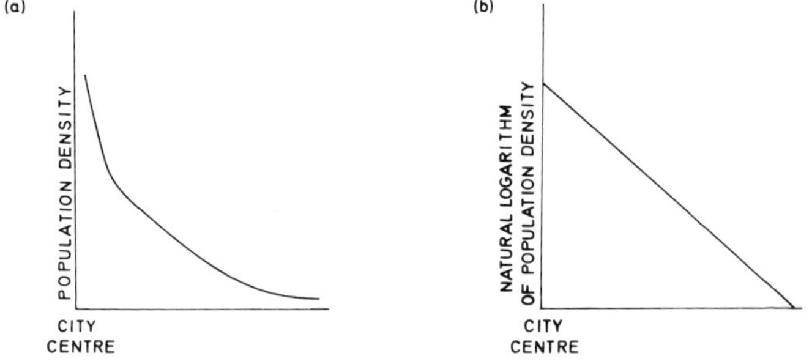


Figure 3.5 POPULATION DENSITY GRADIENTS

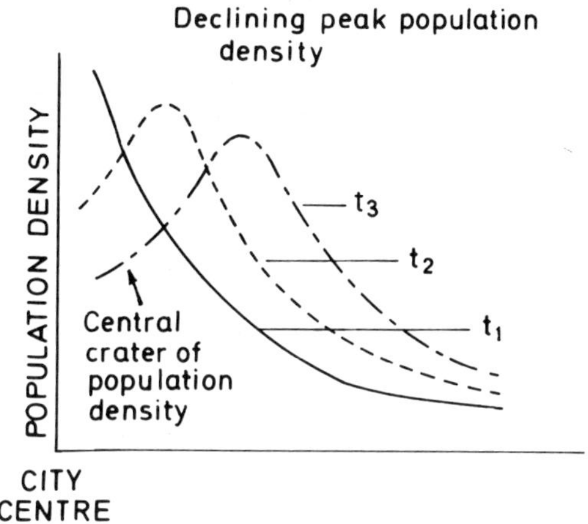


Figure 3.6 POPULATION DENSITY GRADIENTS: CHANGE OVER TIME

Two reasons have been suggested for this decline:

(a) Urban renewal has resulted in a reduction in the density of population in inner city areas.

(b) An increase in higher density housing estates in the suburbs as a consequence of a steep rise in land values. This latter feature is a function of a greater demand for lower density housing and a restriction on the amount of land available for building.

*Project Outline: Population Density Gradients*

Objective

(1) To examine the population density gradient of a particular town or city.

(2) To relate this to other gradient theories based on the idea of bid-rent.

Data

(1) Population density (persons per hectare) for each ward within the administrative boundary of the urban area. This information can be obtained from the county planning

department population censuses.

(2) The distance from the centre of each ward to the city/town centre.

Method

(1) A useful preliminary is to draw a choropleth map of the population densities.

(2) On semi-logarithmic graph paper construct a graph of the population density gradient. A negative exponential relationship will appear as a straight line.

(3) Another possible method to examine the relationship is regression analysis, although the usefulness of this method will depend on the ability of the student. However, the application of this technique will provide valuable additional information.

(4) Time and data permitting examine population density gradients for other year's (e.g. 1961, 1951, 1971).

Discussion

How does the observed pattern of population densities compare with what might be expected?
Account for any differences.
How does the population density gradient relate to ideas of bid-rent?

### 3.3.2 *Models of Residential Differentiation*

Recent analyses of the internal structure of cities have been increasingly focused on the way in which areas in the city are differentiated on the basis of residential characteristics. This field of study has become known as social area analysis. In general three basic factors have been identified:

Socio-economic status

Family life-cycle

Ethnic character.

Socio-economic status

This is related to a household's income, occupation and education of the head of the household. When these variables are mapped in the modern city a sectoral pattern emerges (see Figure 3.7).

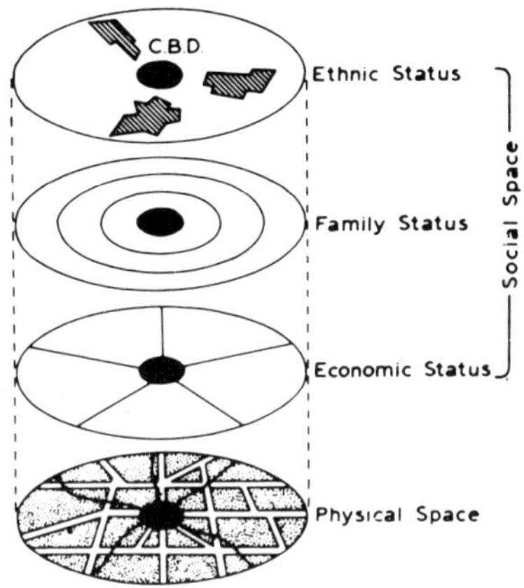


Figure 3.7 SOCIO-ECONOMIC STRUCTURE OF A CITY
(Source: Carter, 1981)

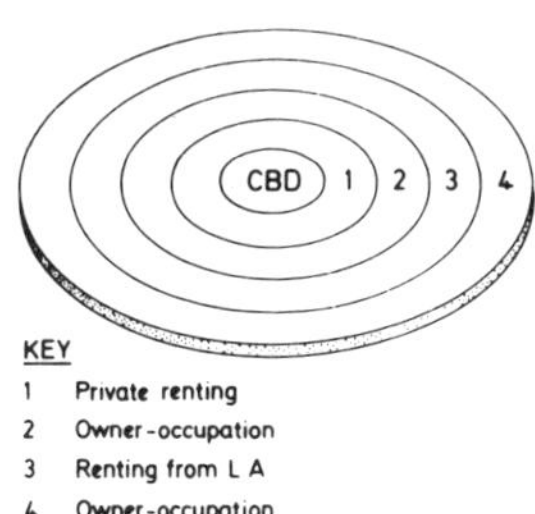


Figure 3.8 TENURE TYPES IN BRITISH CITIES
(Source: Short, 1980)

Family life-cycle

Different stages in this cycle include a young married couple without children, one with a young child, one with teenage children, one where some of the children have left home, and finally a retired couple. In general, family life-cycle varies zonally with the young families living more on the periphery of the city.

Ethnic character

This is related to race, nationality and religion. The tendency is for ethnically distinctive households to cluster in one or more areas near the city centre rather than zonally or sectorally. However, these ghettoes tend to grow outwards along one sector.

In the British context this spatial patterning is somewhat distorted by a number of factors. The most important of these is the influence of public sector housing particularly after World War II. Figure 3.8 is a generalized model of tenure types in British cities. The tendency is for pre-World War II owner-occupied properties to be sandwiched between private renting and public housing areas. Post-war owner-occupiers are found in small areas in the city and, more importantly, in suburban locations.

## *Project Outline: Residential Differentiation*

Objective

To determine whether residential areas in a town/city are spatially differentiated on the basis of socio-economic status, age and ethnic character.

Data

The data required for this project can be obtained from County Planning Authority. The relevant information is:
1971 or 1981 census:
small area statistics (Ward library) 100% population, 100% households and 10% sample.

A map showing the location of wards will also be required.

Method

(1) Select indicators representative of the three differentiating factors, the following are suggested:

a. Age - % of the population aged 0 - 14.

b. Distribution of socio-economic groups 1 - 5 (percentages)

c. Ethnic - % of the residents born in the New

Commonwealth.

(2) Map each variable separately and examine the resultant pattern.

(3) Test whether or not there is a significant difference between identifiable sectors and zones in towns of socio-economic status and age respectively. The appropriate statistical test is the one-way analysis of variance or F - ratio. It is not usually possible to test the distribution of ethnic groups in the same way, however, it should be immediately apparent if they are clustered or restricted to one particular sector or zone.

(4) Attempt a simple classification of Wards based on each pair of variables (bivariate scatter graph) or the three variables acting jointly (triangular graph).

Discussion

Relate the results to models of residential differentiation.

## 3.4 Perception Studies: The Image of the Urban Environment

### 3.4.1 *Introduction*

One of the major developments in human geography during the 1960's was the emergence of a variety of approaches which may be grouped under the term 'behavioural geography'. Initially, the behavioural approach represented a reaction against theories based on the notion of 'economic man' (profit maximization, and so on). Geographers who adopted the behavioural perspective regarded such notions as simple and mechanistic, and wished to incorporate into their explanations the complexities of human behaviour. The widespread adoption of behavioural approaches in geography led some writers to suggest that a 'behavioural revolution' had occured in the subject. However, it is doubtful whether the behavioural approach represented a revolution in the true sense of the term, rather it was complementary to existing approaches.

### 3.4.2 *Perception Studies in Human Geography*

The first developments in human geography involved a series of investigations into the human response to natural hazards, such as floods, drought, snow. The emphasis in these studies was on the examination of how an individual or group of individuals perceived the natural hazard and how this affected their ability to deal with the hazard. Another avenue of research involved the comparison of actual behaviour with the expected behaviour of 'economic man'. In a sample of farms in Central Sweden, for example, it was found that profit maximization was not achieved by farmers and it was concluded that the farmers were 'satisficers' rather than 'optimizers'. Graves (1980) summarised the main areas as:

> '...perception and environmental quality, hazard perception, urban images, perception from certain routes, perception of barriers, micro-areas and personal space, perception of far places, and lastly preferential perception.'

In a widely available text, Gould and White (1974) consider mental maps at a different spatial scale; students ranked the counties of Britain in terms of their locational preferences.

The central feature of the studies noted above is the distinction between the 'objective' environment and the sub-jective' environment. The former term referring to the real world which exists outside the heads of individuals, while the latter is the environment as perceived by man. The central focus of attention is the image, insofar as behaviour depends on the picture of the world we carry in our heads.

Figure 3.9 summarizes the major stages in the formation of the image. The real world sends out a vast number of signals which are perceived by the individual through his senses. However, what is actually perceived is highly selective and depends on a number of personal and cultural factors. Thus only a small part of the information is assimilated. The

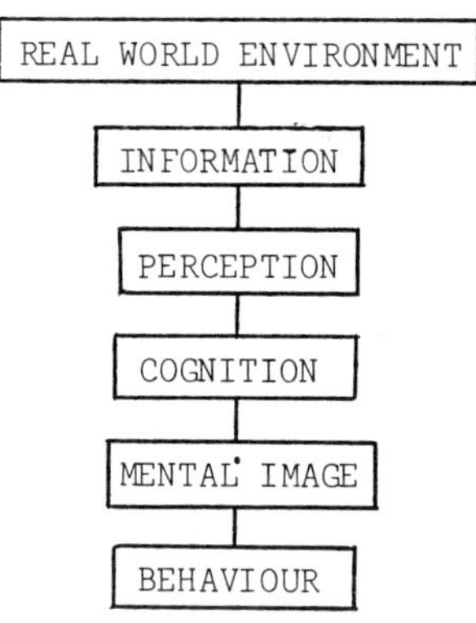


Figure 3.9 ELEMENTS IN THE FORMATION OF THE IMAGE

cognitive stage of the process means that once the information is received it is sorted and organized in the brain so that it fits in with other knowledge and values of the person. The final product is the image which is the mental impression of the world developed over time by individuals through their every day contacts with the environment. Behaviour, therefore, does not depend on the real world as such but on what people think it is; as a result the location of human activities is greatly influenced by geographical images and 'mental maps'.

Although each individual creates his own image, people of the same age, sex, culture perceive reality through similar filters of socialization and experience; there exists, therefore, a broad measure of agreement within particular social groups. As one writer has noted: 'These group images are vital if an individual is to operate successfully in his environment and co-operate with others.'

### 3.4.3 *Perception Studies in the Urban Context*

A growing body of work in urban geography is concerned specifically with how individuals and groups establish images of the city. The translation of these images into a spatial or geographic framework is known as geographical space perception or 'mental maps'. Three broad approaches to geographical space perception have been proposed:

Structural Approach

This focuses on the way in which information about a place is perceived. Clearly, not all the information which the environment sends out can be remembered. People, therefore, structure this information by a process which involves selection and ordering. Mental maps are constructed in which useless detail is omitted, leaving that information which is only necessary for the purposes of the individual, such as finding the way from A to B.

Evaluative Approach

This is not only concerned with the way the environment is structured, but also with the way this affects behaviour and decision-making. For example, the places to which people can migrate are limited to their perception and knowledge of alternative locations.

Preference Approach

If individuals and groups have some particular objective in mind, it is likely that part of the evaluation process will involve them in assessing the relative attractiveness of spatial phenomena, that is they will rank them on a scale of preference. Typical studies include the ranking of counties in terms of residential desirability and towns for various shopping purposes.

### 3.4.4 *Images of the Urban Environment*

In the remainder of this section attention will be focused on the structural approach. This approach is more easily adapted than either the preference or evaluative approach, and would provide the student with a feasible and manageable project.

Methods of Data Collection

The most commonly employed include the following:

Use of sketch maps
Verbal lists of distinctive features
Directions for making specific trips in cities
Informal questions about orientation in the city

The use of sketch maps is the obvious method, however, there are certain drawbacks. The most important is that the investigator can never be certain whether the sketch map reflects the respondent's cartographic skill rather than a true representation of the mental image. In an actual study it is wise to use a variety of methods.

The earliest and most influential structural study was by Kevin Lynch in a book entitled *Image of the City* (1960). The aim of this study was the examination of the visual quality of three American cities (Jersey City, Boston and Los Angeles). Lynch's central theme involved the idea of 'imageability, studying the maps drawn by individuals from memory. This simply describes that quality of cities which makes them memorable, and evokes strong images in most observers. On the basis of field studies and interviews Lynch suggested that five elements formed the structural basis of the mental image of the physical townscape (which could be compared with reality):

1. Paths These are the predominant image, and are the channels along which people move within the city, for example, main streets and pedestrian walkways.

2. Nodes These refer to certain points in the city which stand out as nodes of foci, for example, significant road junctions, squares, roundabouts and crossings. Piccadilly Circus, Times Square and Red Square are all nodes. Nodes represent easily identifiable stages of movement within the city.

3. Landmarks These are also observable point references. Unlike nodes, however, they cannot normally be entered or passed through. Characteristic landmarks are distinctive buildings, statues, signs or shops (for example, Eiffel Tower, Big Ben, Harrods, Statue of Liberty).

4. Districts These are areas with a common character which are immediately identifiable to the inhabitants and often

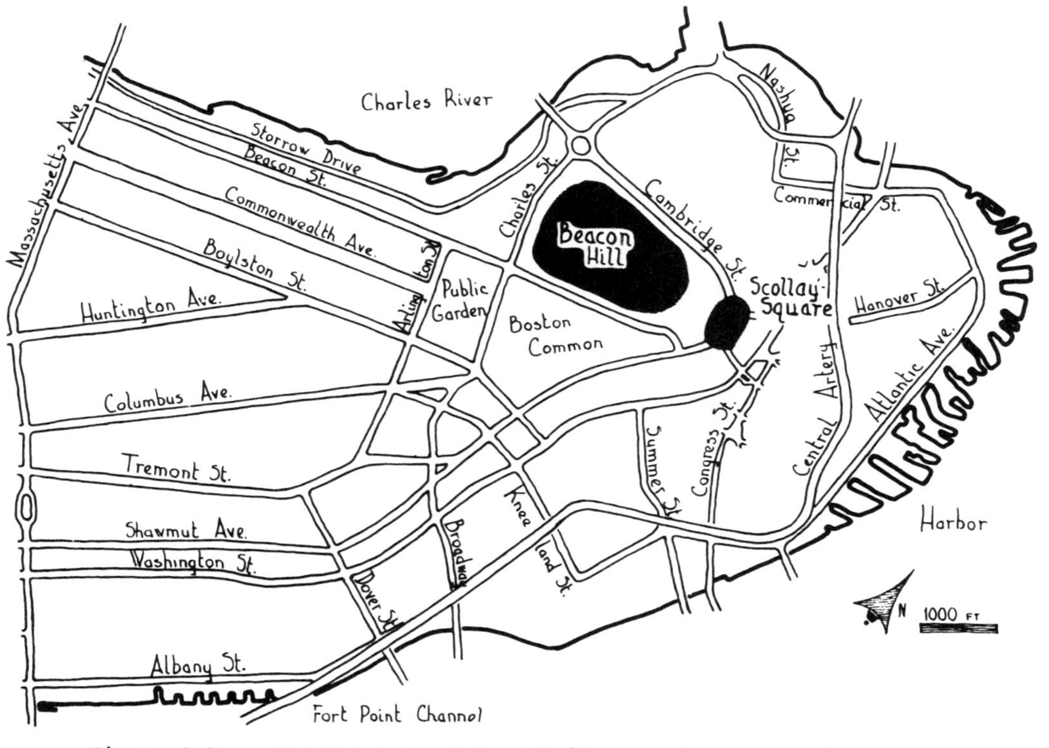


Figure 3.10a OUTLINE PLAN OF BOSTON (Source: Lynch, 1960)

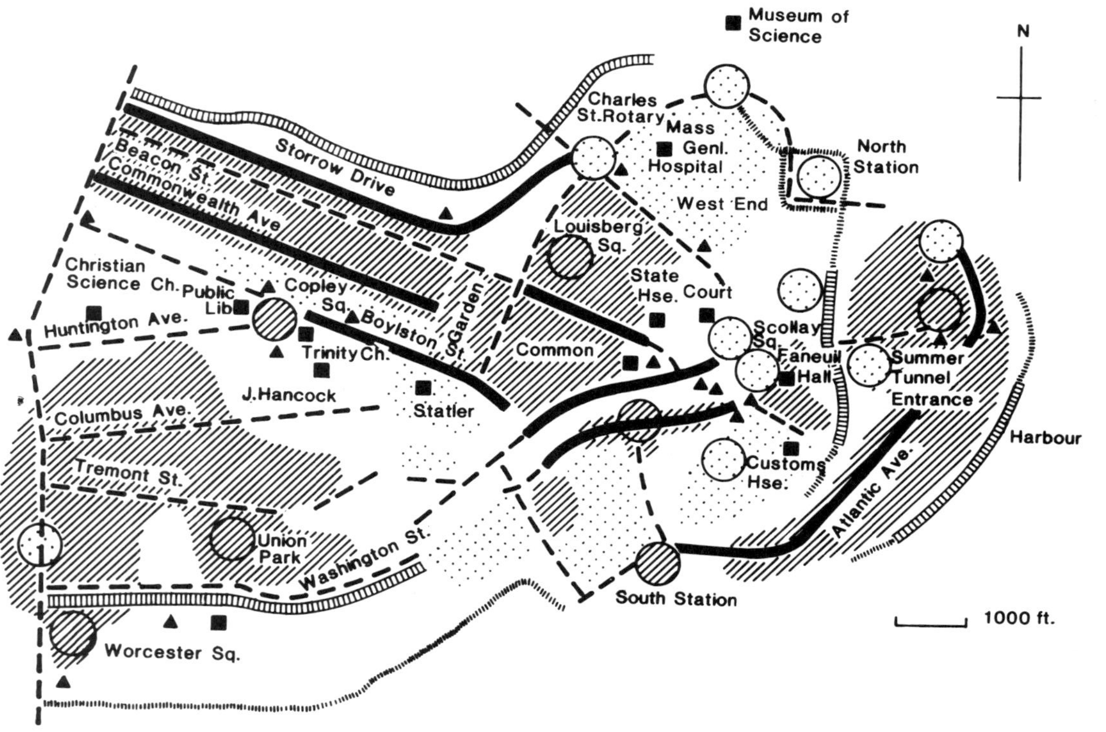


Figure 3.10b VISUAL FORM OF BOSTON (Source: Lynch, 1960)

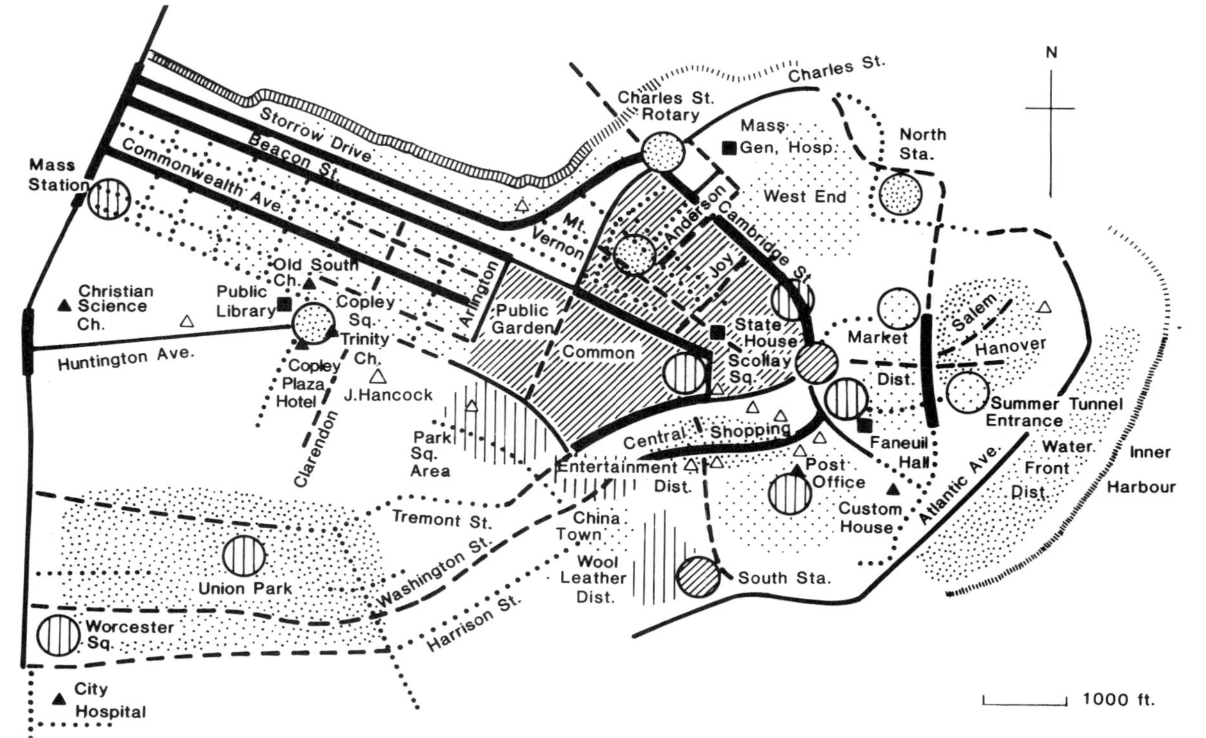


Figure 3.10c IMAGE OF RESPONDENTS BY VERBAL INTERVIEW (Source: Lynch, 1960)

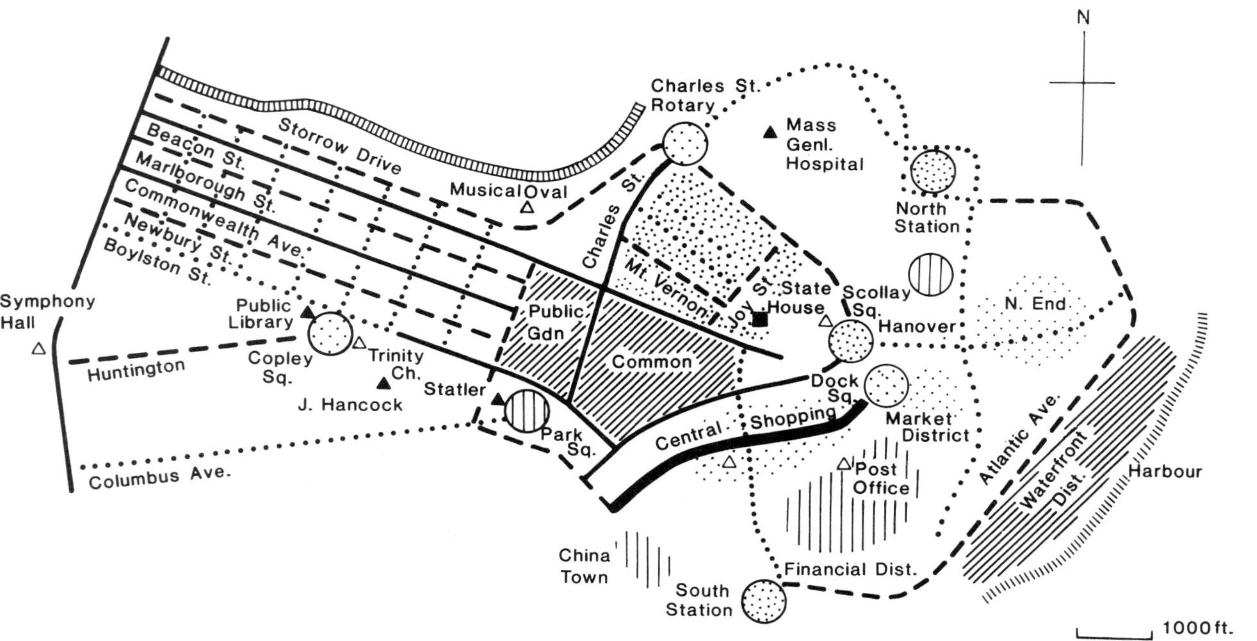


Figure 3.10d IMAGE OF RESPONDENTS BY SKETCH MAPS (Source: Lynch, 1960)

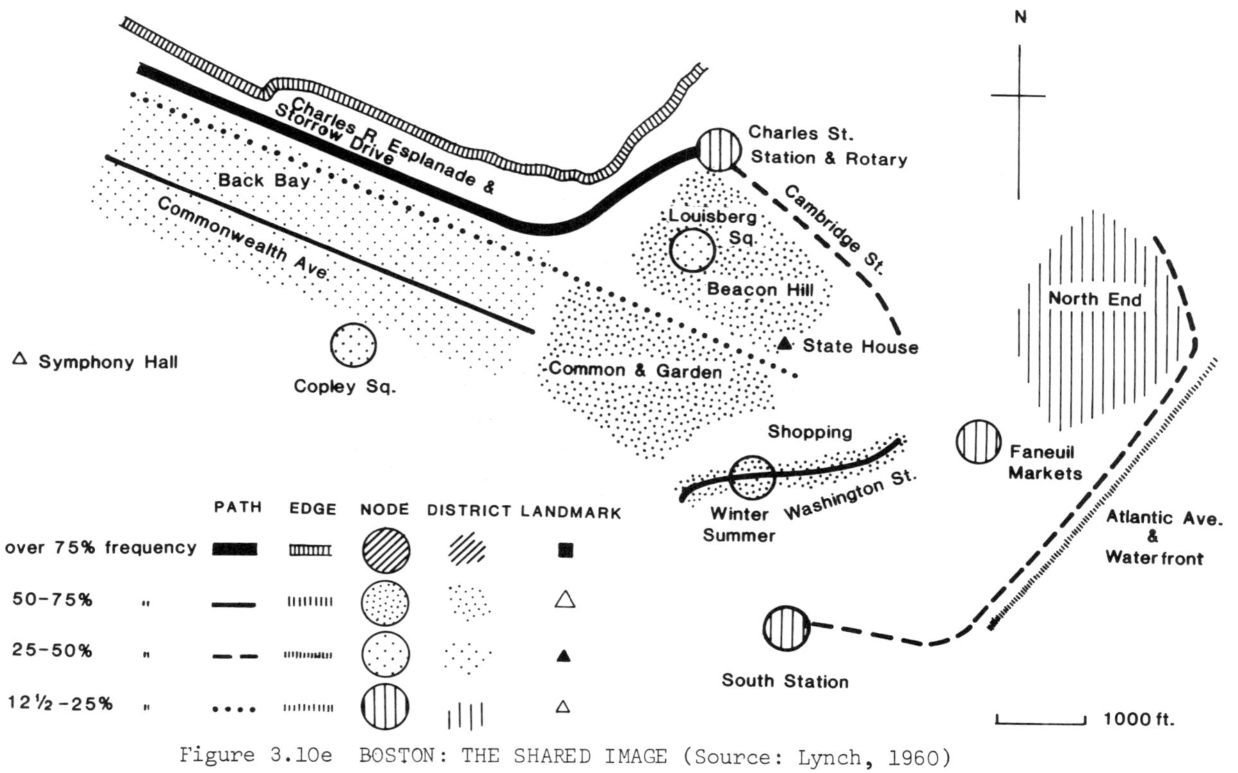


Figure 3.10e BOSTON: THE SHARED IMAGE (Source: Lynch, 1960)

have local names, for instance, an old decaying industrial area, a 'red light' district, a city park.

5. Edges These are linear elements which represent distinctive physical breaks within the city. These can restrict movement or mark the edges of known territory. They can be natural features (for example, breaks of slope, sea or lake shorelines, views) or man-made (railways, canals, urban motorways).

Figure 3.10a - 3.10e show the results of Lynch's work in Boston:

3.10a Outline plan of Boston

3.10b Visual form of Boston as recorded by Lynch

3.10c Image of respondents by verbal interview

3.10d Image of respondents by sketch maps

3.10e The Boston known by most respondents - the shared image.

Recently, Lynch (1977) has developed such work further in an international study of the way in which adolescents use and value their spatial environment.

Some writers have questioned the validity of the five elements proposed by Lynch and suggest that they are dependent on the sort of questions asked. In their perception, do individuals use sequential features (such as roads) and/or spatial features (such as landmarks) for structural referencing in their maps? Pocock feels that a simplified classification may be more useful, consisting of three types of element:

Point features
Linear features

Areas (see Table 3.1)

### 3.4.5 *Example: Durham: Image of a Cathedral City*

Pocock's original study was a detailed and wide ranging examination which focused not only on the physical but also the symbolic aspect of the image. Some of the questions contained in the sample questionnaire (Table 3.2) are therefore not immediately relevant in terms of the physical image (question 8, 9, 10, 11). However, they do provide additional information which the student could use in discussing the general nature of the image.

| | | | | | | | |
|---|---|---|---|---|---|---|---|
| Point Features | A. | Buildings | Ecclesiastical | Linear Features | C. | Linear | Streets |
| | | | Historical | | | | Rivers |
| | | | Civic | | | | Paths |
| | | | Education | | | | |
| | | | Social | Areas | D. | Areas | Nodes |
| | | | Business-commercial | | | | Districts |
| | | | Transport | | | | Open Spaces |
| | B. | Landmarks | Viaducts | | | | |
| | | | Bridges | | | | |
| | | | Monuments | | | | |

TABLE 3.1 THE STRUCTURAL FEATURES OF THE URBAN IMAGE (Source: Pocock, 1975)

TABLE 3.2. THE IMAGE OF THE CITY: A SAMPLE QUESTIONNAIRE

1. County and city of birth:

2. Years lived in (name of County):

3. Sex: Male/female

4. Age group:

5. Occupation (or, if more relevant, last occupation or husband's occupation):

6. If you close your eyes and think of the (name of town or city) what particular picture comes into your head first of all?

7. What other sights or landmarks (sounds, smells or feelings) do you consider characteristic of (name of town or city)? (Good or bad).

8. How attractive as a place to live do you rate the town/city?

| | |
|---|---|
| Very Attractive | Unattractive |
| Attractive | Very Unattractive |
| Don't Know | |

9. How attractive as a place to live would you generally rate the County?

| | |
|---|---|
| Very Attractive | Unattractive |
| Attractive | Very Unattractive |
| Don't Know | |

10. If you had to leave (name of town or city) and live elsewhere, how would you feel?

Very Sad
Sad
Neutral
Happy
Very Happy

11. If you left, what things about the town/city would you particularly miss?

12. Do you think (name of town or city) can be divided into any separate or distinctive parts? If so, could you give a list of the names, or of the name of a street in each area, or their rough position - anything to help me know where each is?

13. Mental map of town/city. Imagine you have a friend about to visit (name of town/city) for the first time, and is anxious to make the most of his/her limited time. You are therefore requested to make a drawing or sketch map of the town/city showing both:

a) Those things which will help him get around.

b) Those things which you think he ought to see.

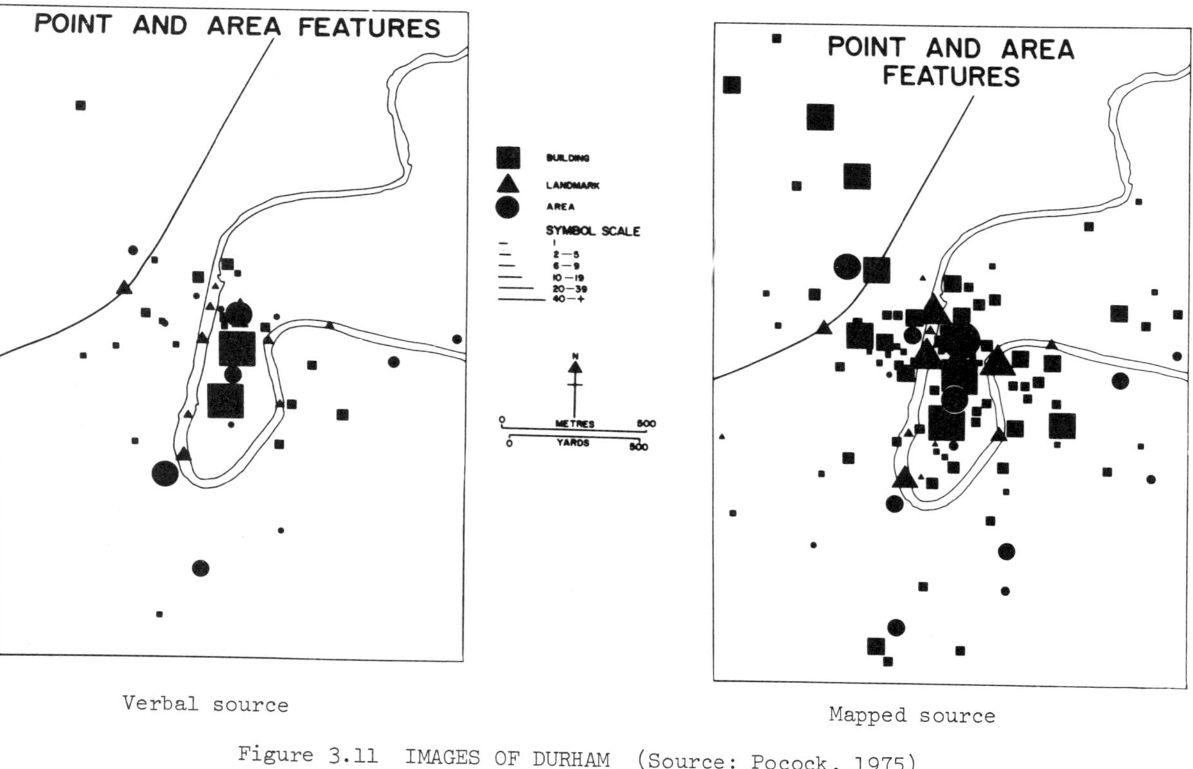


Verbal source

Mapped source

Figure 3.11 IMAGES OF DURHAM (Source: Pocock, 1975)

| | | | Verbal Features | | Mapped Features | |
|---|---|---|---|---|---|---|
| | | | No. of Different Places | Total Responses | No. of Different Places | Total Responses |
| Point Features | A. Buildings | Ecclesiastical | 4 | 70 | 12 | 124 |
| | | Historical | 1 | 57 | 7 | 88 |
| | | Civic | 4 | 12 | 10 | 111 |
| | | Education | 3 | 6 | 21 | 92 |
| | | Social | 4 | 11 | 12 | 128 |
| | | Business-commercial | 4 | 9 | 29 | 102 |
| | | Transport | 2 | 4 | 3 | 77 |
| | | | 22 | 169 | 94 | 722 |
| | B. Landmarks | Viaducts | 1 | 19 | 1 | 14 |
| | | Bridges | 6 | 31 | 6 | 228 |
| | | Monuments | 3 | 31 | 9 | 38 |
| | | | 10 | 81 | 16 | 280 |
| Linear Features | C. Linear | Streets | 12 | 57 | 62 | 1234 |
| | | Paths | 1 | 1 | 4 | 25 |
| | | River | 1 | 54 | 1 | 65 |
| | | Railway | - | - | 1 | 20 |
| | | | 14 | 112 | 68 | 1344 |
| Areas | D. Areas | Nodes | 2 | 46 | 2 | 98 |
| | | Districts | 8 | 17 | 8 | 44 |
| | | Open spaces | 5 | 50 | 10 | 63 |
| | | | 15 | 113 | 20 | 205 |

TABLE 3.3 DURHAM: VERBAL AND MAPPED IMAGES
(Source: Pocock, 1975)

| | Number of Subjects | Mean No. of Elements | Percentage Distribution of Major Categories | | | |
|---|---|---|---|---|---|---|
| | | | Buildings | Landmarks | Linear Places | Areas |
| Males | 42 | 36.5 | 25 | 12 | 56 | 7 |
| Females | 39 | 32.9 | 26 | 9 | 58 | 7 |
| Age < 40 | 27 | 37.9 | 26 | 10 | 56 | 8 |
| Age > 40 | 54 | 33.2 | 25 | 10 | 58 | 7 |
| Resident < 10 years | 20 | 41.9 | 22 | 11 | 60 | 7 |
| Resident > 10 years | 61 | 32.5 | 27 | 10 | 56 | 7 |
| Durham origin | 52 | 32.3 | 27 | 10 | 56 | 7 |
| Non-Durham | 29 | 39.2 | 23 | 11 | 59 | 7 |
| Middle class | 50 | 39.1 | 26 | 10 | 57 | 7 |
| Working class | 31 | 25.7 | 24 | 8 | 58 | 7 |
| Attracted to city | 70 | 36.6 | 26 | 10 | 67 | 7 |
| Not attracted | 11 | 23.3 | 26 | 9 | 60 | 6 |

TABLE 3.4 MAPPED IMAGE : GROUP DIFFERENCES (Source: Pocock 1975)

As noted above the location of Pocock's study was the Cathedral City of Durham. One hundred people were selected from the electoral register to take part in the survey, 94 completed questionnaires were returned. Two methods were used in the collection of data:

1) Verbal lists of distinctive features (questions 6, 7, 12).

2) Mapping technique (questions 13).

Either or both of these techniques could be used as the basis for a project.

Table 3.3 summarizes the responses to both the verbal and mapping technique. The figures in the table refer to the number of different places mentioned and the total responses. For example, 4 ecclesiastical buildings were recorded and these were mentioned 70 times by the respondents. In terms of both the number of places recalled and frequency of occurrence the map technique is far more detailed. The contrast is due, essentially, to the greater emphasis on streets and buildings in the mapped image and as a result mainly of the technique and the wording of instructions. The image of Durham as derived from the two techniques is cartographically portrayed in Figure 3.11. The verbal image is clearly dominated by the peninsula, with the river and its bridges, cathedral and castle facing on to Palace Green, and the Market Place with its landmarks. The only significant extension from the core is North Road leading to the viaduct. In the mapped image the peninsula is still the focal point, however, the extensions are both more numerous and developed particularly to the north and south. Finally, it might be of interest to examine the differences between various social groups. The type of information that may be useful is given in Table 3.4. This table presents the main differences in the mapped image of Durham for a number of different groups. In general the extent of the image (mean number of elements) shows clear differences with class, attraction, length of residence and to a lesser extent, origin. In terms of specific elements, origin and length of residence are related to differences in buildings and linear features. While the sex variable is associated with the greatest contrast in the relative importance accorded to landmarks.

## 3.5 An Introduction to Time - Geography

### 3.5.1 *Background*

The human geographer's traditional concern has been with location in space. Attention has focused not only on absolute location (e.g. latitude and longitude, geographical territory) but also relative location (in terms of cost, accessibility, social interaction). Until recently however, little attention has been paid to location in time, except in a purely static or 'snapshot' sense. This is somewhat surprising insofar as time and space form the backcloth of all our activities (for example, times of lessons, shop opening hours, working hours, etc). Time, like space, can be treated in absolute or relative terms. Absolute time refers to time as measured objectively by a calendar or clock. While relative time is determined in a subjective manner ('too soon' or 'too long', for example). Intuitively, we frequently combine space and time in everyday occurences. For example, in travelling from place A to place B we often substitute time for distance - which varies with the mode of transport.

The main objective of time-geography is the explicit incorporation of the time dimension into human geography. Time-geography, dates from approximately 1970 and is associated with a Swedish geographer called Torsten Hägerstrand. This development has been considered to be by some the major innovative development which occurred in human geography during the last decade. As with any new field of study, time-geography has its own vocabulary, and, although some of the terms are unfamiliar, they should not take long to master. (For a graphic illustration of some of the ideas see Figure 3.12).

### 3.5.2 *Some basic concepts*

1. Time-geography employs an absolute time and space framework. That is, location in space is based on grid coordinates while location in time involves the use of clock or calendar time.

2. The focus of study is the individual or group of individuals in a particular environment context (as defined by space and time).

Figure 3.13 illustrates that space-time is derived by the combination of two-dimensional space with the time dimension.

3. Every individual has a goal. The attainment of this goal involves a series of tasks. When these tasks are added up we have a project.

4. Any behaviour requiring movement involves the individual or group traversing a path through space and time.

5. The ability to pursue a project is not a simple procedure because of the existence of constraints. In this sense time and space are seen as scarce resources which can constrain activity. Three types of constraint have been

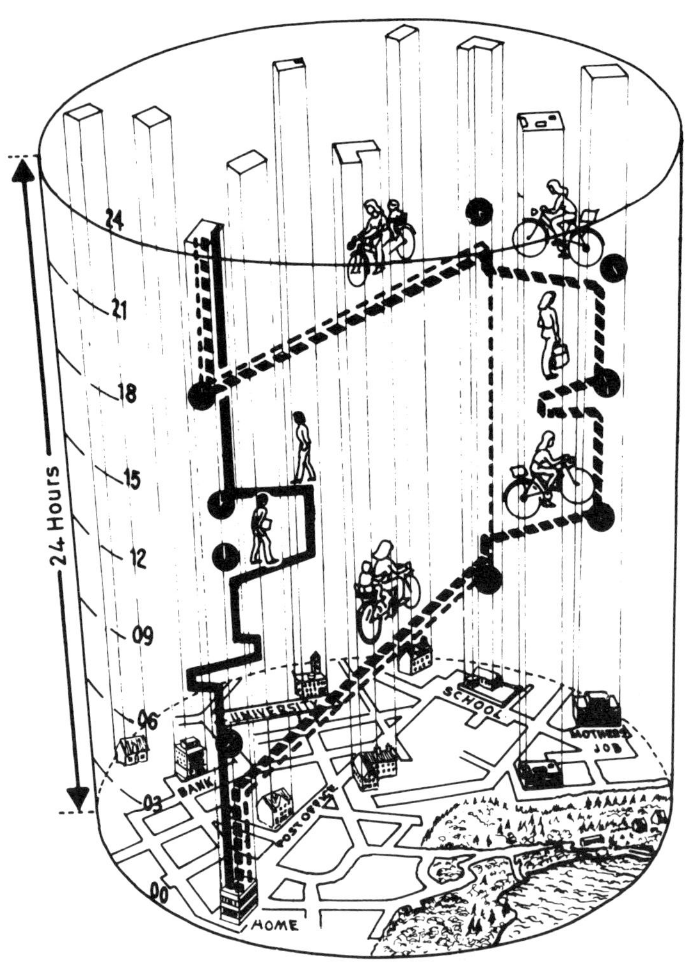


Figure 3.12 A TIME-GEOGRAPHY PERSPECTIVE (from Parkes and Thrift, 1980)

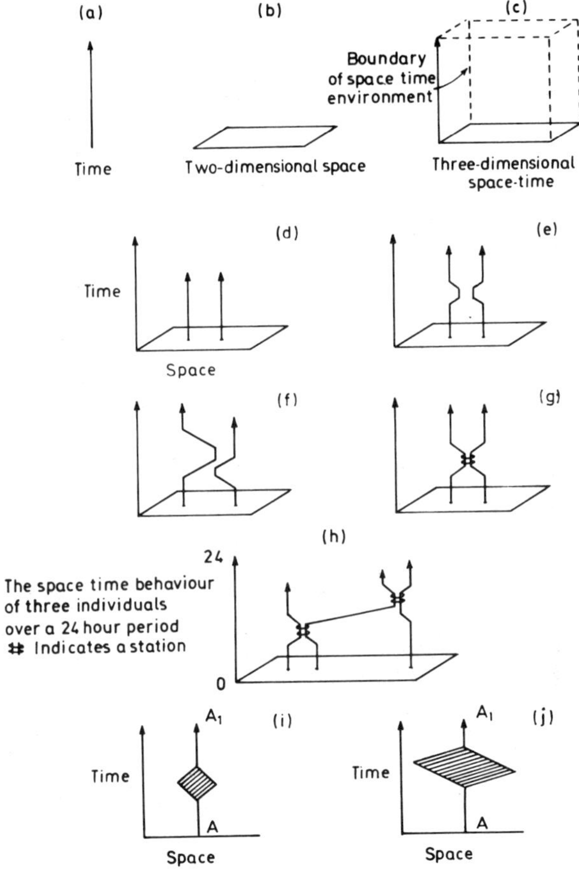


Figure 3.13 TIME-GEOGRAPHY: SOME BASIC CONCEPTS
(Source: Thrift, 1977)

identified:

a) Capability constraints restrict movement in time and space. In time, our movement is constrained by the biological need for approximately eight hours sleep out of every twenty four hours. In space, for example, the extent to which movement can occur is governed by the available means of transport.

b) Coupling constraints require certain individuals and groups to be in particular places at particular times (for example, teachers and pupils in schools).

c) Authority constraints preclude individuals from being in certain places at set times. For example, public houses are 'out of bounds' to certain age groups at all times and 'out of bounds' to all age groups at certain times.

The basic problem for individuals and society is to pack projects into the limited resources of space and time subject to the constraints noted above.

6. Movement of a population in a particular region can be seen as a mass of paths which flow through a set of space-time locations or stations.

A number of these points are illustrated diagramatically in Figure 3.13.

In Figure 3.13d no constraints operate and the paths of the two individuals are simple and straightforward. In Figure 3.13e constraints operate and the two individuals occupy the same location in time but not space. Co-location in space but not in time is indicated in Figure 3.13f, while in Figure 3.13g the existence of coupling constraints means that the individuals occupy the same location in space and time, that is, a station. A more complex example is illustrated in Figure 3.13h.

7. Paths of individuals do not occur in isolation but come in contact as bundles depending on the three constraints noted above and on the location of stations. A household and a school are both examples of bundles, involving individuals congregating at set times and places.

8. Finally, the idea of an individual's daily prism should be introduced. That is the volume of space and time which is within easy reach of an individual within a day. Figure 3.13i refers to an individual who has to return to the same location by a particular time. The mode of transport

is by foot; the steep sides of the prism indicates that the spatial range of the individual is narrow. In Figure 3.13j the mode of transport is by car; the shallow slope of the prism indicates a much wider spatial range.

3.5.3 *Outline Project: Space-time budgets*

Although the application of time-geographic techniques can be quite complex, it is possible to design a relatively straightforward project based on the idea of space-time budgets. A time budget is concerned with the description of activities in time; a space-time budget requires additional information on the location of activities. The activities may be located in territorial space (x and y grid coordinates) or functional space (for example the home, place of work etc.).

The objective of such a project could be simply descriptive, alternatively the focus of attention could be the examination of differences in the space-time budgets of various groups.

The relevant data should be collected in the form of a diary - see Figure 3.14. Information is recorded for a 24 hour period. As illustrated by Figure 3.14 the diary is subdivided into intervals of one hour, but within these intervals specific times of activities should be recorded (preferably to the nearest minute). In summary the diary contains the following information:

1. starting time of a particular activity
2. finishing time of a particular activity
3. location of an activity
4. was the activity carried out with someone else
5. whether a secondary activity was taking place
6. means of transport
7. distance (in miles) from home

In addition a number of background variables should be collected:

1. age
2. male/female
3. marital status - married, married with children, single
4. occupation - note occupation of head of household
   - but also the respondent (include here full-time housewives)
5. length of residence
6. mobility - does the household have a car
   - does the respondent have access to private transport during the day (for example, car, bicycle)

The sample should contain between 30-50 respondents (this

| Time | What did you do? | Time Began | Time Ended | Where | With Whom | Doing Anything Else? | Activity is Away from Home: Means Of Transport | Activity is Away from Home: Distance From Home (Miles) |
|---|---|---|---|---|---|---|---|---|
| Midnight | Sleeping | 12:00 | 7.00 | Home | 0 | 0 | | |
| 1 AM | | | | | | | | |
| 2 AM | | | | | | | | |
| 3 AM | | | | | | | | |
| 4 AM | | | | | | | | |
| 5 AM | | | | | | | | |
| 6 AM | | | | | | | | |
| 7 AM | Got up - let dog in & fed dog | 7.00 | 7.10 | Home | 0 | 0 | | |
| | Showered and dressed | 7.10 | 7.30 | " | 0 | listened to Radio | | |
| | Ate breakfast | 7.30 | 7.45 | " | Wife & 2 sons | " | | |
| | Drove child to bus-stop | 7.45 | 7.50 | In transit | 1 son | " | car | ½ |
| | Drove back from bus-stop | 7.50 | 7.55 | " " | 0 | " | " | ½ |
| 8 AM | Read paper | 7.55 | 8.10 | Home | 0 | " | | |
| | Travelled to Railway Station | 8.10 | 8.15 | In transit | neighbour | talking | car | ½ |
| | Waited for train | 8.15 | 8.22 | " | 0 | 0 | | |
| | Train to City station | 8.22 | 8.55 | " | Work colleague | talking | train | 12 |
| | Walked to office | 8.55 | 9.00 | " | 0 | 0 | walk | 12 |

Figure 3.14 A SPACE-TIME DIARY (Source: Parkes and Thrift, 1980)

would depend on the time available). Although it is difficult to specify the characteristics of the sample in advance, approximately half of the sample should be over the age of 45, similarly the sample should not be disproportionately male or female in composition. In addition it is desirable that the sample contains respondents from a variety of contexts (for example, council estate, private residential estate).

Only those respondents who will be resident in the area during the survey period should be included. It is suggested that respondents are requested to keep a diary of their activities for two 24 hour periods during a week - one on a weekday, and one on a Saturday. Respondents should be reassurred that nothing of a personal or business nature will have to be noted.

The actual activities recorded by the respondents can be reduced in number by classifying them according to the broad categories in Table 3.5. The location of activities is best recorded in terms of functional category space. The latter is preferable to territorial location in that it is easier to equate it with a time location. The following classification is suggested:

1. place of work
2. place of retail services
3. non-retail services (barber, Post Office, bank, doctor, etc.)
4. home
5. entertainment and recreation places
6. other peoples homes
7. in transit

### 3.5.4 *Suggested methods of analysis*

1. The simplest method of analysis is the use of graphs of various types. For example, a bar graph could represent 24 hours. This could then be subdivided according to the length of time each of the activities occupies in the 24 hour period. On this basis it would be possible to compare the bar graphs of different groups of people. Similarly, bar graphs could be constructed illustrating the amount of time spent in specific functional category spaces. Line graphs could be used to illustrate the percentage of times of the day and how this varies for different sub-groups in the sample. Another mode of analysis is illustrated in Figure 3.15. However, instead of focussing on a family, the different 'life lines' could represent selected individuals in different age, occupational or mobility groups for example.

2. Statistical analysis - one of the simplest techniques

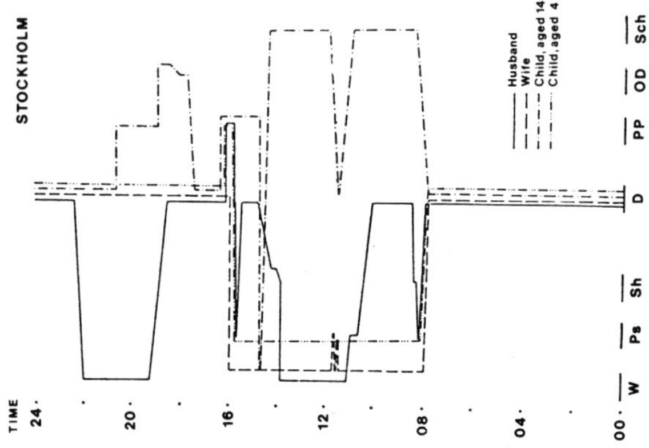


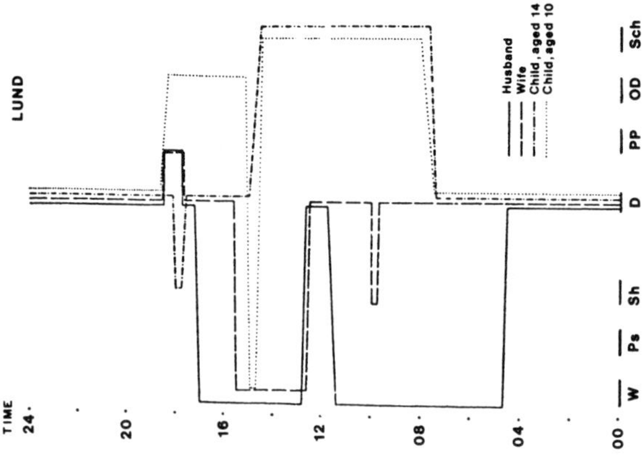


Figure 3.15 LIFELINES: LUND AND STOCKHOLM (Source: Thrift, 1977)

## KINDS OF ACTIVITIES YOU MAY DO DURING THE DAY

(but please use your own words to describe what you are doing)

| | |
|---|---|
| TRAVEL: | All the trips you make, both at home and at work. |
| WORK: | Actual work; work breaks; delays or sitting around at work; work meetings or instruction periods; meals at work; overtime; work brought home. |
| HOUSEWORK: | Preparing meals and snacks; doing dishes; arranging and straightening things; laundry and mending; cleaning house (inside and outside); care of yard and animals; repairs. |
| CHILD CARE: | Baby care; dressing; helping with homework; reading to; playing with; supervising; medical care. |
| SHOPPING: | Groceries,clothes,appliances,or home furnishings; repair shops; other services (for example: barber, hairdresser,doctor,post office). |
| PERSONAL LIFE: | Eating meals and snacks; dressing; care of health or appearance; helping neighbour or friends; sleep or naps. |
| EDUCATION: | Attending classes or lectures; training and correspondence courses; homework; reading for the job. |
| ORGANIZATIONS: | Club meetings or activity; volunteer work; going to church services; other church work. |
| GOING OUT: | Visiting (or dinner with) friends, neighbour or relatives; parties, dances, nightclubs or bars; sports events and fairs; concerts, movies, plays, or museums. |
| ACTIVE LEISURE: | Sports or exercise; playing cards or other games; pleasure trips and walking; hobbies, knitting, painting, or playing music. |
| PASSIVE LEISURE: | Conversations; radio, TV, records; reading books, magazines or newspapers; writing letters; planning, thinking or relaxing. |

TABLE 3.5 CLASSIFICATION OF ACTIVITIES (Source: Parkes and Thrift, 1980)

involves the idea of elasticity. Elasticity is measured by using the coefficient of variation

$$\frac{\text{Standard deviation}}{\text{arithmetic mean}} \times 100 = \frac{Sx}{\bar{x}} \times 100$$

that is, the standard deviation is expressed as a percentage of the mean; the higher the value the greater the variability. Thus, if one wanted to calculate the elasticity of a particular activity, for example, time spent in work, we would add up all the individual times the respondents spend in work (preferably in minutes) and divide through by the number of respondents involved in that activity (not the total number in the sample). The next stage would involve the calculation of the standard deviation, and finally the calculation of the elasticity. An example of such an analysis is presented in Tables 3.6 and 3.7.

TABLE 3.6

Elasticity Values for personal needs activities (obligatory)

| | Week 1 | Week 2 |
|---|---|---|
| Per week | 0.13 (13%) | 0.16 (16%) |
| Per weekdays | 0.15 (15%) | 0.15 (15%) |
| Per weekend | 0.02 (2%) | 0.19 (19%) |

TABLE 3.7

Elasticity Values for leisure activities (discretionary)

| | Week 1 | Week 2 |
|---|---|---|
| Per week day | 1.24 (124%) | 0.87 (87%) |
| Per weekend | 3.39 (339%) | 1.19 (119%) |

The above results are taken from a study which extended over a two week period. Table 3.6 refers to activities which are essentially obligatory (sleeping, eating, etc.). Table 3.7 relates to discretionary activities (entertainment, recreation and so on). Clearly the elasticity values for obligatory activities are much lower than for discretionary activities, as one might expect. Also in Table 3.7 the elasticities vary between weekdays and weekend days indicating the greater variability in time spent in leisure activities on Saturdays and

Sundays. Table 3.7 also reveals an interesting feature, in that Week 1 was the last week of the school holidays which is reflected in the higher elasticities for Week 1 compared with Week 2. Using this simple measure it would be instructive to calculate and compare elasticities for different subgroups in the sample, for example the elderly compared with the young, high social status compared with low social status. It might also be possible to calculate elasticities for distance travelled to different activities.

In discussing the results, the following points may be useful:

1. How does the utilisation of space and time vary between groups?

2. Problems associated with specific groups, the elderly, the young, the unemployed, the immobile, for example.

3. How might these differences relate to time-geography ideas in an intuitive sense? (For example, coupling, capability and authority, constraints, the idea of a 'daily prism').

4. How might these ideas be of use to planners?

## 3.6 Welfare approaches in human geography

### 3.6.1 *Introduction*

During the late 1960s and early 1970s, human geography experienced a change of direction in the guise of social relevance. Within human geography the major stimulus for change was a feeling of disquiet with the methods and techniques associated with the 'quantitative revolution' (which in recent years has increasingly influenced secondary level geography curricula). In particular, the view was expressed that the 'new geography' emphasized methods rather than subject matter and that the overall approach was impersonal and mechanistic in character. Furthermore, many events occurring in the external world (intellectual, social, political and economic) helped to generate a general climate of social concern. Thus, for example, there was a disallusion with the Viet Nam war; a recognition that the prosperity of the previous decades had not been shared by all members of society; a concern for the degradation of the physical environment; and an increased awareness of the extent of social inequality and injustice.

The relevance movement, however, was not a seamless garment, and in a short time a clear distinction between liberal and radical approaches became evident. For the liberals the principal concern is that all members of society should not fall below certain levels of well-being, and are willing to work within the Capitalist system to achieve this end. The radicals, on the other hand, inspired to a greater or lesser extent by the writings of Karl Marx, argue that social justice can only be achieved by the complete restructuring of society, which will also entail a restructuring of the concerns of geography (see, for example, Harvey, 1982).

In the present context, the emphasis is on the liberal contribution. This is not to deny that certain marxist approaches have provided valuable insights. However, it would be difficult to devise a project based on marxist geography which would fit in comfortably with existing 'A' level syllabi.

One of the major contributions of the liberal approach has involved the mapping of patterns of welfare at various spatial scales ( see, for example, Coates, Johnston and Knox, 1977; Smith, 1979). These maps may be used as inputs into planning procedures and as a means of monitoring policy outputs. In general, the geographical perspective

> '...stresses the way in which the <u>spatial</u> structure of the economy and society may work to the advantage of some people in some places and to the disadvantage of others'
>
> (Smith, 1979, p. 18).

However, in examining welfare and inequality there is always the temptation to overemphasise the spatial component. It should always be remembered that the study of welfare and inequality is a multi-disciplinary activity involving sociological, political, economic and geographical perspectives. Indeed, the causes of inequality are principally non-geographical; the geographical contribution rests

> '...on the primary emphasis on where people live as a contributor to differential life experience. The basic descriptive task is to identify these place-to-place variations, which are often overlooked by economists, sociologists and others who are more interested in equality among classes, races or other groups in society.
>
> (Smith, 1979, p. 18).

In the following section, the concern with place-to-place variations is pursued in greater detail through the examination of deprivation at the urban scale.

### 3.6.2 *Urban deprivation*

Urban deprivation is intimately related to the inner-city problem, which has attracted considerable attention from both social scientists and politicians in recent years, and, obviously it has direct spatial characteristics. The essence of urban deprivation can perhaps be best illustrated by referring to the so-called 'Cycle of poverty' (see Figure 3.16). The diagram illustrates the interdependence which exists between unskilled work, low incomes, poor living conditions and poor educational opportunities. This would mean, for example, that people with poor educational opportunities tend to go into unskilled (and perhaps, irregular) work, which consequently affects their income, housing and their children's educational opportunities. In geographical terms, the important feature is that many aspects of poverty and social malaise are spatially concentrated:

> '...people who live in areas characterised by some of these variables, which tend to be the inner areas of cities, have social and physical environments operating against their chances of improving their quality of life'
> (Coates, Johnston and Knox, 1977, p. 179).

It should be noted, however, that the identification of these features is by no means a recent phenomenon. Sixty years ago Burgess referred to the area surrounding the Central Business District as the zone in transition, characterising it as the 'zone of deterioration' or 'zone in decay'. It is precisely in these areas that the inner-city problem manifests itself today.

### 3.6.3 *Measuring Urban Deprivation*

The preceding discussion underlines the important point that the elements of deprivation are very closely interrelated. This situation is frequently referred to as one of multiple deprivation,

> '...the important concept is not that of simple multiplication, but of the functional interrelation of all the aspects of disadvantage'
> (Carter, 1981, p. 308).

The principal concern of urban social goegraphers and administrators has been the operationalisation of the concept of

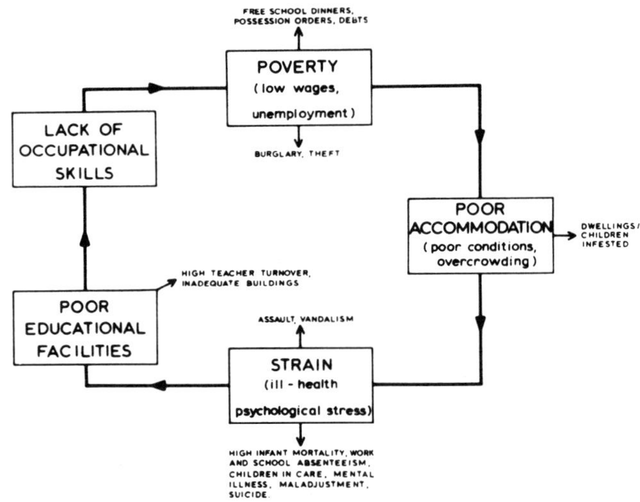


Figure 3.16 CYCLE OF POVERTY
(Source: Coates, Kox and Johnston, 1977)

urban (multiple) deprivation. The problem is essentially twofold. The first problem involves the selection of an appropriate set of indicators and variables. Second, the derivation of a suitable method which combines a number of variables into a single measure of deprivation.

1. Data These studies employ data contained in the Population Census (see Chapter seven), and it is a relatively easy task to define a number of variables which may be used as indicators of deprivation. (See Table 3.8). Clearly, the initial selection of variables determines the characteristics of deprivation that are highlighted.

Variable definition (percentages)

1. Households with more than 1.5 persons per room
2. Households lacking basic amenities (hot water, bath, W.C.)
3. Economically active males unemployed but seeking work or sick.
4. Households without a car
5. Economically active and retired males in socio-economic group 11 (unskilled)
6. Population aged 0-14
7. Population of pensionable age
8. Population of New Commonwealth origin

TABLE 3.8 SELECTED INDICATORS OF DEPRIVATION

The first five items in the Table are fairly straightforward measures of deprivation. Items 6, 7 and 8, considered in isolation, are not indicators of deprivation. However, if the groups defined by these variables are present in significant numbers in a particular area, it is likely that their existence will exacerbate conditions of deprivation.

In many studies, the above type of data is collected for the smallest unit available, that is the enumeration district. Needless to say, this is a very time-consuming process; in terms of school-based projects ward-level data would be sufficiently detailed for the identification of broad patterns of deprivation; and it is usually readily available from the local, main library or local planning offices.

2. Methodology The most straightforward and versatile composite measure of deprivation involves the use of standard scores or standard deviations. In essense, these scores measure the departure of individual observations from the arithmetic mean, expressed in a comparable form. The calculation involves the following

stages (see Appendix I for greater detail on the statistical methods):

a) Calculation of the mean of the set of observations.

$$\bar{x} = \sum_{i=1}^{n} \frac{x_i}{n}$$

were $x_i$ is magnitude of the variable x for observation i

b) The calculation of the standard deviation (Sx) measures the deviation of individual observations from the mean

$$Sx = \sqrt{\sum_{i=1}^{n} \frac{(x_i - \bar{x})^2}{n}}$$

c) The corresponding standard score for any variable's observation $x_i$ is calculated as follows

$$Z_i = \frac{x_i - \bar{x}}{Sx}$$

This gives the result that when $x_i = \bar{x}$, $Z_i = 0$ and when $x_i = Sx$, $Z_i = 1.0$, regardless of the original units of measurement. This means that even if the variables and indicators are measured in different units, they are placed on a comparable scale with a zero mean and unit standard deviation.

d) Finally, if there are m indicators of deprivation, (eight variables were suggested above) then a composite index can be derived for any territory i, $I_i$, as follows

$$I_i = \sum_{i=1}^{m} Z_{ij}$$

where $Z_{ij}$ is the standard score of area i or indicator j.

If this calculation is repeated for all territories a set of composit indicators is derived and could be mapped directly to give a visual impression of spatial variations in the level of deprivation. The application of this simple composite index could provide the basis for an interesting and, potentially, useful project.

In the following example these variables were collected for the seven metropolitan districts which comprise the West Midlands conurbation (see Figure 3.17). In the final table the metropolitan districts were ranked on the basis of their composite scores. High positive scores indicate that an area is deprived in terms of the variables selected. Compare, for example, the scores for Birmingham and Solihull.

*Project Outline: Urban deprivation*

Objective

To identify and analyse spatial patterns of urban deprivation using a simple composite index.

Data

1971 and 1981 census (Small Area Statistics)
Map of wards.
Calculation of Variables suggested in table.

Method

Calculation of the Composite Index
(I) for each ward.

Discussion

1) General comments on the spatial distribution of deprivation.
2. Comparison of the pattern of deprivation with other distributions, for example, population density, ratio of persons to residential land, rateable values.
3. How do the standard models of urban structure (Burgess, Hoyt etc.) help to explain the distribution of deprived areas?
4. How do the identified patterns relate to general patterns of inner-city decline and suburban prosperity?
5. It might be profitable for the student to carry out some primary fieldwork in an area identified as being deprived. For example, a brief examination of the proximity to entertainment facilities, supermarkets, health centres, bus routes and so on may be instructive. While a brief questionnaire survey of residents' attitudes could provide valuable additional information.
6. Finally, some comments should be included on the types of policies which might improve the situation.

| Raw Data (1971 Census) | West Midlands - Metropolitan Districts | | | | | | |
|---|---|---|---|---|---|---|---|
| | 1 | 2 | 3 | 4 | 5 | 6 | 7 |
| 1. % of households with more than 1 persons per room | 13.1 | 12.9 | 8.2 | 11.8 | 7.6 | 11.0 | 12.1 |
| 2. % of households who lack a shower or bath/shower | 12.2 | 8.7 | 6.4 | 9.6 | 1.5 | 8.4 | 9.2 |
| 3. % Econ. active males unemployed | 7.9 | 6.6 | 4.3 | 5.6 | 4.1 | 5.7 | 6.7 |

TABLE 3.9 WEST MIDLANDS METROPOLITAN DISTRICTS - RAW DATA 1971

| Metropolitan District | $x$ | $x - \bar{x}$ | $(x - \bar{x})^2$ | $\frac{x - \bar{x}}{Sx} = Z$ |
|---|---|---|---|---|
| 1 | 13.1 | 2.2 | 4.8 | 1.1 |
| 2 | 12.9 | 2.0 | 4.0 | 1.0 |
| 3 | 8.2 | -2.7 | 7.3 | -1.3 |
| 4 | 11.8 | 0.9 | 0.8 | 0.4 |
| 5 | 7.6 | -3.3 | 10.9 | -1.6 |
| 6 | 11.0 | 0.1 | 0.01 | 0.1 |
| 7 | 12.0 | 1.1 | 1.2 | 0.5 |

$$\Sigma x = 76.6 \qquad \Sigma(x - \bar{x})^2 = 29.02$$

$$\bar{x} = 10.9 \qquad Sx = \sqrt{\frac{\Sigma(x - \bar{x})^2}{n}} = \sqrt{\frac{29.02}{7}} = 2.04$$

TABLE 3.10 CALCULATION OF THE STANDARD SCORE (Z) FOR VARIABLE 1

The same procedure is carried out for variables 2 and 3. This gives the following table of results.

| Metropolitan District | $Z_1$ | $Z_2$ | $Z_3$ | $\Sigma Z$ | Rank |
|---|---|---|---|---|---|
| 1 | 1.1 | 1.3 | 1.8 | 4.2 | 7 |
| 2 | 1.0 | 0.1 | 0.8 | 1.9 | 6 |
| 3 | -1.3 | -0.5 | -0.9 | -2.7 | 2 |
| 4 | 0.4 | 0.6 | 0.1 | 1.1 | 5 |
| 5 | -1.6 | -2.1 | -1.1 | -4.8 | 1 |
| 6 | 0.1 | 0.2 | 0.2 | 0.5 | 4 |
| 7 | 0.5 | 0.4 | -1.1 | -0.4 | 3 |

TABLE 3.11 MATRIX OF STANDARD SCORES

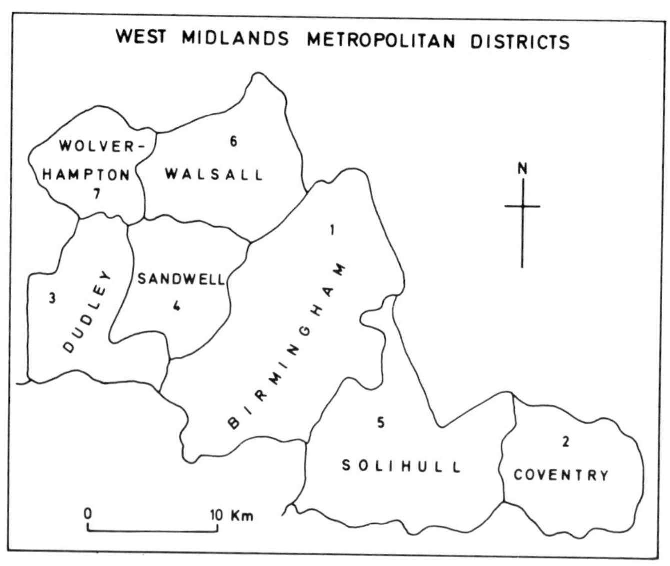


Figure 3.17 WEST MIDLANDS CONURBATION

## 3.7 Some Concluding Thoughts

In this chapter, a range of urban projects have been presented, although it should be appreciated that many individual aspects of specific projects could be applied directly or in a modified form as parts of other non-urban projects. The projects discussed were selected to illustrate a variety of different features of subject matter and approach, and no rigidity in analysis is implied.

Chapter Four

# RURAL SPATIAL STRUCTURE

## 4.1 Introduction

Given the fact that a large and increasing proportion of people live in urban areas, it is not surprising that post-war geographic research has exhibited an urban bias. However, there is evidence in the last few years to suggest an increased interest in rural topics, topics which would often provide stimulating and manageable student projects. The renewed interest can be associated largely with four major issues: the rapid decline in the overall provision of services which is highlighted by the closure of village post offices, schools, shops and so on; the growing concern about the accessibility of inhabitants of rural areas to various facilities, which has been termed *'The Rural Challenge'* by Moseley (1979); the lack of long-term employment opportunities in rural areas and its associated effect on the population structure; and the continued interest in agricultural land-use patterns, which can be related directly to government policies, such as subsidies and so on.

In this chapter, it is demonstrated that such problem-orientated aspects offer enormous scope for investigations by students. In the following four sections, each of the topics listed above are reviewed in detail to indicate the range of potential issues that could be studied. For completeness, in the last section, a general list guide of project areas based on Bowler's (1975) categories for rural research is presented.

First, however, it is necessary to discuss what is meant by 'rural' studies, particularly in relation to 'urban' studies. In general terms, people know what they mean by the term 'rural' but when questioned closely, they find it difficult to define it precisely. Whilst differences obviously exist between rural areas, a broad definition of rural geography is '... the study of recent social, economic, land-use, and spatial changes that have taken place in less-densely populated areas which are commonly recognised by virtue of their visual components as 'countryside'' (Clout, 1972, p. 1). Simply stated, two general kinds of rural area can

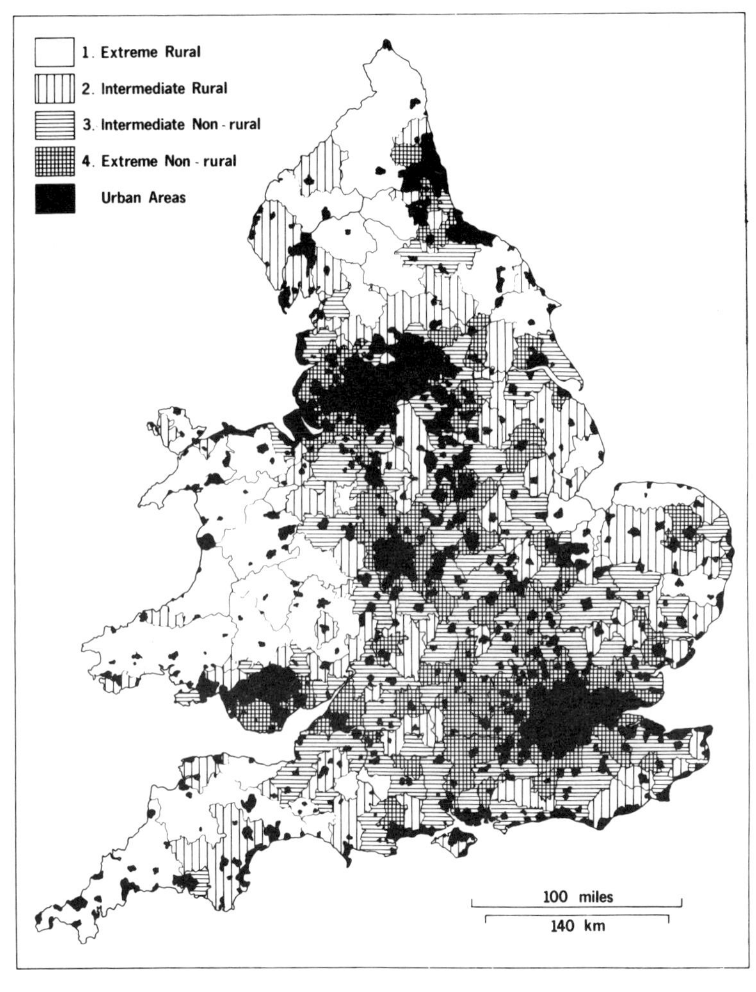


Figure 4.1 INDEX OF RURALITY (Source: Cloke, 1979)

| VARIABLE NAME | CENSUS DATA |
|---|---|
| Population density | Population/acre |
| Population change | % change 1951-61, 1961-71 |
| Population over age 65 | % total population |
| Population men age 15-45 | % total population |
| Population women age 15-45 | % total population |
| Occupancy rate | % population at 1½ per room |
| Occupancy rate | Household/dwelling |
| Household amenities | % households with exclusive use of: hot water; fixed bath; inside W.C. (1971) |
| Occupational structure | % in socio-economic groups: (13) farmers - employers and managers; (14) farmers - own account; (15) agricultural workers |
| Commuting-out pattern | % residents in employment working outside the rural district |
| In-migration | % population resident for less than 5 years |
| Out-migration | % population moved out in last year |
| In/Out migration balance | % in/out migrants |
| Distance from nearest urban centre of 50,000 population | - |
| Distance from nearest urban centre of 100,000 population | - |
| Distance from nearest urban centre of 200,000 population | - |

TABLE 4.1 ORIGINAL VARIABLES IN CLOKE'S (1979) INDEX OF RURALITY

be recognised - the area which is remote from any urban area and the area which is a 'dormitory' of an urban area, although Cloke (1979) has proposed an inductive approach to capture the variety of rural areas. In his description of England and Wales, he derived an index of rurality based on sixteen variables, which were mainly obtained from census data (see table 4.1). As with any multivariate technique, the results are dependent directly on the definitions of the original variables and, in any project, careful selection of a set of variables is essential.

As figure 4.1 portrays, in England and Wales, '...the overall picture of rurality ... is one of a rather fragmented gradation from non-rural to rural' (Cloke, 1979, p. 10). However, it should be appreciated the dichotomy, either rural or urban, is inadequate. In a theoretical discussion of rural-urban relationships, for instance, Cloke and Griffiths (1980) outline the structure of a fluid and complex cyclical model (see Figure 4.2). By recognising a 'dichotomy', 'rural' and

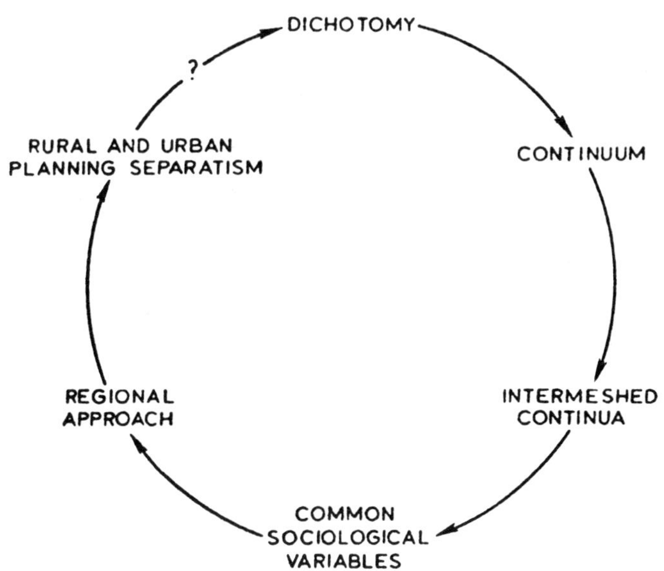


Figure 4.2 A CYCLICAL CONCEPTUALISATION OF THE RURAL-URBAN RELATIONSHIP (Cloke and Griffiths, 1980)

'urban' are viewed as distinct phenomena with different attributes, whereas the 'continuum' perspective suggests a gradation between these two extremes. Pahl (1966), however, argued that the concept of a continuum was too simple and proposed an alternative, 'intermeshed continua' in which a series of continua for different variables represent the gradations between urban and rural. Whilst more realistic, this theoretical framework is very difficult to operationalise in practice. By suggesting that a problem-solving approach would be apposite for rural-based projects, it is important to appreciate that, conceptually urban and rural issues, such as immobility, inequality and socio-economic deprivation, are very similar. This is reflected by the 'common sociological variables' perspective. Developing the rural geography and planning links, Green (1971) argued for a 'regional approach', although, recently, there has been an increasing tendency to see the solution of urban and rural problems as being politically and administratively separate.

Whatever perspective is adopted, at least a cursory mention of the impact of the process of urbanisation on rural areas is required. (see Figure 4.3). Moreover, geographic

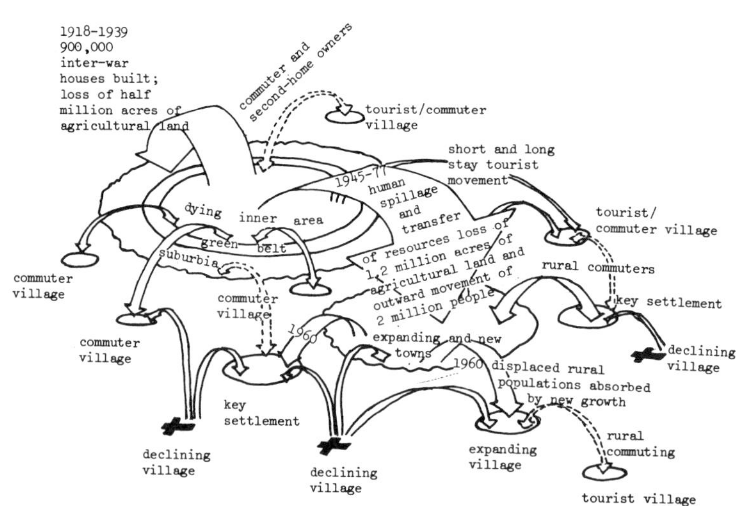


Figure 4.3 A MODEL OF RURAL CHANGE
(Source: Moss, 1978)

studies, such as the patterns and impact of rural-urban migration, are obviously important, feasible student projects.

Finally, at this stage, two points are worth stressing. As with the majority of, (if not all) projects, specific case studies, rather than simply a general discussion of a problem, are required. Secondly, two potential spatial scales of analysis can be recognised: rural settlement patterns within a region or a particular village or hamlet. In relation to the former, the application of a centrality index, which was considered in the preceding chapter, could be appropriate, and, in relation to the latter, it would be useful if a brief description of the settlement's relative location to other centres was presented as a contextual background.

## 4.2 Aspects of service provision in rural areas

A major problem facing many rural settlements is their loss of valuable local services. The publication of the Standing Conference on Rural Community Council's (1978) report, *The Decline of Rural Services*, has resulted in the production of a large number of local studies on this theme. For certain disadvantaged groups, such as the poor and elderly, this problem is exacerbated by the reduction in the quantity and quality of public transport provision. For commercial services, the potential of obtaining economies of scale from larger facilities and the reduction in the rural population or market potential are the basic, underlying reasons for this situation. Similar reasons underpin the planned reduction in public services, particularly education and health services (see, for example, Haynes and Bentham (1979) for a discussion of health care provision), as part of an overall rationalisation programme. However, the closure of such local facilities for economic reasons ignores potential adverse social effects.

In terms of project work, many facets of this general decline of services merit detailed local investigation. Clearly, in any introduction, it would be necessary to outline briefly the general tendency for such alterations in post-war rural Britain. In terms of fieldwork, a study of the implications of such changes in specific locations could be undertaken by a questionnaire survey. It is suggested that particular attention should be given to the impacts on different groups of people. Given the problem-orientated approach, an evaluation of relevant planning policies would be essential, and students should be expected to offer some solutions to the particular issues which they address. As a general classification to possible solutions, planning policies could be directed towards either taking the people to the services or taking the services to the people. Obviously, the former requires an efficient transportation system (which is discussed in the next section), and examples of the latter include mobile shops and libraries which already serve many rural areas.

## 4.3 Rural public transportation and accessibility

In his book, *Accessibility: the rural challenge*, Moseley (1979) provides a broad ranging examination of accessibility problems in rural areas. Basically, the relationship between the levels of public transport provision and private car ownership has changed enormously in the post-war period. Whilst, in general, the greater car ownership has resulted in an increase in household mobility, it has also caused the decline in the provision of public transport. Again, unfortunately, this change has had a differential impact on people living in rural areas; the poor, the elderly, the housewives and the teenagers have been disadvantaged by the withdrawal of these services.

In terms of project work, whilst aspects of this issue would provide well-defined and manageable topics, it is important to appreciate that, as with all the rural problems, it is only part of an integrated rural complex. The level of public transportation affects, and also is affected by, the level and pattern of service provision, the employment opportunities and the demographic structure. Moreover, in any such study, it would be a fundamental requisite to provide an unambiguous and operational definition of the concept of accessibility (see, for example, the discussion of potential models in Chapter seven).

Thus, although transport is only one component of the rural problem, it is explicitly spatial in nature and would link easily with other parts of a geographic curriculum.

## 4.4 Aspects of rural employment

Simply stated, there has been a long history of the decline in rural employment opportunities, which is partly a result of the changing economic structure of the country. Problems of rural employment are expressed in terms of the continued, but expected, dominance of primary industrial activities, the relatively high unemployment rates especially for women, the relatively low level of skills and the relatively low wage levels. Gilg (1980) presents a recent review of planning for rural employment, and Hodge and Whitby (1981) consider trends, options and choices.

In terms of potential projects, it would be appropriate to evaluate the success of various planning response to rural unemployment in specific local areas. Although, in the present economic climate, it should be appreciated that rural areas are competing with urban areas to attract new enterprises, various grant and subsidies are available nationally (such as the Council for Small Industries in Rural Areas and the Rural Development Boards), and through the European Economic Community for development in rural areas. Recent interest has been given to small-scale, village-based workshops, and, if these are present in a local area, an evaluation of their success and possible future development would be an appropriate student project.

From a purely geographical perspective, some evaluation must be given to whether it is appropriate to try to generate a dispersed pattern of a relatively large number of small economic developments or to concentrate development in a few settlements (forgetting the needs of the other settlements). In Britain, rural settlement planning has followed a so-called 'key settlement policy' (see Cloke (1979) for a detailed, wide-ranging discussion), although, recently, there has been some indication of dissatisfaction with this inequitable approach.

## 4.5 Rural land-use patterns

For student projects, detailed land-use mapping, whether it is in urban or in rural areas, can provide a useful foundation (see Rhind and Hudson (1980) for an overview of land use issues). Whilst maps are available with this data, if the classifications employed or the spatial scale of analysis are inappropriate, valuable experience can be gained by actual mapping in the field. Whilst such an exercise is of practical importance, it is suggested that a student project would have to go beyong a simple description of the patterns (and a discussion of the data collection procedure and the associated difficulties). Various land use models are now taught in a geographic course, and their applicability could be tested.

For instance, in a rural land-use project, Von Thunen's 'isolated state', in which a series of different agricultural land use zones are arranged concentrically around a central city, could be examined (see also Chisholm's (1968) book, '*Rural Settlement and Land Use*'). Clearly, for completeness the major features of the model should be explained, but attention must be given to the range of underlying, simplified assumptions which are often forgotten. Are the assumptions satisfied? If the observed land use patterns are different from those predicted by the model, it is essential to suggest explanations for the discrepancies. Can the model be extended? In terms of Von Thunen's model for instance, it is possible to indicate the implications of relaxing the following restrictive assumptions: competing market centres can replace the assumption of a single nodal point; empirical evidence of transport costs for different products would indicate that they are not linearly related to distance; detailed fieldwork, using a suitable spatial sampling frame, would provide an indication of spatial variations in soil fertility; and, by questionnaire survey, it should be possible to obtain insights into potential economies of scale. (For reference, attention is drawn to Griffen's (1973) testing of the Von Thunen model in Uruguay).

4.6 Bowler's (1975) categories for rural research

By systematically treating the following seven categories (which are, obviously, neither mutually exclusive nor exhaustive), a large number of possible student projects should come to mind: agriculture; forestry; rural settlement; rural population; rural transport; recreation and tourism; and rural (development) planning. Clearly, aspects of these topics have been discussed in the preceding sections, but, for completeness, additional ideas are listed in this final section on rural-based projects.

In relation to agricultural studies, attention has been given to Von Thunen's work, although, by incorporating the temporal dimension, it would be possible to examine alterations in land use patterns. In such an analysis, links with changing land quality and market conditions must be considered; land tenure would also be significant.

One theme that is common to a number of the specified categories relates to rural areas as both an economic and a natural resource (and often the trade-off is a political decision); useful background volumes are by Blacksell and Gilg (1981) and by Green (1981). For example, in terms of forestry, its recent rapid use as a valuable economic commodity has led to depletion problems, and, consequently, there has been a need to introduce conservation measures. Forests are also one type of landscape that is attractive for recreation and tourism. Specific issues that could be considered within this category would be an examination of second/holiday homes in an area, particularly the spatial pattern of their owners place of residence. Davies and O'Farrell (1981), for instance, have completed a spatial and temporal analysis of second home ownership in west Wales. Suitable areas of such studies would be National Parks, although, obviously, a large number of different topics would be suitable for these specially designated areas. For a recent, incisive assessment of National Parks, the MacEwans (1982) consider conservation or cosmetics. With regard to rural (development) planning, whilst the National Parks present a range of special problems, the general environment-resource crisis is also found (as illustrated by the recent Vale of Belvoir inquiry).

Sufficient background details have been given already to demonstrate the scope for student projects on aspects of rural settlement, rural population and rural transport. It should be noted that the methodological discussion in Appendix II is appropriate for a pattern analysis of the distribution of settlements, which could be linked directly to central place theory. As suggested in the earlier sections, whilst a description of the demographic structure of rural populations in itself would be insufficient for a project, it does provide the necessary background for many topics and would be a major part of any study of the social and economic impacts of rural-urban migration.

Chapter Five

# DEMOGRAPHIC STUDIES WITH SPECIAL REFERENCE TO THE BRITISH CENSUS

## 5.1 Introduction

In this chapter, rather than examining a range of possible projects directly, attention is focused on a specific, important data set which could provide the basis for a variety of projects. This data set is the British Census and is of particular importance to human geographers. Indeed, the Population Census is the largest data set likely to be encountered by the student of British geography, and, importantly, it is widely available.

The Concise Oxford Dictionary defines a population census as '...an official numbering of population with various statistics'. The modern census implies an official government activity, involving a comprehensive coverage of the entire population (in a country or parts of a country) and a set of data (demographic, social and economic) referring to a specific point in time. It is desirable that a census should form part of a regular series, thus allowing temporal comparisons to be made. The modern census, therefore, is not just a 'counting of heads', but involves the collection and publication of the characteristics (as attributes) of individuals, such as age, sex, marital status, economic activity and place of birth.

## 5.2 A brief history of the modern census

The modern census is a product of the nineteenth and twentieth centuries; the first modern census was held in the U.S.A. in 1790, and was followed quickly by England and France in 1801. In Britain a complete census has been held every ten years since 1801 (except for 1941). In addition, a 10% sample census was held in 1966 (but not in 1976), giving a total of nineteen censuses.

The characteristics of, and problems associated with, early censuses are summarized in table 5.1. The information collected in the earliest censuses (1801-1831) was very

| CENSUS | VARIABLES | AREAS | COMMENTS |
|---|---|---|---|
| 1801 | Number of houses, families males/females; number of persons engaged in agriculture, trade/manufacturing/crafts, others | Collected for parishes but only published for ancient geographical counties | |
| 1811 | As above, but information on occupation referred to families rather than persons, therefore, difficult to compare with 1801 | | |
| 1821 | Age question introduced but not compulsory | | |
| 1831 | Age question deleted | | |
| 1841 | Age question reintroduced; name, sex, occupation, also place of birth is first recorded. | 1841-1911 a hierarchical system was introduced consisting of registration counties, districts and sub-districts. | 1841 - first census based on a system of civil registration. Information was collected on the basis of the household with each individual recorded separately. Information first recorded in enumerators' books (available up to 1881 at present). |
| 1851<br>1861<br>1871<br>1881 | Name, occupation, age, place of birth (based on county and parish). Also information on the relationship of each individual to the head of household and marital status. Details on disabled elements of the population (deaf/mute; blind; imbecile/idiot; lunatic). | | 1851 - a religious census was also held but published separately.<br><br>After 1851 changes in occupational categories frequently occur making comparison difficult. |
| 1891 | Additional questions: Welsh speaking (wales only), number of rooms in each household. Employment in trade and industry, division into employers, employed and self employed. | | |

TABLE 5.1 CHARACTERISTICS OF EARLY CENSUSES

rudimentary in character and amounted to little more than head counts. This data, however, does allow a simple analysis of the rate of increases in population and its general spatial distribution to be made. The only innovation during this period was the introduction of a non-compulsory age question in 1821. Major developments in the census were reflected in the 1841 enumeration. This was the first census based on a system of civil registration; the enumeration was carried out on a household basis and the names of individual members were recorded. In addition, the enumerator transferred the information entered in the census forms to a book (for a particular enumeration district) which then formed the basis for the aggregate summaries contained in the printed volumes. After a period of one hundred years, the enumerators' books become generally available and they provide valuable information on the age, sex, place of birth, and occupation of individuals together with their household arrangements. This information allows detailed analyses of the social and demographic structure of nineteenth century communities to be made. (In 1841 a new spatial framework was also adopted which involved a system of registration counties, districts and sub-districts). In general, however, the 1851 census is accepted as the first census on which detailed local studies can be based. For example, although migration studies can be undertaken for 1841, the 1851 census provides information on migrants at the level of the county and parish.

Two cautionary points need to be made in relation to the early censuses. First, although detailed studies of occupations and trades can be made after 1851, frequent changes in categories make it almost impossible to study occupational changes over any long period of time. Second, most County Record Offices and some local libraries in larger towns possess microfilm copies (or, in some cases, transcriptions) of enumeration books relating to their immediate area. However, photostat copies cannot be made from these sources because they are copyright of the Public Record Office. (Copies of the originals can be requested from the Public Record Office, Chancery Lane, London WC2).

From the mid-nineteenth century to the mid-twentieth century, the British census underwent relatively few changes. In 1911, the areal basis changed from Registration Counties to that of Local Authority Areas. While an attempt was made to develop occupation-based definitions of social class. Other changes have included the production of County Reports and the provision of a greater range of data on household characteristics (for a detailed discussion of the 1971 census, see Dewdney, 1981).

## 5.3 The 1981 Census

The modern census is compulsary and backed by law, and must be held at intervals of not less than five years. The questions to be asked must fall under the following topic headings (Census Act, 1920):

(a) Name, sex, age

(b) Occupation, profession, trade or employment

(c) Nationality, race, birthplace, language

(d) Place of abode, character of dwelling

(e) Condition as to marriage, relationship to head of family, issue born in marriage

(f) 'Any other matter with respect to which it is desirable to obtain statistical information with a view to ascertaining the social or civil conditions of the population'.

It is important to note that this latter point does not give carte blanche to the authorities, for example, the proposed ethnic question for the 1981 census was eventually abandoned because of adverse public opinion.

### 5.3.1 *Organization of the census*

The 1981 census retained the four-tier hierarchy introduced in England and Wales in 1971 (a similar structure was used in Scotland):

1. 110 census supervisors each responsible for the recruiting, training and overseeing of twenty census officers

2. 2200 census officers responsible for a census district containing on average fifty Emuneration Districts (EDs)

3. 6400 assistant census officers - trained and recruited by the census officers

4. 103100 enumerators trained by a census officer. Each enumerator is responsible for one ED (in some cases more than one ED) containing on average 165 households.

### 5.3.2 *Census areas*

The basic unit is the Enumeration District. In 1981, there were 112300 EDs in England and Wales. Wherever possible ED boundaries must follow significant physical features such as railways and roads. Although an average urban ED contains a population of five hundred people, local conditions result in variations in both areal size and population. ED boundaries also vary from census to census, for example, during the period 1971-1981 approximately 40% of ED boundaries changed.

Using the ED as the basic building brick, data can be produced for areal units at various levels of aggregation. Pre-local government organization (pre-1974) typical aggregations were: wards/parishes, urban districts, rural districts, counties and standard regions. Post-1974 a new geographical framework was introduced consisting of wards/parishes, districts, counties and standard regions (see Figure 5.1). It is noted that the geocoding of the 1971 census by 1 km grid square provided an extremely flexible base for aggregation. In addition, statistics are produced for administrative areas in England and Wales such as health regions, parliamentary constituencies, towns as constituted before local government reorganization, city centres, New Towns and National Parks. Figure 5.2 and Table 5.2 provide details of the standard regions of Great Britain, pre- and post-1974.

### 5.3.3 *The census schedule*

The 1981 census schedule was the simplest since 1931 and consisted of two main sections (see Table 5.3, the actual census schedule is reproduced in Appendix VI ).

1. Questions H1-H5. These refer to household accommodation and include household tenure, accommodation and amenities. This information is of particular interest to social geographers and planners. In fact these questions represent, essentially, a housing census.

2. Questions 1-17. These questions refer to the attributes of each individual, and few censuses outside the UK cover so wide a range.

   Some of the major reasons for the inclusion of specific questions are provided below.

1. Five questions are basic to the population count (questions 1, 2, 3, 6, 7)

2. Two questions are used for analysis of demographic trends as well as for resource allocation, for example, Rate Support Grant, allocations to Health Authorities (questions 4, 5)

3. Two questions provide basic information on immigration

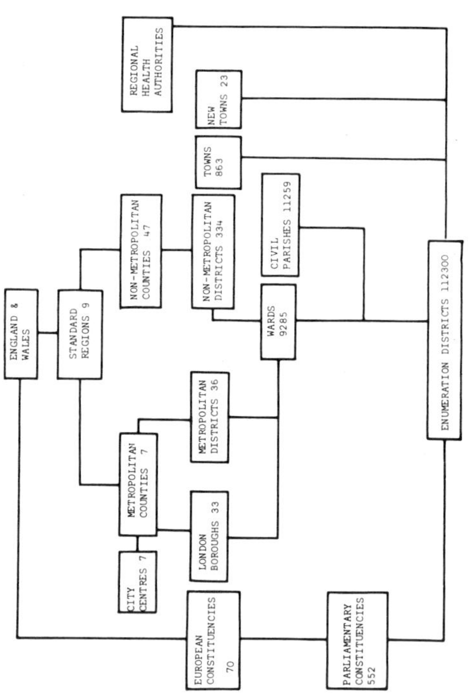


Figure 5.1 1981 CENSUS: GEOGRAPHICAL FRAMEWORK

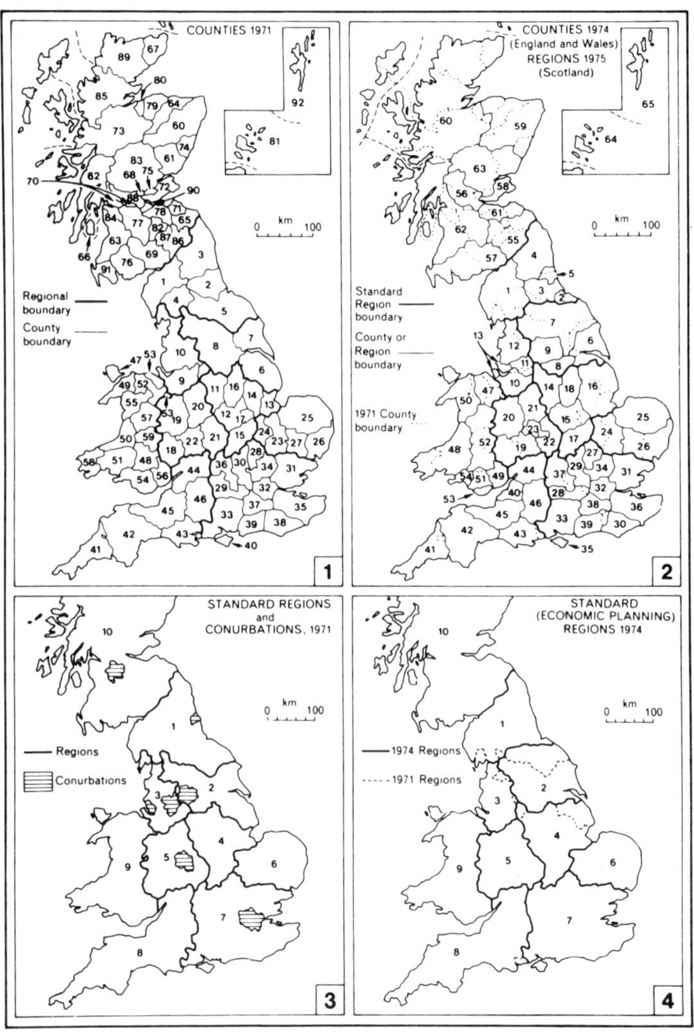


Figure 5.2 STANDARD REGIONS: 1971 AND 1974
(Source: Dewdney, 1981)

| | 1981 |
|---|---|
| 1. NORTH | |
| 1. Cumbria | 483427 |
| 2. Cleveland | 865775 |
| 3. Durham | 604728 |
| 4. Northumberland | 299905 |
| 5. Tyne and Wear (Metropolitan County) | 1143245 |
| 2. YORKSHIRE and HUMBERSIDE | |
| 6. Humberside | 84766 |
| 7. North Yorkshire | 666610 |
| 8. South Yorkshire (Metropolitan County) | 1301813 |
| 9. West Yorkshire (Metropolitan County) | 2037510 |
| 3. NORTH WEST | |
| 10. Cheshire | 926293 |
| 11. Greater Manchester (Metropolitan County) | 2594778 |
| 12. Lancashire | 1372118 |
| 13. Merseyside (Metropolitan County | 151307 |
| 4. EAST MIDLANDS | |
| 14. Derbyshire | 906929 |
| 15. Leicestershire | 842577 |
| 16. Lincolnshire | 547560 |
| 17. Northamptonshire | 527532 |
| 18. Nottinghamshire | 982631 |
| 5. WEST MIDLANDS | |
| 19. Hereford and Worcester | 630218 |
| 20. Salop | 375610 |
| 21. Staffordshire | 1012320 |
| 22. Warwickshire | 473620 |
| 23. West Midlands (Metropolitan County) | 2644634 |
| 6. EAST ANGLIA | |
| 24. Cambridgeshire | 575177 |
| 25. Norfolk | 693490 |
| 26. Suffolk | 596354 |
| 7. SOUTH EAST | |
| 27. Bedfordshire | 504986 |
| 28. Berkshire | 675153 |
| 29. Buckinghamshire | 565992 |
| 7. SOUTH EAST (cont.) | |
| 30. East Sussex | 652568 |
| 31. Essex | 1459065 |
| 32. Greater London (Metropolitan County) | 6696008 |
| 33. Hampshire | 1456267 |
| 34. Hertfordshire | 954535 |
| 35. Isle of Wight | 118192 |
| 36. Kent | 1453055 |
| 37. Oxfordshire | 515079 |
| 38. Surrey | 999393 |
| 39. West Sussex | 658562 |
| 8. SOUTH WEST | |
| 40. Avon | 909408 |
| 41. Cornwall | 430506 |
| 42. Devon | 952000 |
| 43. Dorset | 591990 |
| 44. Gloucestershire | 499351 |
| 45. Somerset | 424988 |
| 46. Wiltshire | 518167 |
| 9. WALES | |
| 47. Clwyd | 390173 |
| 48. Dyfed | 329977 |
| 49. Gwent | 439684 |
| 50. Gwynedd | 230468 |
| 51. Mid Glamorgan | 537866 |
| 52. Powys | 110467 |
| 53. South Glamorgan | 384633 |
| 10. SCOTLAND | |
| 55. Borders | 99248 |
| 56. Central | 273078 |
| 57. Dunfries and Galloway | 145078 |
| 58. Fife | 326480 |
| 59. Grampian | 470596 |
| 60. Highland | 200030 |
| 61. Lothian | 735892 |
| 62. Strathclyde | 2397827 |
| 63. Tayside | 391529 |
| 64. Orkney (Island Area) | 18906 |
| 65. Shetland (Island Area) | 26716 |
| 66. Western Isles (Island Area) | 31766 |

TABLE 5.2 STANDARD REGIONS AND COUNTIES: 1981 POPULATION TOTALS

Household-housing and amenities

H1 Number of rooms
H2 Tenure
H3 Availability of amenities
H4 (and panel A) Nature of building and whether sharing entrance to building.
H5 Availability of cars and vans

General population information

1. Full name
2. Sex
3. Date of birth
4. Marital status
5.* Relationship in household
6. Whereabouts on census night
7. Usual address

Migration

8. Address one year ago
9. Country of birth

Employment

10. Activity last week (e.g. full-time job, housewife, retired, part-time job, student).
11.*Name and business of employer (industry)
12.*Occupation.
13. Employment status (whether employed, self-employed, etc.)

Work place and transport to work

14.*Address of place of work
15.*Means of transport used on journey to work
16.*Higher educational, professional and vocational qualifications attained after age 18.

Welsh language (in Wales only)

17. Whether speaking, reading and writing Welsh

* 10% sample questions.

TABLE 5.3 TOPICS COVERED IN THE 1981 CENSUS SCHEDULE

and internal migration (questions 8, 9)

4. Four questions are 'benchmarks' on economic activity. They provide a standard against which regularly collected statistical returns and sample surveys can be checked. The questions in this category cover the working population, industry, occupations and social classes, and help in measuring occupational mortality (questions 10, 11, 12, 13).

5. Question 14 on a persons's place of work is used to determine the boundaries of travel-to-work areas and assisted areas and for planning employment services.

6. Question 15 on means of travel to work is used, in conjunction with question 14 (and H5) in planning roads and transport services, in structure plans and in planning homes and jobs.

7. Question 16 provides information about the availability of qualified manpower and (in conjunction with questions 11, 12) its deployment.

8. Four questions measure the stock and characteristics of the nation's housing and are also used for resource allocation, for example Rate Support (questions H1, H2, H3, H4).

The questions themselves are of different types. The simplest are the 'tick-box' variety (H2, H3 and H5, for example):

H5 Cars and Vans

Please tick the appropriate box to indicate the number of cars and vans normally available for use by you or members of your household (other than visitors)

0 ☐ None

1 ☐ One

2 ☐ Two

3 ☐ Three or more

Include any car or van provided by employers if normally available for use by you or members of your household but exclude vans used solely for the carriage of goods.

Some questions require very simple numerical answers (for example, 'number of rooms' (H1)):

H1 Rooms

Please count the rooms in your household's accommodation.

Do not count:

small kitchens, that is those under 2 metres (6ft.6ins.) wide, bathrooms, W.C's.

Number of rooms ............................

However, obviously, correct answers depend on careful adherence to the definitions set out on the schedule.

## 5.4 Census output

Before the census data is actually made available a vast amount of processing of the raw data is required. This includes coding, punching and storage of data on computer tapes from which the final tabulation are produced. Census output can be divided into three classes according to the form in which it is made available. First, there are the published census volumes which contain data at local authority area level and above (for example, the County Reports and volumes on particular topics such as Migration or Household Composition). Second, there is the standard unpublished output which has a standard format determined before the processing of results. The most heavily used unpublished output are the Small Area Statistics (SAS) which consist of a set of tables giving data for areas as small as the enumeration district. Finally, there is the unpublished special output designed to the specification of the user and is, therefore, more costly to produce.

In the following sub-sections a more detailed examination is made of the SAS and the published census volumes.

### 5.4.1 *Small Area Statistics (SAS)*

The SAS are a relatively recent development in the history of the British census, dating from 1951 when the General Register Office agreed to make data available for thirty-five census tracts (amalgamation of EDs) for the city of Oxford. In 1961, SAS were available on demand and thus coverage was incomplete. Although information for civil parishes, wards and enumeration districts were made available for the 1966 sample census, 1971 was the first 100% census for which SAS data were available as standard unpublished output.

In addition to the standard administrative divisions, two non-census areal bases (that is, independent of the processes of census-taking) are currently in use. The first system utilises the lines of the National Grid; the enumerator assigns a grid reference to each address at which households and persons were enumerated. In 1971 this information enabled standard SAS to be provided for each 1 km square and in areas for which maps at a scale of 1:2500 or better were available (in general the urban areas) for each 100 m square. For the 1981 census SAS data for 100 m squares will only be prepared on request, making the cost of individual purchases prohibitive.

The second non-census areal base used in the 1981 census is the post code unit. A full post code zone (for example BR2 6AX) covers about fifteen postal delivery points, unless it is assigned to a single large user. In Scotland these post code zones have been aggregated to form enumeration districts. As post codes build hierarchically (BR2 6A, BR26 and BR2) combinations of EDs (approximately ten) can be made to form post code sectors. The General Register Office Scotland have produced data for such sectors from the 1981 census. In England and Wales post codes have a more limited role, being used in coding addresses other than the place of enumeration, such as the place of work and the place of residence one year before census day.

Finally, reference should be made to the availability of 1971-1981 comparable small areas (census tracts) and statistics. These census tracts attempt to overcome the problem of boundary changes and topic coverage from census to census. Each tract is defined in terms of the 1971 and 1981 EDs and consists of one or more EDs from each census. EDs unchanged between the censuses are included as tracts. In rural areas problems of comparison are not as acute and civil parishes form a suitable base for 1971-81 comparisons.

#### 5.4.1.2 *Structure of the SAS*

The SAS are computed from the original census forms and cover a full range of statistics. The information is available in (computer) machine-readable and hard-copy forms. In addition, maps of the basic areas (EDs and wards) can also be obtained.

For England and Wales, the 1981 SAS consists of over 5500 pieces of information contained in eight pages of tables, (see Figure 5.3). These can be divided into three main sections.

1. Pages 1, 2, 3 and 4, consist of population and household tables on a 100% basis. Page 1 provides information on the demographic and employment characteristics of the population, while page 2 tabulates data relating to tenure characteristics, basic amenities, number of rooms and persons, for example, at the level of the household. Page 3 contains detailed information on household composition. Data on groups in the population with particular needs such as lone parent and pensioner households

are contained on page 4.

2. Population tables on a 10% sample basis, available throughout Great Britain (pages 5 and 6).

3. Additional tables for areas in England , or these tables and the Welsh language table for areas in Wales only.

Information is not published which allows the identification of an individual household or person. Each cell in all tables in sections one and three except tables 1 and 17 is modified by +1, 0, or -1. In addition, if an ED or ward has less than twenty-five residents, tables counting persons in sections one and three (with the exception of tables 1 and 17) are not released. If an ED or ward has less than eight resident households, tables counting households in sections one and three are not released (except for table 17). In section 2, person and household tables are suppressed if the person and/or household tables have been suppressed in sections one and three. The tables are amalgamated with those of adjoining areas.

As can be seen from Figure 5.3, few of the tables are one-dimensional in character, most are cross-tabulations showing the relationship between two or more variables. In addition, there is considerable scope for aggregating counts and deriving new variables, extracting key indicators and calculating multivariate measures, and for aggregating statistics to particular geographical zones of interest to the user.

SAS can be used for a number of purposes:

1. Local authorities use SAS for the planning and management of local services, and for monitoring the effect of policies.

2. In commerce, SAS are used to look at the characteristics of the populations in catchment areas of retail outlets or to pinpoint localities with market potential.

3. In central government, SAS analysed over the country as a whole, have been used to locate areas of multiple deprivation.

4. SAS are also a source of information for individuals, local businessmen, members of parliament and local interest groups with questions about their communities.

5. Finally, in schools, students use SAS to study their local communities, while the SAS are the largest body of data on the population available for academic research in general.

**CENSUS 1981 SMALL AREA STATISTICS**

**PAGE 1** **100%**

ED No

Map Reference

Note: $ = ED same as in 1971
$$ = Special Enumeration District

Crown copyright reserved | Frame No:

Separate explanatory notes are available

**1 All persons present; plus absent residents * in private households**

| | TOTAL PERSONS | In private households | | Not in private households | |
|---|---|---|---|---|---|
| | | Males | Fmles | Males | Fmles |
| 1 All present res | 1 | 3 | 4 | 6 | 7 |
| 2 All absent res | 8 | 10 | 11 | xxx | xxx |
| 3 All visitors | 15 | 17 | 18 | 20 | 21 |
| Res in UK | 22 | 24 | 25 | 27 | 28 |
| Res outside UK | 29 | 31 | 32 | 34 | 35 |
| ALL PRESENT 1981 | 36 | 38 | 39 | 41 | 42 |
| 1971 BASE (1+3) | | | | | |
| ALL RESIDENT 1981 | 43 | 45 | 46 | 48 | 49 |
| 1981 BASE (1+2) | | | | | |

**2 All residents**

| Age | TOTAL PERSONS | Males | | Females | | Ret'd Males |
|---|---|---|---|---|---|---|
| | | SWD | Mrr'd | SWD | Mrr'd | |
| TOTAL | 50 | 52 | 53 | 55 | 56 | 190 |
| 0-4 | 57 | 59 | xxx | 62 | xxx | xxx |
| 5-9 | 64 | 66 | xxx | 69 | xxx | xxx |
| 10-14 | 71 | 73 | xxx | 76 | xxx | xxx |
| 15 | 78 | 80 | xxx | 83 | xxx | xxx |
| 16-19 | 85 | 87 | 88 | 90 | 91 | xxx |
| 20-24 | 92 | 94 | 95 | 97 | 98 | xxx |
| 25-29 | 99 | 101 | 102 | 104 | 105 | xxx |
| 30-34 | 106 | 108 | 109 | 111 | 112 | xxx |
| 35-39 | 113 | 115 | 116 | 118 | 119 | 199 |
| 40-44 | 120 | 122 | 123 | 125 | 126 | 200 |
| 45-49 | 127 | 129 | 130 | 132 | 133 | 201 |
| 50-54 | 134 | 136 | 137 | 139 | 140 | 202 |
| 55-59 | 141 | 143 | 144 | 146 | 147 | 203 |
| 60-64 | 148 | 150 | 151 | 153 | 154 | 204 |
| 65-69 | 155 | 157 | 158 | 160 | 161 | 205 |
| 70-74 | 162 | 164 | 165 | 167 | 168 | 206 |
| 75-79 | 169 | 171 | 172 | 174 | 175 | 207 |
| 80-84 | 176 | 178 | 179 | 181 | 182 | 208 |
| 85+ | 183 | 185 | 186 | 188 | 189 | 209 |

**3 Persons present not in private households**

| Establishments | TOTAL PERS | Males (M) | Fmles (F) | Not usually resident | | Residents: Staff ## | | Residents: Other | |
|---|---|---|---|---|---|---|---|---|---|
| | | | | M | F | M | F | M | F |
| TOTAL # | 210 | 211 | 212 | 214 | 215 | 217 | 218 | 219 | 220 |
| Hotels/boarding houses | 221 | 222 | 223 | 225 | 226 | 228 | 229 | 230 | 231 |
| Children's homes | 232 | 233 | 234 | 236 | 237 | 239 | 240 | 241 | 242 |
| Old people's homes | 243 | 244 | 245 | 247 | 248 | 250 | 251 | 252 | 253 |
| Psychiatric hospitals | 254 | 255 | 256 | 258 | 259 | 261 | 262 | 263 | 264 |
| Other hospitals | 265 | 266 | 267 | 269 | 270 | 272 | 273 | 274 | 275 |
| Schools and colleges | 276 | 277 | 278 | 280 | 281 | 283 | 284 | 285 | 286 |
| Prison dept estabs | 287 | 288 | 289 | 291 | 292 | 294 | 295 | 296 | 297 |
| Hostels/lodging houses | 298 | 299 | 300 | 302 | 303 | 305 | 306 | 307 | 308 |
| Other establishments | 309 | 310 | 311 | 313 | 314 | 316 | 317 | 318 | 319 |

**4 All residents**

| Country of birth | Males | Fmles |
|---|---|---|
| TOTAL | 321 | 322 |
| England | 324 | 325 |
| Scotland | 327 | 328 |
| Wales | 330 | 331 |
| Rest of UK | 333 | 334 |
| Irish Rep | 336 | 337 |
| Old Comm'th | 339 | 340 |
| New Comm'th | 342 | 343 |
| East Africa | 345 | 346 |
| Africa Rem | 348 | 349 |
| Caribbean | 351 | 352 |
| India | 354 | 355 |
| Bangladesh | 357 | 358 |
| Far East | 360 | 361 |
| Mediterr | 363 | 364 |

**5 All residents aged 16 or over**

| Economic Position | TOTAL PERS | Males | | Females | |
|---|---|---|---|---|---|
| | | SWD | Mrr'd | SWD | Mrr'd |
| ALL PERSONS 16+ | 380 | 382 | 383 | 385 | 386 |
| Total econ active | 387 | 389 | 390 | 392 | 393 |
| Working | 394 | 396 | 397 | 399 | 400 |
| Seeking work | 401 | 403 | 404 | 406 | 407 |
| Temp sick | 408 | 410 | 411 | 413 | 414 |
| Total econ inact | 415 | 417 | 418 | 420 | 421 |

| | | |
|---|---|---|
| Remainder | 366 | 367 |
| Pakistan | 369 | 370 |
| Other E.C. | 372 | 373 |
| Other Europe | 375 | 376 |
| Rest of World | 378 | 379 |

| | | | | | |
|---|---|---|---|---|---|
| Perm sick | 422 | 424 | 425 | 427 | 428 |
| Retired | 429 | 431 | 432 | 434 | 435 |
| Student | 436 | 438 | 439 | 441 | 442 |
| Other inactive | 443 | 445 | 446 | 448 | 449 |

**NOTES**

* Persons returned as usually resident but absent on census night in private households with one or more other person(s) present (Table 1)

\# Includes campers vagrants etc (Table 3)

\#\# Includes relatives of staff (Tables 3 and 6)

**6 All persons present**

| Age | TOTAL PERSONS | In private households | | Not in private households | | | | | |
|---|---|---|---|---|---|---|---|---|---|
| | | | | All present | | Residents | | | |
| | | | | | | Staff ## | | Other | |
| | | M | F | M | F | M | F | M | F |
| TOTAL | 450 | 452 | 453 | 455 | 456 | 457 | 458 | 459 | 460 |
| 0-4 | 461 | 463 | 464 | 466 | 467 | 468 | 469 | 470 | 471 |
| 5-15 | 472 | 474 | 475 | 477 | 478 | 479 | 480 | 481 | 482 |
| 16-24 | 483 | 485 | 486 | 488 | 489 | 490 | 491 | 492 | 493 |
| 25-34 | 494 | 496 | 497 | 499 | 500 | 501 | 502 | 503 | 504 |
| 35-44 | 505 | 507 | 508 | 510 | 511 | 512 | 513 | 514 | 515 |
| 45-54 | 516 | 518 | 519 | 521 | 522 | 523 | 524 | 525 | 526 |
| 55-59 | 527 | 529 | 530 | 532 | 533 | 534 | 535 | 536 | 537 |
| 60-64 | 538 | 540 | 541 | 543 | 544 | 545 | 546 | 547 | 548 |
| 65-69 | 549 | 551 | 552 | 554 | 555 | 556 | 557 | 558 | 559 |
| 70-74 | 560 | 562 | 563 | 565 | 566 | 567 | 568 | 569 | 570 |
| 75+ | 571 | 573 | 574 | 576 | 577 | 578 | 579 | 580 | 581 |
| Single | 582 | 584 | 585 | 587 | 588 | 589 | 590 | 591 | 592 |
| Married | 593 | 595 | 596 | 598 | 599 | 600 | 601 | 602 | 603 |
| Students aged 16 or over | 604 | 606 | 607 | 609 | 610 | 611 | 612 | 613 | 614 |

**7 All residents aged 16 or over in employment**

| Employment status | Males | Females | |
|---|---|---|---|
| | | SWD | Mrr'd |
| ALL IN EMPLOYMENT | 615 | 616 | 617 |
| Apprentices and trainees | 618 | 619 | 620 |
| Employees supervising others | 621 | 622 | 623 |
| Other employees | 624 | 625 | 626 |
| Self-empl without employees | 627 | 628 | 629 |
| Self-empl with employees | 630 | 631 | 632 |
| ALL EMPLOYEES | 633 | 634 | 635 |
| Working full-time | 636 | 637 | 638 |
| Working part-time | 639 | 640 | 641 |

**8 All residents aged 1 or over with a usual address 1 year before census different from present usual address**

| Age | TOTAL PERS | Males | | Females | |
|---|---|---|---|---|---|
| | | SWD | Mrr'd | SWD | Mrr'd |
| TOTAL | 642 | 644 | 645 | 647 | 648 |
| 1-4 | 649 | 651 | xxx | 654 | xxx |
| 5-15 | 656 | 658 | xxx | 661 | xxx |
| 16-24 | 663 | 665 | 666 | 668 | 669 |
| 25-34 | 670 | 672 | 673 | 675 | 676 |
| 35-44 | 677 | 679 | 680 | 682 | 683 |
| 45-59 | 684 | 686 | 687 | 689 | 690 |
| 60-64 | 691 | 693 | 694 | 696 | 697 |
| 65+ | 698 | 700 | 701 | 703 | 704 |

**9 All Economically Active (EA) residents**

| Age | TOTAL PERSONS EA | Males EA | Females EA | | In emp working f/t | | | Not in employment | | | Self empl pers |
|---|---|---|---|---|---|---|---|---|---|---|---|
| | | | | | Males | Females | | Males | Females | | |
| | | | SWD | Mrr'd | | SWD | Mrr'd | | SWD | Mrr'd | |
| TOTAL | 719 | 720 | 722 | 723 | 790 | 792 | 793 | 860 | 862 | 863 | 705 |
| 16-19 | 724 | 725 | 727 | 728 | 795 | 797 | 798 | 865 | 867 | 868 | 706 |
| 20-24 | 729 | 730 | 732 | 733 | 800 | 802 | 803 | 870 | 872 | 873 | 707 |
| 25-29 | 734 | 735 | 737 | 738 | 805 | 807 | 808 | 875 | 877 | 878 | 708 |
| 30-34 | 739 | 740 | 742 | 743 | 810 | 812 | 813 | 880 | 882 | 883 | 709 |
| 35-39 | 744 | 745 | 747 | 748 | 815 | 817 | 818 | 885 | 887 | 888 | 710 |
| 40-44 | 749 | 750 | 752 | 753 | 820 | 822 | 823 | 890 | 892 | 893 | 711 |
| 45-49 | 754 | 755 | 757 | 758 | 825 | 827 | 828 | 895 | 897 | 898 | 712 |
| 50-54 | 759 | 760 | 762 | 763 | 830 | 832 | 833 | 900 | 902 | 903 | 713 |
| 55-59 | 764 | 765 | 767 | 768 | 835 | 837 | 838 | 905 | 907 | 908 | 714 |
| 60-64 | 769 | 770 | 772 | 773 | 840 | 842 | 843 | 910 | 912 | 913 | 715 |
| 65-69 | 774 | 775 | 777 | 778 | 845 | 847 | 848 | 915 | 917 | 918 | 716 |
| 70-74 | 779 | 780 | 782 | 783 | 850 | 852 | 853 | 920 | 922 | 923 | 717 |
| 75+ | 784 | 785 | 787 | 788 | 855 | 857 | 858 | 925 | 927 | 928 | 718 |

Figure 5.3 1981 CENSUS: SAS (Crown Copyright. Reproduced with the permission of the Controller of HMSO)

**CENSUS 1981 SMALL AREA STATISTICS**

**PAGE 2** **100%**

**ED No**

**Map Reference**

**Note:** $ = ED same as in 1971
$$ = Special Enumeration District

 | Frame No:

Separate explanatory notes are available

**11 Household spaces; rooms in household spaces; rooms in hotels and boarding houses**

| Occupancy type | TOTAL H/HOLD SPACES | TOTAL ROOMS |
|---|---|---|
| ALL TYPES OF OCCUPANCY | 1149 | 1159 |
| H/hold enum with usual resident(s) | 1150 | 1160 |
| Absent household | 1151 | 1161 |
| H/hold enum with no usual resident(s) | | |
| Owner occupied | 1152 | 1162 |
| not owner occupied | 1153 | 1163 |
| Second residence (unoccupied at Census) | 1154 | 1164 |
| Holiday accomm (unoccupied at Census) | 1155 | 1165 |
| Vacant (new, never occupied) | 1156 | 1166 |
| Vacant (under improvement) | 1157 | 1167 |
| Vacant (other) | 1158 | 1168 |
| Hotels and boarding houses | xxx | 1169 |

**12 Private households with residents; residents; cars in households**

| | TOTALS | No car | 1 car | 2 cars | 3 or more cars | TOTAL CARS* |
|---|---|---|---|---|---|---|
| Households | 1170 | 1171 | 1172 | 1173 | 1174 | 1175 |
| Persons in households | 1176 | 1177 | 1178 | 1179 | 1180 | xxx |

* H/hlds with 3 or more cars are counted as having 3 cars

**10 Private households (H) with residents; residents (P)**

| Tenure | | TOTALS | Bath and inside WC present: Both excl | Bath and inside WC present: One/both shared | Lack bath or inside WC | Neither bath nor inside WC | Lack bath | Lack inside WC | Share inside WC | Persons per room: >1.5 | Persons per room: >1-1.5 | No car |
|---|---|---|---|---|---|---|---|---|---|---|---|---|
| ALL HOUSEHOLDS | H | 929 | 930 | 931 | 932 | 933 | 934 | 935 | 936 | 945 | 946 | 949 |
| | P | 937 | 938 | 939 | 940 | 941 | 942 | 943 | 944 | 947 | 948 | 950 |
| All permanent | H | 951 | 952 | 953 | 954 | 955 | 956 | 957 | 958 | 1079 | 1080 | 1111 |
| | P | 959 | 960 | 961 | 962 | 963 | 964 | 965 | 966 | 1081 | 1082 | 1112 |
| Owner occupied | H | 967 | 968 | 969 | 970 | 971 | 972 | 973 | 974 | 1083 | 1084 | 1113 |
| | P | 975 | 976 | 977 | 978 | 979 | 980 | 981 | 982 | 1085 | 1086 | 1114 |
| Council etc | H | 983 | 984 | 985 | 986 | 987 | 988 | 989 | 990 | 1087 | 1088 | 1115 |
| | P | 991 | 992 | 993 | 994 | 995 | 996 | 997 | 998 | 1089 | 1090 | 1116 |
| Housing association | H | 999 | 1000 | 1001 | 1002 | 1003 | 1004 | 1005 | 1006 | 1091 | 1092 | 1117 |
| | P | 1007 | 1008 | 1009 | 1010 | 1011 | 1012 | 1013 | 1014 | 1093 | 1094 | 1118 |
| Rented with business | H | 1015 | 1016 | 1017 | 1018 | 1019 | 1020 | 1021 | 1022 | 1095 | 1096 | 1119 |
| | P | 1023 | 1024 | 1025 | 1026 | 1027 | 1028 | 1029 | 1030 | 1097 | 1098 | 1120 |
| By virtue of employment | H | 1031 | 1032 | 1033 | 1034 | 1035 | 1036 | 1037 | 1038 | 1099 | 1100 | 1121 |
| | P | 1039 | 1040 | 1041 | 1042 | 1043 | 1044 | 1045 | 1046 | 1101 | 1102 | 1122 |
| Other rented unfurnished | H | 1047 | 1048 | 1049 | 1050 | 1051 | 1052 | 1053 | 1054 | 1103 | 1104 | 1123 |
| | P | 1055 | 1056 | 1057 | 1058 | 1059 | 1060 | 1061 | 1062 | 1105 | 1106 | 1124 |
| Other rented furnished | H | 1063 | 1064 | 1065 | 1066 | 1067 | 1068 | 1069 | 1070 | 1107 | 1108 | 1125 |
| | P | 1071 | 1072 | 1073 | 1074 | 1075 | 1076 | 1077 | 1078 | 1109 | 1110 | 1126 |
| Non-permanent | H | 1127 | 1128 | 1129 | 1130 | 1131 | 1132 | 1133 | 1134 | 1143 | 1144 | 1147 |
| | P | 1135 | 1136 | 1137 | 1138 | 1139 | 1140 | 1141 | 1142 | 1145 | 1146 | 1148 |

**13 Private households (H) with residents; residents (P); rooms in household spaces**

| Tenure | | Households with the following rooms: 1 | 2 | 3 | 4 | 5 | 6 | 7+ | TOTALS | TOTAL ROOMS |
|---|---|---|---|---|---|---|---|---|---|---|
| All permanent | H | 1182 | 1183 | 1184 | 1185 | 1186 | 1187 | 1188 | 1189 | 1190 |
| | P | 1191 | 1192 | 1193 | 1194 | 1195 | 1196 | 1197 | 1198 | xxx |
| Owner occupied | H | 1200 | 1201 | 1202 | 1203 | 1204 | 1205 | 1206 | 1207 | 1208 |
| | P | 1209 | 1210 | 1211 | 1212 | 1213 | 1214 | 1215 | 1216 | xxx |
| Council etc | H | 1218 | 1219 | 1220 | 1221 | 1222 | 1223 | 1224 | 1225 | 1226 |
| | P | 1227 | 1228 | 1229 | 1230 | 1231 | 1232 | 1233 | 1234 | xxx |
| Housing association | H | 1236 | 1237 | 1238 | 1239 | 1240 | 1241 | 1242 | 1243 | 1244 |
| | P | 1245 | 1246 | 1247 | 1248 | 1249 | 1250 | 1251 | 1252 | xxx |
| Rented with business | H | 1254 | 1255 | 1256 | 1257 | 1258 | 1259 | 1260 | 1261 | 1262 |
| | P | 1263 | 1264 | 1265 | 1266 | 1267 | 1268 | 1269 | 1270 | xxx |
| By virtue of employment | H | 1272 | 1273 | 1274 | 1275 | 1276 | 1277 | 1278 | 1279 | 1280 |
| | P | 1281 | 1282 | 1283 | 1284 | 1285 | 1286 | 1287 | 1288 | xxx |
| Other rented unfurnished | H | 1290 | 1291 | 1292 | 1293 | 1294 | 1295 | 1296 | 1297 | 1298 |
| | P | 1299 | 1300 | 1301 | 1302 | 1303 | 1304 | 1305 | 1306 | xxx |
| Other rented furnished | H | 1308 | 1309 | 1310 | 1311 | 1312 | 1313 | 1314 | 1315 | 1316 |
| | P | 1317 | 1318 | 1319 | 1320 | 1321 | 1322 | 1323 | 1324 | xxx |
| Non-permanent | H | 1326 | 1327 | 1328 | 1329 | 1330 | xxx | xxx | 1333 | 1334 |
| | P | 1335 | 1336 | 1337 | 1338 | 1339 | xxx | xxx | 1342 | xxx |

**14 Private h/holds with residents; residents; rooms in h/hold spaces**

| H/holds with the following persons | Households with the following rooms | | | | | | | TOTAL H/HLDS | TOTAL ROOMS |
|---|---|---|---|---|---|---|---|---|---|
| | 1 | 2 | 3 | 4 | 5 | 6 | 7+ | | |
| ALL H/HLDS | 1344 | 1345 | 1346 | 1347 | 1348 | 1349 | 1350 | 1351 | 1352 |
| 1 | 1353 | 1354 | 1355 | 1356 | 1357 | 1358 | 1359 | 1360 | 1361 |
| 2 | 1362 | 1363 | 1364 | 1365 | 1366 | 1367 | 1368 | 1369 | 1370 |
| 3 | 1371 | 1372 | 1373 | 1374 | 1375 | 1376 | 1377 | 1378 | 1379 |
| 4 | 1380 | 1381 | 1382 | 1383 | 1384 | 1385 | 1386 | 1387 | 1388 |
| 5 | 1389 | 1390 | 1391 | 1392 | 1393 | 1394 | 1395 | 1396 | 1397 |
| 6 | 1398 | 1399 | 1400 | 1401 | 1402 | 1403 | 1404 | 1405 | 1406 |
| 7+ | 1407 | 1408 | 1409 | 1410 | 1411 | 1412 | 1413 | 1414 | 1415 |
| TOTAL PERS | 1416 | 1417 | 1418 | 1419 | 1420 | 1421 | 1422 | 1423 | xxx |

**15 Private households with residents**

| H/holds with the following persons | Tenure of households in permanent buildings | | | | | | | | Households in non-perm accom |
|---|---|---|---|---|---|---|---|---|---|
| | All perm-anent | Owner occ | Council etc | House assoc | Rented with business | By virtue of emp | Other rented | | |
| | | | | | | | Unfurn | Furn | |
| ALL H/HOLDS | 1425 | 1426 | 1427 | 1428 | 1429 | 1430 | 1431 | 1432 | 1489 |
| 1 | 1433 | 1434 | 1435 | 1436 | 1437 | 1438 | 1439 | 1440 | 1490 |
| 2 | 1441 | 1442 | 1443 | 1444 | 1445 | 1446 | 1447 | 1448 | 1491 |
| 3 | 1449 | 1450 | 1451 | 1452 | 1453 | 1454 | 1455 | 1456 | 1492 |
| 4 | 1457 | 1458 | 1459 | 1460 | 1461 | 1462 | 1463 | 1464 | 1493 |
| 5 | 1465 | 1466 | 1467 | 1468 | 1469 | 1470 | 1471 | 1472 | 1494 |
| 6 | 1473 | 1474 | 1475 | 1476 | 1477 | 1478 | 1479 | 1480 | 1495 |
| 7+ | 1481 | 1482 | 1483 | 1484 | 1485 | 1486 | 1487 | 1488 | 1496 |

**16 Priv h/holds with pers present but no residents; pers present; rooms and cars in such h/holds**

| | TOTAL H/HOLDS | Bath + inside wc excl | TOTAL PERS PRES | Stdt aged 16+ | TOTAL ROOMS | TOTAL CARS* |
|---|---|---|---|---|---|---|
| H/holds with no usually res persons | 1497 | 1498 | 1499 | 1500 | 1501 | 1502 |

* H/holds with 3+ cars are counted as having 3 cars

**17 Line 1: 1981 private h/holds with pers present (1971 pop base); present residents and visitors; rooms**
**Line 2: 1981 private households* (1981 pop base); present and absent residents; rooms**

| | H/holds (1981) with the following persons | | | | | | | | TOTAL HOUSE-HOLDS | TOTAL PERS (1981) | TOTAL ROOMS (1981) |
|---|---|---|---|---|---|---|---|---|---|---|---|
| | 0 | 1 | 2 | 3 | 4 | 5 | 6 | 7+ | | | |
| 1: 1971 pop base | xxx | 1504 | 1505 | 1506 | 1507 | 1508 | 1509 | 1510 | 1511 | 1512 | 1513 |
| 2: 1981 pop base | 1514 | 1515 | 1516 | 1517 | 1518 | 1519 | 1520 | 1521 | 1522 | 1523 | 1524 |

*Private households with 1 or more usual residents with at least 1 person (a resident or a visitor) present, or with a visitor or visitors present but no usual residents ie a household with '0 persons'

Figure 5.3 (continued)

**CENSUS 1981 SMALL AREA STATISTICS**

**PAGE 3** 100%

ED No

Map Reference

Note: $ = ED same as in 1971
$$ = Special Enumeration District

Crown copyright reserved | Frame No:

Separate explanatory notes are available

**NOTES**

* Persons aged 16 or over (Tables 18 and 22)
** Includes households with no persons usually resident aged 16 or over (Table 18)
\# Includes a small number of heads aged under 16 counted in the 16-29 row (Table 26)
\# Includes residents in households with heads aged under 16 (Table 26)

**18 Private households with residents; residents**

| Households with the following adults * | Households with no person aged 0-15 | Households with person(s) aged 0-15 | | | | | | | Persons in households | | | | |
|---|---|---|---|---|---|---|---|---|---|---|---|---|---|
| | | Households | With one person aged 0-15 | | With two or more aged 0-15 | | | With no person aged 0-15 | | With person(s) aged 0-15 | | |
| | | | Aged 0-4 | Aged 5-15 | All Aged 0-4 | All Aged 5-15 | Others | TOTAL PERS | Pers econ act | Pers aged 0-15 | Pers aged 16+ | Pers econ act |
| ALL HOUSEHOLDS ** | 1525 | 1526 | 1527 | 1528 | 1529 | 1530 | 1531 | 1574 | 1575 | 1576 | 1577 | 1578 |
| 1 Male | 1532 | 1533 | 1534 | 1535 | 1536 | 1537 | 1538 | 1579 | 1580 | 1581 | 1582 | 1583 |
| 1 Female | 1539 | 1540 | 1541 | 1542 | 1543 | 1544 | 1545 | 1584 | 1585 | 1586 | 1587 | 1588 |
| 2 (Married male + married female) | 1546 | 1547 | 1548 | 1549 | 1550 | 1551 | 1552 | 1589 | 1590 | 1591 | 1592 | 1593 |
| 2 (Other) | 1553 | 1554 | 1555 | 1556 | 1557 | 1558 | 1559 | 1594 | 1595 | 1596 | 1597 | 1598 |
| 3+ (Married male(s) + married female(s) with or without others) | 1560 | 1561 | 1562 | 1563 | 1564 | 1565 | 1566 | 1599 | 1600 | 1601 | 1602 | 1603 |
| 3+ (Other) | 1567 | 1568 | 1569 | 1570 | 1571 | 1572 | 1573 | 1604 | 1605 | 1606 | 1607 | 1608 |

**19 Married women resident in private households of married male plus one married female with or without others; number of persons aged 0-15 in such households**

| In households with: | TOTAL MRR'D WOMEN | Mrr'd women econ active | Mrr'd women in employment | | |
|---|---|---|---|---|---|
| | | | TOTAL | Working full-time | Working pt-time |
| No person aged 0-15 | 1609 | 1610 | 1617 | 1618 | 1619 |
| Person(s) aged 0-4 with or without any aged 5-15 | 1611 | 1612 | 1620 | 1621 | 1622 |
| Person(s) aged | 1613 | 1614 | 1623 | 1624 | 1625 |

**20 Residents aged 16 or over in private households**

| Economic position | TOTAL PERS | Males | | Females | |
|---|---|---|---|---|---|
| | | SWD | Mrr'd | SWD | Mrr'd |
| ALL PERSONS 16+ | 1629 | 1630 | 1631 | 1632 | 1633 |
| Total econ act | 1634 | 1635 | 1636 | 1637 | 1638 |
| Working f/time | 1639 | 1640 | 1641 | 1642 | 1643 |
| Working p/time | 1644 | 1645 | 1646 | 1647 | 1648 |
| Seeking work | 1649 | 1650 | 1651 | 1652 | 1653 |
| Temp sick | 1654 | 1655 | 1656 | 1657 | 1658 |
| Total econ inact | 1659 | 1660 | 1661 | 1662 | 1663 |
| Perm sick | 1664 | 1665 | 1666 | 1667 | 1668 |
| Retired | 1669 | 1670 | 1671 | 1672 | 1673 |
| Student | 1674 | 1675 | 1676 | 1677 | 1678 |
| Other inactive | 1679 | 1680 | 1681 | 1682 | 1683 |

**21 Residents in private households**

| Age | TOTAL PERSONS | Males | | Females | |
|---|---|---|---|---|---|
| | | SWD | Mrr'd | SWD | Mrr'd |
| TOTAL | 1684 | 1686 | 1687 | 1689 | 1690 |
| 0-4 | 1691 | 1693 | xxx | 1696 | xxx |
| 5-9 | 1698 | 1700 | xxx | 1703 | xxx |
| 10-14 | 1705 | 1707 | xxx | 1710 | xxx |
| 15 | 1712 | 1714 | xxx | 1717 | xxx |
| 16-19 | 1719 | 1721 | 1722 | 1724 | 1725 |
| 20-24 | 1726 | 1728 | 1729 | 1731 | 1732 |
| 25-29 | 1733 | 1735 | 1736 | 1738 | 1739 |
| 30-34 | 1740 | 1742 | 1743 | 1745 | 1746 |
| 35-39 | 1747 | 1749 | 1750 | 1752 | 1753 |

| | | | | | | |
|---|---|---|---|---|---|---|
| 5-15 only | | | | | | |
| TOTAL PERSONS AGED 0-15 | 1615 | 1616 | 1626 | 1627 | 1628 | |

| Age | | | | | |
|---|---|---|---|---|---|
| 40-44 | 1754 | 1756 | 1757 | 1759 | 1760 |
| 45-49 | 1761 | 1763 | 1764 | 1766 | 1767 |
| 50-54 | 1768 | 1770 | 1771 | 1773 | 1774 |
| 55-59 | 1775 | 1777 | 1778 | 1780 | 1781 |
| 60-64 | 1782 | 1784 | 1785 | 1787 | 1788 |
| 65-69 | 1789 | 1791 | 1792 | 1794 | 1795 |
| 70-74 | 1796 | 1798 | 1799 | 1801 | 1802 |
| 75-79 | 1803 | 1805 | 1806 | 1808 | 1809 |
| 80-84 | 1810 | 1812 | 1813 | 1815 | 1816 |
| 85+ | 1817 | 1819 | 1820 | 1822 | 1823 |

**22 Private households with residents; residents aged 16 or over**

| Households with the following adults * | Households with the following persons aged 0-15 | | | | | |
|---|---|---|---|---|---|---|
| | TOTAL | 0 | 1 | 2 | 3 | 4+ |
| TOTAL HOUSEHOLDS | 1824 | 1825 | 1826 | 1827 | 1828 | 1829 |
| 1 econ inactive | 1830 | 1831 | 1832 | 1833 | 1834 | 1835 |
| 1 econ active | 1836 | 1837 | 1838 | 1839 | 1840 | 1841 |
| 2+, all econ inactive | 1842 | 1843 | 1844 | 1845 | 1846 | 1847 |
| 2+, 1 econ active | 1848 | 1849 | 1850 | 1851 | 1852 | 1853 |
| 2+, 2+ econ active | 1854 | 1855 | 1856 | 1857 | 1858 | 1859 |
| TOTAL ADULTS* | 1860 | 1861 | 1862 | 1863 | 1864 | 1865 |
| Persons econ active | 1866 | 1867 | 1868 | 1869 | 1870 | 1871 |
| Pers econ act out of employmnt (inc above) | 1872 | 1873 | 1874 | 1875 | 1876 | 1877 |

**23 Married women resident in priv households**

| Age | TOTAL MARRIED WOMEN | Married women econ active | Married women in empl | | |
|---|---|---|---|---|---|
| | | | TOTAL | Work f/time | Work p/time |
| TOTAL | 1878 | 1879 | 1890 | 1891 | 1892 |
| 16-24 | 1880 | 1881 | 1893 | 1894 | 1895 |
| 25-34 | 1882 | 1883 | 1896 | 1897 | 1898 |
| 35-44 | 1884 | 1885 | 1899 | 1900 | 1901 |
| 45-59 | 1886 | 1887 | 1902 | 1903 | 1904 |
| 60+ | 1888 | 1889 | 1905 | 1906 | 1907 |

**24 Residents aged 16-24 in private households**

| Age | Persons | | Married | | Student | | Econ active | | EA out of empl | |
|---|---|---|---|---|---|---|---|---|---|---|
| | Males | Fmles | Males | Fmles | Males | Fmles | Males | Fmles | Males | Fmles |
| TOTAL | 1908 | 1909 | 1928 | 1929 | 1948 | 1949 | 1968 | 1969 | 1988 | 1989 |
| 16 | 1910 | 1911 | 1930 | 1931 | 1950 | 1951 | 1970 | 1971 | 1990 | 1991 |
| 17 | 1912 | 1913 | 1932 | 1933 | 1952 | 1953 | 1972 | 1973 | 1992 | 1993 |
| 18 | 1914 | 1915 | 1934 | 1935 | 1954 | 1955 | 1974 | 1975 | 1994 | 1995 |
| 19 | 1916 | 1917 | 1936 | 1937 | 1956 | 1957 | 1976 | 1977 | 1996 | 1997 |
| 20 | 1918 | 1919 | 1938 | 1939 | 1958 | 1959 | 1978 | 1979 | 1998 | 1999 |
| 21 | 1920 | 1921 | 1940 | 1941 | 1960 | 1961 | 1980 | 1981 | 2000 | 2001 |
| 22 | 1922 | 1923 | 1942 | 1943 | 1962 | 1963 | 1982 | 1983 | 2002 | 2003 |
| 23 | 1924 | 1925 | 1944 | 1945 | 1964 | 1965 | 1984 | 1985 | 2004 | 2005 |
| 24 | 1926 | 1927 | 1946 | 1947 | 1966 | 1967 | 1986 | 1987 | 2006 | 2007 |

**25 Residents aged 0-15 in private households**

| Age | Males | Females |
|---|---|---|
| TOTAL | 2009 | 2010 |
| 0 | 2012 | 2013 |
| 1 | 2015 | 2016 |
| 2 | 2018 | 2019 |
| 3 | 2021 | 2022 |
| 4 | 2024 | 2025 |
| 5 | 2027 | 2028 |
| 6 | 2030 | 2031 |
| 7 | 2033 | 2034 |
| 8 | 2036 | 2037 |
| 9 | 2039 | 2040 |
| 10 | 2042 | 2043 |
| 11 | 2045 | 2046 |
| 12 | 2048 | 2049 |
| 13 | 2051 | 2052 |
| 14 | 2054 | 2055 |
| 15 | 2057 | 2058 |

**26 Residents in private households**

| Age | TOTAL | Males | | | Females | | |
|---|---|---|---|---|---|---|---|
| | | Mrr'd | S | W/D | Mrr'd | S | W/D |
| Persons age | PERSONS AGED 16 OR OVER IN PRIVATE HOUSEHOLDS | | | | | | |
| ALL 16+ | 2059 | 2061 | 2062 | 2063 | 2065 | 2066 | 2067 |
| 16-29 | 2068 | 2070 | 2071 | 2072 | 2074 | 2075 | 2076 |
| 30-44 | 2077 | 2079 | 2080 | 2081 | 2083 | 2084 | 2085 |
| 45-64/59 | 2086 | 2088 | 2089 | 2090 | 2092 | 2093 | 2094 |
| Pensioners | 2095 | 2097 | 2098 | 2099 | 2101 | 2102 | 2103 |
| Heads age | HEADS IN PRIVATE HOUSEHOLDS ## | | | | | | |
| ALL 16+ | 2104 | 2106 | 2107 | 2108 | 2110 | 2111 | 2112 |
| 16-29 | 2113 | 2115 | 2116 | 2117 | 2119 | 2120 | 2121 |
| 30-44 | 2122 | 2124 | 2125 | 2126 | 2128 | 2129 | 2130 |
| 45-64/59 | 2131 | 2133 | 2134 | 2135 | 2137 | 2138 | 2139 |
| Pensioners | 2140 | 2142 | 2143 | 2144 | 2146 | 2147 | 2148 |
| Heads age | ALL PERS IN H/HLDS BY HEADS AGE, SEX, MAR/STAT # | | | | | | |
| ALL 16+ | 2149 | 2151 | 2152 | 2153 | 2155 | 2156 | 2157 |
| 16-29 | 2158 | 2160 | 2161 | 2162 | 2164 | 2165 | 2166 |
| 30-44 | 2167 | 2169 | 2170 | 2171 | 2173 | 2174 | 2175 |
| 45-64/59 | 2176 | 2178 | 2179 | 2180 | 2182 | 2183 | 2184 |
| Pensioners | 2185 | 2187 | 2188 | 2189 | 2191 | 2192 | 2193 |

Figure 5.3 (continued)

**CENSUS 1981 SMALL AREA STATISTICS**

**PAGE 4** 100%

ED No

Map Reference

Note: $ = ED same as in 1971
$$ = Special Enumeration District

Crown copyright reserved | Frame No:

Separate explanatory notes are available

**NOTES**

#Persons aged 16 and over (Tables 27 + 31)

≠Inc households with no persons usually resident aged 16 and over (Table 29)

*Inc renting from LAs, New Town Corporations and Scottish Special Housing Assoc (Table 29)

**27** Lone adults # resident in private households of one adult with residents aged 0-15, number of persons aged 0-15 in such households

| In households with child(ren): | Male lone 'parents' | | | | | Female lone 'parents' | | | | |
|---|---|---|---|---|---|---|---|---|---|---|
| | TOTAL | Econ active | In employment | | | TOTAL | Econ active | In employment | | |
| | | | TOTAL | f/time | p/time | | | TOTAL | f/time | p/time |
| Aged 0-4 with or w/out any aged 5-15 | 2194 | 2195 | 2200 | 2201 | 2202 | 2209 | 2210 | 2215 | 2216 | 2217 |
| Aged 5-15 only | 2196 | 2197 | 2203 | 2204 | 2205 | 2211 | 2212 | 2218 | 2219 | 2220 |
| TOTAL PERSONS AGED 0-15 | 2198 | 2199 | 2206 | 2207 | 2208 | 2213 | 2214 | 2221 | 2222 | 2223 |

**28** Private households with residents not in self-contained accommodation; rooms in such households

| Households with the following persons | H/holds not in self-contained accom | | | | | | |
|---|---|---|---|---|---|---|---|
| | TOTAL | One or more persons per rm | Bath + inside WC excl | Lack bath | Lack inside WC | No car | TOTAL ROOMS |
| TOTAL | 2224 | 2228 | 2232 | 2233 | 2234 | 2248 | 2244 |
| 1 person | 2225 | 2229 | 2235 | 2236 | 2237 | 2249 | 2245 |
| 2 persons | 2226 | 2230 | 2238 | 2239 | 2240 | 2250 | 2246 |
| 3+ persons | 2227 | 2231 | 2241 | 2242 | 2243 | 2251 | 2247 |

**29** Private h/holds with residents; residents aged 0-15, and aged 60+(Fmles) and 65+ (males)

| Household type: Households with persons Aged 16+ | Aged 0-15 | TOTAL HOUSE-HOLDS | Tenure of households in permanent buildings: TOTAL | Owner occ | Coun-cil etc * | Hous-ing assoc | Rented with bus-iness | Rented by virtue of emp | Other rented: Unfurn | Other rented: Furn | House-holds in non-perm accom |
|---|---|---|---|---|---|---|---|---|---|---|---|
| ALL HOUSEHOLDS ≠ | Any | 2252 | 2262 | 2263 | 2264 | 2265 | 2266 | 2267 | 2268 | 2269 | 2342 |
| 1 pensioner | 0 | 2253 | 2270 | 2271 | 2272 | 2273 | 2274 | 2275 | 2276 | 2277 | 2343 |
| 1 adult# under pensionable age | 0 | 2254 | 2278 | 2279 | 2280 | 2281 | 2282 | 2283 | 2284 | 2285 | 2344 |
| 1 adult any age | 1+ | 2255 | 2286 | 2287 | 2288 | 2289 | 2290 | 2291 | 2292 | 2293 | 2345 |
| Mrr'd male with mrr'd female without others | 0 | 2256 | 2294 | 2295 | 2296 | 2297 | 2298 | 2299 | 2300 | 2301 | 2346 |
| | 1+ | 2257 | 2302 | 2303 | 2304 | 2305 | 2306 | 2307 | 2308 | 2309 | 2347 |

| | | | | | | | | | | |
|---|---|---|---|---|---|---|---|---|---|---|
| 3+, mrr'd male(s) with mrr'd female(s) with/without others | 0 | 2258 | 2310 | 2311 | 2312 | 2313 | 2314 | 2315 | 2316 | 2317 | 2348 |
| | 1+ | 2259 | 2318 | 2319 | 2320 | 2321 | 2322 | 2323 | 2324 | 2325 | 2349 |
| 2+ Others | 0 | 2260 | 2326 | 2327 | 2328 | 2329 | 2330 | 2331 | 2332 | 2333 | 2350 |
| | 1+ | 2261 | 2334 | 2335 | 2336 | 2337 | 2338 | 2339 | 2340 | 2341 | 2351 |
| Households containing pers of pens age only (any number) | 0 | 2352 | 2353 | 2354 | 2355 | 2356 | 2357 | 2358 | 2359 | 2360 | 2361 |
| Persons aged 0-4 | | 2362 | 2366 | 2367 | 2368 | 2369 | 2370 | 2371 | 2372 | 2373 | 2398 |
| Persons aged 5-15 | | 2363 | 2374 | 2375 | 2376 | 2377 | 2378 | 2379 | 2380 | 2381 | 2399 |
| Pers of pensionable age up to and including 74 | | 2364 | 2382 | 2383 | 2384 | 2385 | 2386 | 2387 | 2388 | 2389 | 2400 |
| Persons aged 75+ | | 2365 | 2390 | 2391 | 2392 | 2393 | 2394 | 2395 | 2396 | 2397 | 2401 |

**30** **Private h/holds with resident head with different address 1 year before census; residents in such h/holds; all residents in private h/holds with different address 1 yr before census**

| | Households with migrants | | | | | | |
|---|---|---|---|---|---|---|---|
| | TOTAL | One or more pers per rm | Bath + inside WC excl | Lack bath | Lack inside WC | Not self-cont accom | No car |
| | Households with migrant heads | | | | | | |
| Households | 2402 | 2404 | 2406 | 2407 | 2408 | 2412 | 2414 |
| Persons | 2403 | 2405 | 2409 | 2410 | 2411 | 2413 | 2415 |
| | All households with migrants | | | | | | |
| All migrants | 2416 | 2417 | 2418 | 2419 | 2420 | 2421 | 2422 |

**31** **Private households with dependent children; private households with one or more resident(s) aged 0-15; residents in such households**

| | Households with child(ren) | | | | | | |
|---|---|---|---|---|---|---|---|
| | TOTAL | One or more persons per room | Bath + inside WC excl | Lack bath | Lack inside WC | Not self-cont accom | No car |
| ALL H/HOLDS WITH DEPENDENT CHILD(REN) | 2423 | 2424 | 2425 | 2426 | 2427 | 2428 | 2429 |
| H/holds containing at least one one parent family with dep child(ren) | 2430 | 2431 | 2432 | 2433 | 2434 | 2435 | 2436 |
| H/holds with 3 or more dep children | 2437 | 2438 | 2439 | 2440 | 2441 | 2442 | 2443 |
| ALL H/HOLDS WITH ONE OR MORE PERSONS AGED 0-15 | 2444 | 2453 | 2462 | 2463 | 2464 | 2489 | 2498 |
| H/holds of 1 adult# + 1 or more 0-15 | | | | | | | |
| Households | 2445 | 2454 | 2465 | 2466 | 2467 | 2490 | 2499 |
| Persons 0-4 | 2446 | 2455 | 2468 | 2469 | 2470 | 2491 | 2500 |
| Persons 5-15 | 2447 | 2456 | 2471 | 2472 | 2473 | 2492 | 2501 |
| Other h/holds with persons 0-15 | | | | | | | |
| Households | 2448 | 2457 | 2474 | 2475 | 2476 | 2493 | 2502 |
| Adults# | 2449 | 2458 | 2477 | 2478 | 2479 | 2494 | 2503 |
| Persons 0-4 | 2450 | 2459 | 2480 | 2481 | 2482 | 2495 | 2504 |
| Persons 5-15 | 2451 | 2460 | 2483 | 2484 | 2485 | 2496 | 2505 |
| H/holds with 3 or more persons 0-15 | 2452 | 2461 | 2486 | 2487 | 2488 | 2497 | 2506 |

**32** **Private households with one or more residents of pensionable age, residents in such households**

| | Households with pensioners | | | | | | |
|---|---|---|---|---|---|---|---|
| | TOTAL | One or more persons per room | Bath + inside WC excl | Lack bath | Lack inside WC | Not self-cont accom | No car |
| TOTAL H/HOLDS WITH 1 OR MORE PENSIONERS | 2507 | 2516 | 2525 | 2526 | 2527 | 2552 | 2561 |
| Lone male 65-74 | 2508 | 2517 | 2528 | 2529 | 2530 | 2553 | 2562 |
| Lone male 75+ | 2509 | 2518 | 2531 | 2532 | 2533 | 2554 | 2563 |
| Lone female 60-74 | 2510 | 2519 | 2534 | 2535 | 2536 | 2555 | 2564 |
| Lone female 75+ | 2511 | 2520 | 2537 | 2538 | 2539 | 2556 | 2565 |
| 2+ all pens >75 | 2512 | 2521 | 2540 | 2541 | 2542 | 2557 | 2566 |
| 2+, all pens, any 75+ | 2513 | 2522 | 2543 | 2544 | 2545 | 2558 | 2567 |
| 1 or more pensioners with 1 non-pensioner | 2514 | 2523 | 2546 | 2547 | 2548 | 2559 | 2568 |
| 1 or more pens with 2 or more non-pens | 2515 | 2524 | 2549 | 2550 | 2551 | 2560 | 2569 |
| TOTAL PERS IN H/HOLDS WITH PENSIONERS | 2570 | 2574 | 2578 | 2579 | 2580 | 2590 | 2594 |
| Total pens persons | 2571 | 2575 | 2581 | 2582 | 2583 | 2591 | 2595 |
| Total persons 75+ | 2572 | 2576 | 2584 | 2585 | 2586 | 2592 | 2596 |
| Total persons 85+ | 2573 | 2577 | 2587 | 2588 | 2589 | 2593 | 2597 |

Figure 5.3 (Continued)

| CENSUS 1981 SMALL AREA STATISTICS | |
|---|---|
| PAGE 5 10% | |
| ED No | |
| Map Reference | |
| Note: $ = ED same as in 1971<br>$$ = Special Enumeration District<br>† = Importing Enumeration District | |
| Crown copyright reserved | Frame No: |
| Separate explanatory notes are available | |

The figures in these tables are a 10% sample of the Census

**45** Residents; private households with residents (100% + 10% sample)

| | Resident persons | | | PRIVATE H/HOLDS |
|---|---|---|---|---|
| | TOTAL | Not in private h/holds | In private h/holds | |
| 100% | 4442 | 4443 | 4444 | 4445 |
| 10% | 4446 | 4447 | 4448 | 4449 |

**44** Residents aged 16 or over in employment (10% sample)

| Socio-economic group (SEG) | Industry | | | | | | | | Working full time | Working part time | Working outside dist of residence |
|---|---|---|---|---|---|---|---|---|---|---|---|
| | Agric | Energy and Water | Manuf | Constr | Distrib and Catering | Trans-port | Other Services | TOTAL IN EMPL | | | |
| 1 | 4223 | 4224 | 4225 | 4226 | 4227 | 4228 | 4229 | 4230 | 4375 | 4376 | 4413 |
| 2 | 4231 | 4232 | 4233 | 4234 | 4235 | 4236 | 4237 | 4238 | 4377 | 4378 | 4414 |
| 3 | 4239 | 4240 | 4241 | 4242 | 4243 | 4244 | 4245 | 4246 | 4379 | 4380 | 4415 |
| 4 | 4247 | 4248 | 4249 | 4250 | 4251 | 4252 | 4253 | 4254 | 4381 | 4382 | 4416 |
| 5.1 | 4255 | 4256 | 4257 | 4258 | 4259 | 4260 | 4261 | 4262 | 4383 | 4384 | 4417 |
| 5.2 | 4263 | 4264 | 4265 | 4266 | 4267 | 4268 | 4269 | 4270 | 4385 | 4386 | 4418 |
| 6 | 4271 | 4272 | 4273 | 4274 | 4275 | 4276 | 4277 | 4278 | 4387 | 4388 | 4419 |
| 7 | 4279 | 4280 | 4281 | 4282 | 4283 | 4284 | 4285 | 4286 | 4389 | 4390 | 4420 |
| 8 | 4287 | 4288 | 4289 | 4290 | 4291 | 4292 | 4293 | 4294 | 4391 | 4392 | 4421 |
| 9 | 4295 | 4296 | 4297 | 4298 | 4299 | 4300 | 4301 | 4302 | 4393 | 4394 | 4422 |
| 10 | 4303 | 4304 | 4305 | 4306 | 4307 | 4308 | 4309 | 4310 | 4395 | 4396 | 4423 |
| 11 | 4311 | 4312 | 4313 | 4314 | 4315 | 4316 | 4317 | 4318 | 4397 | 4398 | 4424 |
| 12 | 4319 | 4320 | 4321 | 4322 | 4323 | 4324 | 4325 | 4326 | 4399 | 4400 | 4425 |
| 13 | 4327 | 4328 | 4329 | 4330 | 4331 | 4332 | 4333 | 4334 | 4401 | 4402 | 4426 |
| 14 | 4335 | 4336 | 4337 | 4338 | 4339 | 4340 | 4341 | 4342 | 4403 | 4404 | 4427 |
| 15 | 4343 | 4344 | 4345 | 4346 | 4347 | 4348 | 4349 | 4350 | 4405 | 4406 | 4428 |
| 16 | 4351 | 4352 | 4353 | 4354 | 4355 | 4356 | 4357 | 4358 | 4407 | 4408 | 4429 |
| 17 | 4359 | 4360 | 4361 | 4362 | 4363 | 4364 | 4365 | 4366 | 4409 | 4410 | 4430 |
| TOTAL | 4367 | 4368 | 4369 | 4370 | 4371 | 4372 | 4373 | 4374 | 4411 | 4412 | 4431 |
| Working outside distr of residence | 4432 | 4433 | 4434 | 4435 | 4436 | 4437 | 4438 | 4439 | 4440 | 4441 | xxx |

**46 Residents aged 16 or over in employment (10% sample)**

| Sex and age | Industry | | | | | | | |
|---|---|---|---|---|---|---|---|---|
| | TOTAL IN EMPL | Agric | Energy and Water | Manuf | Constr | Distrib and Catering | Trans-port | Other Services |
| Males 16+ | 4450 | 4451 | 4452 | 4453 | 4454 | 4455 | 4456 | 4457 |
| 16-29 | 4458 | 4459 | 4460 | 4461 | 4462 | 4463 | 4464 | 4465 |
| 30-44 | 4466 | 4467 | 4468 | 4469 | 4470 | 4471 | 4472 | 4473 |
| 45-64 | 4474 | 4475 | 4476 | 4477 | 4478 | 4479 | 4480 | 4481 |
| 65+ | 4482 | 4483 | 4484 | 4485 | 4486 | 4487 | 4488 | 4489 |
| Fmls 16+ | 4490 | 4491 | 4492 | 4493 | 4494 | 4495 | 4496 | 4497 |
| 16-29 | 4498 | 4499 | 4500 | 4501 | 4502 | 4503 | 4504 | 4505 |
| 30-44 | 4506 | 4507 | 4508 | 4509 | 4510 | 4511 | 4512 | 4513 |
| 45-59 | 4514 | 4515 | 4516 | 4517 | 4518 | 4519 | 4520 | 4521 |
| 60+ | 4522 | 4523 | 4524 | 4525 | 4526 | 4527 | 4528 | 4529 |

**47 Residents aged 16 or over in employment (10% sample)**

| SEG | Means of travel to work | | | | | | | | | | |
|---|---|---|---|---|---|---|---|---|---|---|---|
| | Car pool | Car psngr | Car driver | Bus | BR train | Under-ground | Motor-cycle | Pedal-cycle | On foot | Other + n/s | Works at home |
| 1 | 4530 | 4531 | 4532 | 4533 | 4534 | 4535 | 4536 | 4537 | 4538 | 4539 | 4540 |
| 2 | 4541 | 4542 | 4543 | 4544 | 4545 | 4546 | 4547 | 4548 | 4549 | 4550 | 4551 |
| 3 | 4552 | 4553 | 4554 | 4555 | 4556 | 4557 | 4558 | 4559 | 4560 | 4561 | 4562 |
| 4 | 4563 | 4564 | 4565 | 4566 | 4567 | 4568 | 4569 | 4570 | 4571 | 4572 | 4573 |
| 5.1 | 4574 | 4575 | 4576 | 4577 | 4578 | 4579 | 4580 | 4581 | 4682 | 4583 | 4584 |
| 5.2 | 4585 | 4586 | 4587 | 4588 | 4589 | 4590 | 4591 | 4592 | 4593 | 4594 | 4595 |
| 6 | 4596 | 4597 | 4598 | 4599 | 4600 | 4601 | 4602 | 4603 | 4604 | 4605 | 4606 |
| 7 | 4607 | 4608 | 4609 | 4610 | 4611 | 4612 | 4613 | 4614 | 4615 | 4616 | 4617 |
| 8 | 4618 | 4619 | 4620 | 4621 | 4622 | 4623 | 4624 | 4625 | 4626 | 4627 | 4628 |
| 9 | 4629 | 4630 | 4631 | 4632 | 4633 | 4634 | 4635 | 4636 | 4637 | 4638 | 4639 |
| 10 | 4640 | 4641 | 4642 | 4643 | 4644 | 4645 | 4646 | 4647 | 4648 | 4649 | 4650 |
| 11 | 4651 | 4652 | 4653 | 4654 | 4655 | 4656 | 4657 | 4658 | 4659 | 4660 | 4661 |
| 12 | 4662 | 4663 | 4664 | 4665 | 4666 | 4667 | 4668 | 4669 | 4670 | 4671 | 4672 |
| 13 | 4673 | 4674 | 4675 | 4676 | 4677 | 4678 | 4679 | 4680 | 4681 | 4682 | 4683 |
| 14 | 4684 | 4685 | 4686 | 4687 | 4688 | 4689 | 4690 | 4691 | 4692 | 4693 | 4694 |
| 15 | 4695 | 4696 | 4697 | 4698 | 4699 | 4700 | 4701 | 4702 | 4703 | 4704 | 4705 |
| 16 | 4706 | 4707 | 4708 | 4709 | 4710 | 4711 | 4712 | 4713 | 4714 | 4715 | 4716 |
| 17 | 4717 | 4718 | 4719 | 4720 | 4721 | 4722 | 4723 | 4724 | 4725 | 4726 | 4727 |
| Total pers | 4728 | 4729 | 4730 | 4731 | 4732 | 4733 | 4734 | 4735 | 4736 | 4737 | 4738 |
| Total mls | 4739 | 4740 | 4741 | 4742 | 4743 | 4744 | 4745 | 4746 | 4747 | 4748 | 4749 |
| Total fmls | 4750 | 4751 | 4752 | 4753 | 4754 | 4755 | 4756 | 4757 | 4758 | 4759 | 4760 |
| Persons working o/s distr of resid | 4761 | 4762 | 4763 | 4764 | 4765 | 4766 | 4767 | 4768 | 4769 | 4770 | xxx |
| | Persons in employment in private households with: | | | | | | | | | | |
| No car | 4772 | 4773 | 4774 | 4775 | 4776 | 4777 | 4778 | 4779 | 4780 | 4781 | 4782 |
| 1 car | 4783 | 4784 | 4785 | 4786 | 4787 | 4788 | 4789 | 4790 | 4791 | 4792 | 4793 |
| 2+ cars | 4794 | 4795 | 4796 | 4797 | 4798 | 4799 | 4800 | 4801 | 4802 | 4803 | 4804 |

**48 Residents aged 18 or over (10% sample)**

| Age | Persons with degrees, professional and vocational qualifications | | |
|---|---|---|---|
| | TOTAL | Males | Females |
| 18-29 | 4805 | 4806 | 4807 |
| 30-44 | 4808 | 4809 | 4810 |
| 45-64/59 | 4811 | 4812 | 4813 |
| Pensioners | 4814 | 4815 | 4816 |
| In employment | 4817 | 4818 | 4819 |

Figure 5.3 (Continued)

**CENSUS 1981 SMALL AREA STATISTICS**

**PAGE 6** **10%**

ED No

Map Reference

Note: $ = ED same as in 1971
$$ = Special Enumeration District
† = Importing Enumeration District

 Frame No:

Separate explanatory notes are available

**The figures in these tables are a 10% sample of the Census**

**50 Residents, economically active or retired (10% sample)**

| SEG | Economic position: All res econ act or retired | EA not in emp: Males | EA not in emp: Females SWD | EA not in emp: Females Mrrd | Retd males | Econ active: Males | Econ active: Females SWD | Econ active: Females Mrrd | Econ active mig-rant |
|---|---|---|---|---|---|---|---|---|---|
| 1 | 5147 | 5166 | 5167 | 5168 | 5223 | 5242 | 5243 | 5244 | 5299 |
| 2 | 5148 | 5169 | 5170 | 5171 | 5224 | 5245 | 5246 | 5247 | 5300 |
| 3 | 5149 | 5172 | 5173 | 5174 | 5225 | 5248 | 5249 | 5250 | 5301 |
| 4 | 5150 | 5175 | 5176 | 5177 | 5226 | 5251 | 5252 | 5253 | 5302 |
| 5.1 | 5151 | 5178 | 5179 | 5180 | 5227 | 5254 | 5255 | 5256 | 5303 |
| 5.2 | 5152 | 5181 | 5182 | 5183 | 5228 | 5257 | 5258 | 5259 | 5304 |
| 6 | 5153 | 5184 | 5185 | 5186 | 5229 | 5260 | 5261 | 5262 | 5305 |
| 7 | 5154 | 5187 | 5188 | 5189 | 5230 | 5263 | 5264 | 5265 | 5306 |
| 8 | 5155 | 5190 | 5191 | 5192 | 5231 | 5266 | 5267 | 5268 | 5307 |
| 9 | 5156 | 5193 | 5194 | 5195 | 5232 | 5269 | 5270 | 5271 | 5308 |
| 10 | 5157 | 5196 | 5197 | 5198 | 5233 | 5272 | 5273 | 5274 | 5309 |
| 11 | 5158 | 5199 | 5200 | 5201 | 5234 | 5275 | 5276 | 5277 | 5310 |

**49 Private households with residents; residents; families of resident persons (10% sample)**

| | Households | | | | | | | Persons | | | Families by family type | |
|---|---|---|---|---|---|---|---|---|---|---|---|---|
| | Selected tenures of h/hlds in perm blgs | | | Migrant head of h/hold | No car | Ret head of h/hold | ALL HOUSE-HOLDS | ALL PERS | Pers EA | Dep child-ren | | |
| | Owner occ | Council etc | Unfurn-ished | | | | | | | | Mrrd couple | Lone parent |
| SEG | By SEG of EA or retired (with previous occupation stated) head of household | | | | | | | | | | head of family | |
| 1 | 4820 | 4821 | 4822 | 4877 | 4896 | 4915 | 4934 | 4953 | 4954 | 4955 | 5010 | 5011 |
| 2 | 4823 | 4824 | 4825 | 4878 | 4897 | 4916 | 4935 | 4956 | 4957 | 4958 | 5012 | 5013 |
| 3 | 4826 | 4827 | 4828 | 4879 | 4898 | 4917 | 4936 | 4959 | 4960 | 4961 | 5014 | 5015 |
| 4 | 4829 | 4830 | 4831 | 4880 | 4899 | 4918 | 4937 | 4962 | 4963 | 4964 | 5016 | 5017 |
| 5.1 | 4832 | 4833 | 4834 | 4881 | 4900 | 4919 | 4938 | 4965 | 4966 | 4967 | 5018 | 5019 |
| 5.2 | 4835 | 4836 | 4837 | 4882 | 4901 | 4920 | 4939 | 4968 | 4969 | 4970 | 5020 | 5021 |
| 6 | 4838 | 4839 | 4840 | 4883 | 4902 | 4921 | 4940 | 4971 | 4972 | 4973 | 5022 | 5023 |
| 7 | 4841 | 4842 | 4843 | 4884 | 4903 | 4922 | 4941 | 4974 | 4975 | 4976 | 5024 | 5025 |
| 8 | 4844 | 4845 | 4846 | 4885 | 4904 | 4923 | 4942 | 4977 | 4978 | 4979 | 5026 | 5027 |
| 9 | 4847 | 4848 | 4849 | 4886 | 4905 | 4924 | 4943 | 4980 | 4981 | 4982 | 5028 | 5029 |
| 10 | 4850 | 4851 | 4852 | 4887 | 4906 | 4925 | 4944 | 4983 | 4984 | 4985 | 5030 | 5031 |
| 11 | 4853 | 4854 | 4855 | 4888 | 4907 | 4926 | 4945 | 4986 | 4987 | 4988 | 5032 | 5033 |
| 12 | 4856 | 4857 | 4858 | 4889 | 4908 | 4927 | 4946 | 4989 | 4990 | 4991 | 5034 | 5035 |
| 13 | 4859 | 4860 | 4861 | 4890 | 4909 | 4928 | 4947 | 4992 | 4993 | 4994 | 5036 | 5037 |
| 14 | 4862 | 4863 | 4864 | 4891 | 4910 | 4929 | 4948 | 4995 | 4996 | 4997 | 5038 | 5039 |
| 15 | 4865 | 4866 | 4867 | 4892 | 4911 | 4930 | 4949 | 4998 | 4999 | 5000 | 5040 | 5041 |
| 16 | 4868 | 4869 | 4870 | 4893 | 4912 | 4931 | 4950 | 5001 | 5002 | 5003 | 5042 | 5043 |
| 17 | 4871 | 4872 | 4873 | 4894 | 4913 | 4932 | 4951 | 5004 | 5005 | 5006 | 5044 | 5045 |
| Total (inc never act) | 4874 | 4875 | 4876 | 4895 | 4914 | 4933 | 4952 | 5007 | 5008 | 5009 | 5046 | 5047 |
| Retired 1 | 5048 | 5049 | 5050 | 5057 | 5060 | 5063 | 5066 | 5069 | 5070 | 5071 | 5078 | 5079 |
| Retired 2 | 5051 | 5052 | 5053 | 5058 | 5061 | 5064 | 5067 | 5072 | 5073 | 5074 | 5080 | 5081 |
| Retired 3 | 5054 | 5055 | 5056 | 5059 | 5062 | 5065 | 5068 | 5075 | xxx | 5077 | 5082 | 5083 |
| TOTAL PERS | 5084 | 5085 | 5086 | 5093 | 5096 | 5099 | 5102 | xxx | xxx | xxx | 5105 | 5106 |
| Pers EA | 5087 | 5088 | 5089 | 5094 | 5097 | 5100 | 5103 | xxx | xxx | xxx | 5107 | 5108 |
| Dep child | 5090 | 5091 | 5092 | 5095 | 5098 | 5101 | 5104 | xxx | xxx | xxx | 5109 | 5110 |
| | Households with | | | | | | | | | | | |
| No family | 5111 | 5112 | 5113 | 5120 | 5123 | 5126 | 5129 | 5132 | 5133 | xxx | xxx | xxx |
| 1 family | 5114 | 5115 | 5116 | 5121 | 5124 | 5127 | 5130 | 5135 | 5136 | 5137 | 5143 | 5144 |
| 2+ families | 5117 | 5118 | 5119 | 5122 | 5125 | 5128 | 5131 | 5138 | 5139 | 5140 | 5145 | 5146 |

| | | | | | | | | | |
|---|---|---|---|---|---|---|---|---|---|
| 12 | 5159 | 5202 | 5203 | 5204 | 5235 | 5278 | 5279 | 5280 | 5311 |
| 13 | 5160 | 5205 | 5206 | 5207 | 5236 | 5281 | 5282 | 5283 | 5312 |
| 14 | 5161 | 5208 | 5209 | 5210 | 5237 | 5284 | 5285 | 5286 | 5313 |
| 15 | 5162 | 5211 | 5212 | 5213 | 5238 | 5287 | 5288 | 5289 | 5314 |
| 16 | 5163 | 5214 | 5215 | 5216 | 5239 | 5290 | 5291 | 5292 | 5315 |
| 17 | 5164 | 5217 | 5218 | 5219 | 5240 | 5293 | 5294 | 5295 | 5316 |
| TOTAL | 5165 | 5220 | 5221 | 5222 | 5241 | 5296 | 5297 | 5298 | 5317 |

**NOTES**

**Retired 1 = retired [head], previous occupation stated (Tables 49 and 52)**
**Retired 2 = retired [head] (no previous occupation stated) + never active [head] - in h/hold/family with at least one econ active person (Tables 49 and 52)**
**Retired 3 = retired [head] (no previous occupation stated) + never active [head] - in household/family without an economically active person (Tables 49 and 52)**

**51 Residents aged 16 or over in employment (10% sample)**

| Employment Status | TOTAL IN EMPL | Selected industries: Agric (ex forest + fishing) | Forestry and fishing | Manuf | Constr | Distrib and catering | Fin-ance | Pub admin + other services |
|---|---|---|---|---|---|---|---|---|
| **Males** | 5318 | 5328 | 5329 | 5330 | 5331 | 5332 | 5333 | 5334 |
| **Self-emp with empl** | 5319 | 5335 | 5336 | 5337 | 5338 | 5339 | 5340 | 5341 |
| **Self-emp w/o empl** | 5320 | 5342 | 5343 | 5344 | 5345 | 5346 | 5347 | 5348 |
| **Employees** | 5321 | 5349 | 5350 | 5351 | 5352 | 5353 | 5354 | 5355 |
| **Females** | 5322 | 5356 | 5357 | 5358 | 5359 | 5360 | 5361 | 5362 |
| **Self-emp with empl** | 5323 | 5363 | 5364 | 5365 | 5366 | 5367 | 5368 | 5369 |
| **Self-emp w/o empl** | 5324 | 5370 | 5371 | 5372 | 5373 | 5374 | 5375 | 5376 |
| **Employees** | 5325 | 5377 | 5378 | 5379 | 5380 | 5381 | 5382 | 5383 |
| **Working full-time** | 5326 | 5384 | 5385 | 5386 | 5387 | 5388 | 5389 | 5390 |
| **Working part-time** | 5327 | 5391 | 5392 | 5393 | 5394 | 5395 | 5396 | 5397 |

**52 Residents in private households, private households (10% sample)**

| Social class | H/hlds and Residents in h/hlds: House-holds | Per-sons | M'rrd fmls | Pers 0-15 | Pers 65+ | Residents 16+ in h/hlds: Males | Males 16-64 | M'rrd fmls | SWD fmls |
|---|---|---|---|---|---|---|---|---|---|
| | By social class of EA head of household | | | | | EA persons social class | | | |
| I | 5398 | 5409 | 5410 | 5411 | 5442 | 5453 | 5454 | 5455 | 5456 |
| II | 5399 | 5412 | 5413 | 5414 | 5443 | 5457 | 5458 | 5459 | 5460 |
| IIIN | 5400 | 5415 | 5416 | 5417 | 5444 | 5461 | 5462 | 5463 | 5464 |
| IIIM | 5401 | 5418 | 5419 | 5420 | 5445 | 5465 | 5466 | 5467 | 5468 |
| IV | 5402 | 5421 | 5422 | 5423 | 5446 | 5469 | 5470 | 5471 | 5472 |
| V | 5403 | 5424 | 5425 | 5426 | 5447 | 5473 | 5474 | 5475 | 5476 |
| Armed forces + inad desc | 5404 | 5427 | 5428 | 5429 | 5448 | 5477 | 5478 | 5479 | 5480 |
| | Households with inactive head | | | | | Inactive persons | | | |
| Retired 1 | 5405 | 5430 | 5431 | 5432 | 5449 | 5481 | 5482 | 5483 | 5484 |
| Retired 2 | 5406 | 5433 | 5434 | 5435 | 5450 | 5485 | 5486 | 5487 | 5488 |
| Retired 3 | 5407 | 5436 | 5437 | 5438 | 5451 | 5489 | 5490 | 5491 | 5492 |
| TOTAL | 5408 | 5439 | 5440 | 5441 | 5452 | 5493 | 5494 | 5495 | 5496 |

**53 Residents economically active but not in employment (10% sample)**

| | Former industry: Agric | Energy and water | Manuf | Constr | Distrib and catering | Trans-port | Other Services |
|---|---|---|---|---|---|---|---|
| Males | 5497 | 5498 | 5499 | 5500 | 5501 | 5502 | 5503 |
| SWD females | 5504 | 5505 | 5506 | 5507 | 5508 | 5509 | 5510 |
| Married females | 5511 | 5512 | 5513 | 5514 | 5515 | 5516 | 5517 |

Figure 5.3 (Continued)

**CENSUS 1981 SMALL AREA STATISTICS**

**PAGE 7** **100%**

ED No

Map Reference

Note: $ = ED same as in 1971
$$ = Special Enumeration District

 | Frame No:

Separate explanatory notes are available

33 Household spaces; permanent buildings; non-permanent accommodation

| | Household spaces in permanent buildings with: | | | | | | | | | | | Non-perm accom | |
|---|---|---|---|---|---|---|---|---|---|---|---|---|---|
| | TOTAL H/HOLD SPACES | Self-contained accommodation | | | | | | Not self-contained accom with shared entrance from o/side the building | | | Cara vans | Other non-perm |
| | | TOTAL | Purp-ose built flats | Sep entr-ance from o/side bldg | Shared entrance from outside the building | | | Total | Bed sits | Other | | |
| | | | | | 2+ rooms bath + inside WC excl | Flat-lets | Other | | | | | |
| TOTAL HOUSEHOLD SPACES | 2598 | 2599 | 2600 | 2601 | xxx | xxx | xxx | 2605 | xxx | xxx | 2608 | 2609 |
| Resident household spaces | 2610 | 2611 | 2612 | 2613 | 2614 | 2615 | 2616 | 2617 | 2618 | 2619 | 2620 | 2621 |
| Other household spaces | 2622 | 2623 | 2624 | 2625 | xxx | xxx | xxx | 2629 | xxx | xxx | 2632 | 2633 |

**34 Resident household spaces; rooms; residents in private households**

| | Household spaces in permanent buildings | | | | | Non-perm accom* |
|---|---|---|---|---|---|---|
| | TOTAL | Purpose built flats | Sep entrance from o/side bldg | Shrd entrance from o/s bldg: Self-cont accom | Shrd entrance from o/s bldg: Not self-cont accom | |
| TOTAL ROOMS | 2634 | 2635 | 2636 | 2637 | 2638 | 2674 |
| Rooms: | | | | | | |
| 1 | 2639 | 2640 | 2641 | 2642 | 2643 | 2675 |
| 2 | 2644 | 2645 | 2646 | 2647 | 2648 | 2676 |
| 3 | 2649 | 2650 | 2651 | 2652 | 2653 | 2677 |
| 4 | 2654 | 2655 | 2656 | 2657 | 2658 | 2678 |
| 5/6 | 2659 | 2660 | 2661 | 2662 | 2663 | 2679 |
| 7/8 | 2664 | 2665 | 2666 | 2667 | 2668 | xxx |
| 9+ | 2669 | 2670 | 2671 | 2672 | 2673 | xxx |
| TOTAL PERSONS | 2682 | 2683 | 2684 | 2685 | 2686 | 2722 |
| Persons usually resident: | | | | | | |
| 1 | 2687 | 2688 | 2689 | 2690 | 2691 | 2723 |
| 2 | 2692 | 2693 | 2694 | 2695 | 2696 | 2724 |
| 3 | 2697 | 2698 | 2699 | 2700 | 2701 | 2725 |
| 4 | 2702 | 2703 | 2704 | 2705 | 2706 | 2726 |
| 5 | 2707 | 2708 | 2709 | 2710 | 2711 | 2727 |
| 6 | 2712 | 2713 | 2714 | 2715 | 2716 | 2728 |
| 7+ | 2717 | 2718 | 2719 | 2720 | 2721 | 2729 |
| Persons per room: | | | | | | |
| Greater than 1.5 | 2730 | 2731 | 2732 | 2733 | 2734 | 2755 |
| >1 - 1.5 | 2735 | 2736 | 2737 | 2738 | 2739 | 2756 |
| >0.5 - 1 | 2740 | 2741 | 2742 | 2743 | 2744 | 2757 |
| 0.5 or under | 2745 | 2746 | 2747 | 2748 | 2749 | 2758 |
| TOTAL H/HOLD SPACES | 2750 | 2751 | 2752 | 2753 | 2754 | 2759 |

* Maximum number of rooms in non-perm accom is five

**35 Resident household spaces**

| | Household spaces in permanent buildings | | | | | Non-perm accom |
|---|---|---|---|---|---|---|
| | TOTAL | Purpose built flats | Separate entrance from outside building | Shared entrance from o/s bldg: Self-contained accom | Shared entrance from o/s bldg: Not self-contained accom | |
| Amenities: | | | | | | |
| Exclusive bath with: | | | | | | |
| exclusive inside WC | 2760 | 2761 | 2762 | 2763 | 2764 | 2805 |
| shared inside WC | 2765 | 2766 | 2767 | 2768 | 2769 | 2806 |
| no inside WC | 2770 | 2771 | 2772 | 2773 | 2774 | 2807 |
| Shared bath with: | | | | | | |
| exclusive inside WC | 2775 | 2776 | 2777 | 2778 | 2779 | 2808 |
| shared inside WC | 2780 | 2781 | 2782 | 2783 | 2784 | 2809 |
| no inside WC | 2785 | 2786 | 2787 | 2788 | 2789 | 2810 |
| No bath with: | | | | | | |
| exclusive inside WC | 2790 | 2791 | 2792 | 2793 | 2794 | 2811 |
| shared inside WC | 2795 | 2796 | 2797 | 2798 | 2799 | 2812 |
| no inside WC | 2800 | 2801 | 2802 | 2803 | 2804 | 2813 |
| Tenure: | | | | | | |
| ALL TENURES | 2814 | 2815 | 2816 | 2817 | 2818 | 2859 |
| Owner occupied freehold | 2819 | 2820 | 2821 | 2822 | 2823 | 2860 |
| Owner occupied leasehold | 2824 | 2825 | 2826 | 2827 | 2828 | 2861 |
| Council, New Town, etc | 2829 | 2830 | 2831 | 2832 | 2833 | 2862 |
| Housing association | 2834 | 2835 | 2836 | 2837 | 2838 | 2863 |
| Rented with business | 2839 | 2840 | 2841 | 2842 | 2843 | 2864 |
| By virtue of employment | 2844 | 2845 | 2846 | 2847 | 2848 | 2865 |
| Other rented unfurnished | 2849 | 2850 | 2851 | 2852 | 2853 | 2866 |
| Other rented furnished | 2854 | 2855 | 2856 | 2857 | 2858 | 2867 |

Figure 5.3 (Continued)

**CENSUS 1981 SMALL AREA STATISTICS**

**PAGE 8** **100%**

**ED No**

**Map Reference**

**Note:** $ = ED same as in 1971
$$ = Special Enumeration District

 | Frame No:

Separate explanatory notes are available

**36 Private households with resident heads born in the New Commonwealth or Pakistan**

| | New Commonwealth or Pakistani headed households | | | | | | |
|---|---|---|---|---|---|---|---|
| | TOTAL | 1 or more persons per room | Bath + inside WC excl | Lack bath | Lack inside WC | Not in self-cont accom | No car |
| Households | 2868 | 2869 | 2870 | 2871 | 2872 | 2873 | 2874 |

**37 Residents in private households**

| Birthplace of household head | TOTAL PERSONS | Age and birthplace of persons | | | | | | | | | | | | | | TOTAL HEADS OF HOUSE-HOLDS |
|---|---|---|---|---|---|---|---|---|---|---|---|---|---|---|---|---|
| | | All ages | | 0-4 | | 5-15 | | 16-29 | | 30-44 | | 45 to pensionable age | | Pensionable age and over | | |
| | | In UK | Outside UK | In UK | Outside UK | In UK | Outside UK | In UK | Outside UK | In UK | Outside UK | In UK | Outside UK | In UK | Outside UK | |
| TOTAL | 2875 | 2876 | 2877 | 2878 | 2879 | 2880 | 2881 | 2882 | 2883 | 2884 | 2885 | 2886 | 2887 | 2888 | 2889 | 2950 |
| UK | 2890 | 2891 | 2892 | 2893 | 2894 | 2895 | 2896 | 2897 | 2898 | 2899 | 2900 | 2901 | 2902 | 2903 | 2904 | 2951 |
| Irish Republic | 2905 | 2906 | 2907 | 2908 | 2909 | 2910 | 2911 | 2912 | 2913 | 2914 | 2915 | 2916 | 2917 | 2918 | 2919 | 2952 |
| New Commonwealth and Pakistan | 2920 | 2921 | 2922 | 2923 | 2924 | 2925 | 2926 | 2927 | 2928 | 2929 | 2930 | 2931 | 2932 | 2933 | 2934 | 2953 |
| Rest of World | 2935 | 2936 | 2937 | 2938 | 2939 | 2940 | 2941 | 2942 | 2943 | 2944 | 2945 | 2946 | 2947 | 2948 | 2949 | 2954 |

**38 Households (H) with residents, rooms and resident persons (P) in owner occupied accommodation in permanent buildings**

| Tenure | TOTAL HOUSEHOLDS | Households with the following persons 1 | 2 | 3 | 4 | 5 | 6 | 7+ |
|---|---|---|---|---|---|---|---|---|
| Freehold | 2955 | 2956 | 2957 | 2958 | 2959 | 2960 | 2961 | 2962 |
| Leasehold | 2963 | 2964 | 2965 | 2966 | 2967 | 2968 | 2969 | 2970 |

| Tenure | | Households with the following rooms 1 | 2 | 3 | 4 | 5 | 6 | 7+ | TOTALS | TOTAL ROOMS |
|---|---|---|---|---|---|---|---|---|---|---|
| F/hold | H | 2971 | 2972 | 2973 | 2974 | 2975 | 2976 | 2977 | 2978 | 2979 |
| | P | 2980 | 2981 | 2982 | 2983 | 2984 | 2985 | 2986 | 2987 | xxx |
| L/hold | H | 2989 | 2990 | 2991 | 2992 | 2993 | 2994 | 2995 | 2996 | 2997 |
| | P | 2998 | 2999 | 3000 | 3001 | 3002 | 3003 | 3004 | 3005 | xxx |

| Tenure | | TOTALS | Bath and inside WC present: Both excl | Bath and inside WC present: One/both shared | Lack inside WC or bath | Neither inside WC nor bath | Lack bath | Lack inside WC | Share inside WC | Persons per room: >1.5 | Persons per room: >1-1.5 | No car |
|---|---|---|---|---|---|---|---|---|---|---|---|---|
| F/hold | H | 3007 | 3008 | 3009 | 3010 | 3011 | 3012 | 3013 | 3014 | 3039 | 3040 | 3047 |
| | P | 3015 | 3016 | 3017 | 3018 | 3019 | 3020 | 3021 | 3022 | 3041 | 3042 | 3048 |
| L/hold | H | 3023 | 3024 | 3025 | 3026 | 3027 | 3028 | 3029 | 3030 | 3043 | 3044 | 3049 |
| | P | 3031 | 3032 | 3033 | 3034 | 3035 | 3036 | 3037 | 3038 | 3045 | 3046 | 3050 |

**39 Persons aged 3 or over: Pres & abs h'hold residents**

| Age | TOTAL PER-SONS | Speaking Welsh: Total | Speaking Welsh: Not Speak-ing Eng-lish | Speaking Welsh, Speaking English: Reads and writes Welsh | Speaking Welsh, Speaking English: Others | Not Speak-ing Welsh |
|---|---|---|---|---|---|---|
| TOTAL RESIDENTS | 3051 | 3052 | 3053 | 3054 | 3055 | 3056 |
| 3-4 | 3057 | 3058 | 3059 | 3060 | 3061 | 3062 |
| 5-15 | 3063 | 3064 | 3065 | 3066 | 3067 | 3068 |
| 16-24 | 3069 | 3070 | 3071 | 3072 | 3073 | 3074 |
| 25-44 | 3075 | 3076 | 3077 | 3078 | 3079 | 3080 |
| 45-64 | 3081 | 3082 | 3083 | 3084 | 3085 | 3086 |
| 65+ | 3087 | 3088 | 3089 | 3090 | 3091 | 3092 |
| Present residents and visitors (1981) (1971 Base) | 3093 | 3094 | 3095 | 3096 | 3097 | 3098 |

Figure 5.3 (Continued)

The data contained in the SAS may be used directly or form the basis for the calculation of new or derived variables. Table 5.4 illustrates how new variables may be derived using the deprivation indicators discussed in Chapter Three. The figures refer to the cell numbers contained in Figure 5.3, and not to actual data. Thus, the first variable, 'households with greater than 1.5 persons per room' is derived by dividing the value contained in cell 945 by the value contained in cell 1522 multiplied by 100.

### 5.4.2 *The published census volumes (England and Wales)*

This source of data is published by Her Majesty's Stationery Office (HMSO), and consists of a number of distinct components.

#### 5.4.2.1 *Preliminary reports*

These are published fairly quickly after enumeration has been completed. Although the figures are generally accurate they are provisional, and, therefore, subject to revision. The two Preliminary Reports are as follows:

1981 Census: Preliminary Report: England and Wales
1981 Census: Preliminary Report on Towns

#### 5.4.2.2 *County Monitors (one for each county)*

These are published in advance of the main county reports and contain basic summary statistics together with a brief commentary and comparisons with previous censuses. In addition, a national and regional summary of the County Monitors is also available.

#### 5.4.2.3 *County Reports (two volumes per county)*

The County Reports are the most important source of published census statistics for counties and local authority districts and consist of two parts.

Part 1 The Part 1 report contains tables derived from answers to the 100% items on the census form (see Table 5.3) and cover the following topics:

| | |
|---|---|
| General tables | 3 tables |
| Demographic Statistics | 8 tables |
| Economic characteristics | 5 tables |
| Housing and Amenities | 17 tables |
| Household composition | 8 tables |
| Welsh language (Wales only) | 1 table |

All tables, except table 1, present statistics for

| | Variable name (all in percentages) | Cell numbers |
|---|---|---|
| 1. | Households with >1.5 persons per room | $\frac{945}{1522} \times 100$ |
| 2. | Households lacking basic amenities (neither bath nor inside W.C.) | $\frac{933}{1522} \times 100$ |
| 3. | Economic active males unemployed but seeking work or sick | $\frac{403+404+410+411}{389+390} \times 100$ |
| 4. | Households without a car | $\frac{949}{1522} \times 100$ |
| 5. | Economic active and retired males in socio-economic group 11 (unskilled) | $\frac{5158}{5165} \times 100$ |
| 6. | Population aged 0-14 | $\frac{57+64+71}{43} \times 100$ |
| 7. | Population of pensionable age | $\frac{153+154+157 \text{ to } 161+164 \text{ to } 168 +171 \text{ to } 175+178 \text{ to } 182+185 \text{ to } 189}{50} \times 100$ |
| 8. | Population of New Commonwealth origin | $\frac{342+343}{43} \times 100$ |

TABLE 5.4 DERIVING VARIABLES FROM THE 1981 SAS - AN EXAMPLE

both the county and its local authority districts. Table 1 gives the population present on census night for every census from 1891-1981 and calculates the amount and percentage of population change from census to census. Where applicable, changes to district and county boundaries since April 1974 are given. The County Monitor is reprinted at the front of each Part 1 Report to provide key indicators. Each Report also contains a map to illustrate the boundaries and position of each tounty and its districts.

Part 2. The Part 2 Reports contain statistics derived from the 10% items. Topics covered include: occupation, industry, relationship in household, address of place of work, means of daily journey to work. (9 tables).

In comparison with earlier censuses, the 1981 tables cover a wider range of topics. The majority of counts contained in the SAS are published in the County Reports so that small areas can be compared with larger ones, both within and between counties. However, the tables have additional information at district and county level, and tables in the County Reports are presented in a different order from their SAS equivalents to allow tables to be grouped into broad sub-sections.

5.4.2.4 *Economic activity leaflets*
These leavelets provide greater detail on occupation, industry and socio-economic groups for each county.

5.4.2.5 *Statistics for wards and parishes*
Leaflets are also published for each county giving basic statistics for District Wards as existing on census day, and for civil parishes and successor parishes in England and Communities in Wales as defined in 1974.

5.4.2.6 *Great Britain: national and regional summaries*
Regional and national versions of the tables in Part 1 and Part 2 of the County Reports are also available in two volumes. Summary versions of the county report tables also form the basis for additional reports on local authorities, towns, rural areas, New Towns and parliamentary constituencies.

5.4.2.7 *National Volumes*
These consist of a number of topic volumes reporting results on a national level. Volumes based on 100% items are listed below:

1. Historical tables - giving comparison of the population with previous censuses.

2. Sex, age and marital status.

3. Birthplaces.

4. Population in communal establishments.

5. Usual residence - giving comparison of different population bases.

6. Housing and households (including availability of cars).

7. Welsh language (in a bilingual volume).

8. Persons of pensionable age.

9. Migration (cross-analysed with 100% items).

The following volumes are based on the 10% sample.

1. Economic Activity (occupation, industry, socio-economic group).

2. Household and family composition - household tables.

3. Household and family composition - family tables.

4. Qualified manpower.

5. Migration and transport.

6. Migration (cross-analysed with 10% items).

5.5 Availability of Census Information

Many of the published volumes discussed above can be consulted at most local main libraries. Copies of the SAS for particular areas within a county can be obtained free of charge from the local authority concerned. Bowever, unrestricted requests for SAS from schools may result in local authorities being unable to cope with demand. In certain authorities, therefore, clearing houses have been established where a single representative for the schools (possibly a teachers' resource centre) obtains a copy of a selection of SAS suitable for educational purposes, then distributes them to the schools. Although the data will be supplied as paper copies of the SAS, tbis information could quite easily be transferred for use on schools' micro-computers.

Further information on the SAS can be obtained from:

Census Customer Services,
Office of Population Censuses and Surveys,
Titchfield,
Fansham,
Hants PO15 5RR

Telephone Titchfield (0329) 42511 Ext. 231 and 296.

Chapter Six

# TRANSPORT STUDIES

> 'Transport is an indispensable feature of modern life and, because of its function and importance, has a profound impact upon our lives.'
>
> (Robinson and Bamford, 1978, p.3)

## 6.1 Introduction

In recent years transport studies have become an increasingly important and distinctive aspect of human geography. In considering the suitability of projects for 'A' level students, attention has been focused on spatial and environmental transport problems which would appear to be relevant at the local level. In this chapter, therefore, the following themes are examined. In section 6.2 the so-called urban transport problem provides the context within which three major issues are examined. First, the environmental effects of traffic on pedestrians (danger, noise, pollution, visual intrusion and delay). Second, the assessment of user-satisfaction with various aspects of public transport. Third, the impact of the urban transport problem on car users. The idea of space-time sampling is introduced in section 6.3 through the examination of both traffic and pedestrian flows. While in section 6.4 alternative measures of distance are considered in terms of time and cost. Finally, attention is focused on the potential of network analysis in the formulation of projects.

In this chapter, outline projects are not included in every section. However, a number of methods and approaches are presented which, either individually or in combination, could provide the basis for a range of interesting projects.

## 6.2 The Urban Transport Problem

At the intra-urban scale it is possible to devise a number of manageable projects which are of local significance. The themes outlined below revolve around the idea of the 'urban

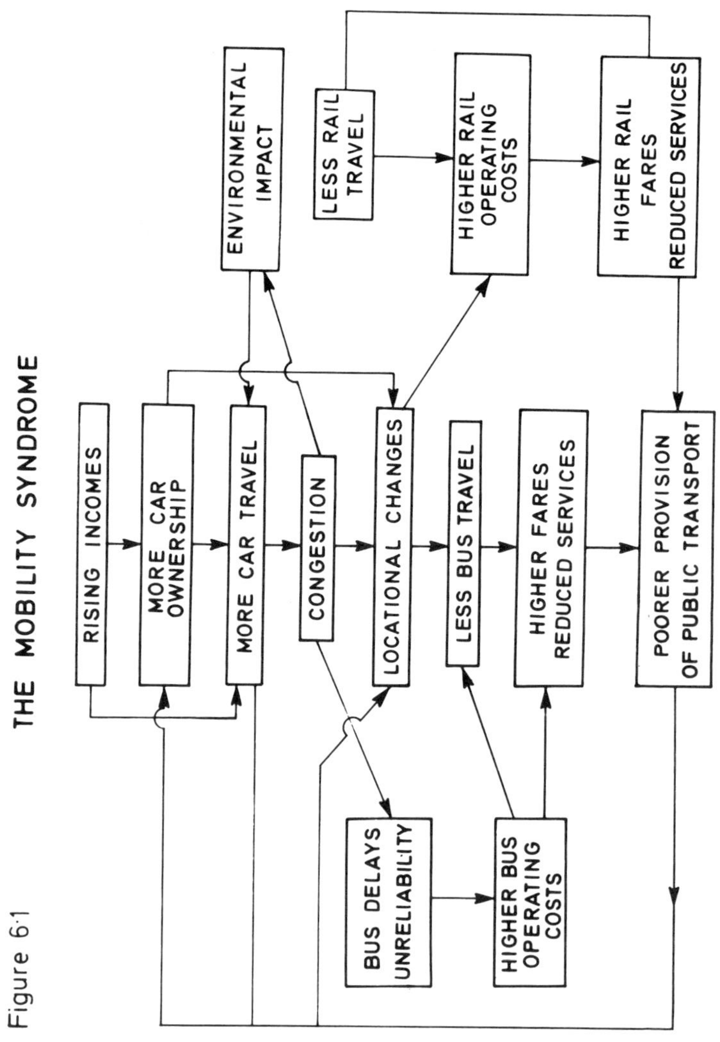


Figure 6·1 THE MOBILITY SYNDROME

transport problem'. The major issues involved are summarized in Figure 6.1. In essence the urban transport problem consists of three distinct, but interrelated, factors:

1. The growth and changing distribution of population and the associated increase in travel demand. Urban public transport developed as a response to this demand firstly with trams and later with motor buses. With increased personal mobility housing developed further away from the central area, which consequently resulted in increased travel demand in and out of the central area. In addition it may be noted that the morphology of many towns and cities in Britain was established prior to the advent of the motor car and thus were unsuited to increased traffic densities.

2. The growth of vehicle numbers. From 1949-75, the number of motor vehicles in use rose by 325%, while road length only increased by 17%. Particularly important is the growth of commercial vehicles both in number and in size/carrying capacity. For example during 1960-70 there was a 5.6% increase in Heavy Goods Vehicles, but in the heaviest category (unladen weight over 8 tonnes) there was a 500% increase. This clearly has important implications for traffic flows and environmental problems in urban areas.

3. Decline in the use of public transport. Since 1965 private transport passenger-km's have experienced growth while Public Service Vehicles (PSV) passenger-km's have decreased (see Table 6.1). During 1960-76 estimated passenger-km increased by 56.3% However private cars increased by 119.5%. While P.S.V. fell by 22.4% The major reasons for this decline can be seen in Figure 6.1.

| | 1961 | 1966 | 1971 | 1976 | 1980 |
|---|---|---|---|---|---|
| Air | 1.0 | 1.8 | 2.0 | 2.3 | 2.8 |
| Rail | 39 | 35 | 36 | 33 | 36 |
| P.S.V. | 69 | 62 | 56 | 53 | 51 |
| Private | 161 | 255 | 330 | 367 | 433 |
| Bicycles | 11 | 7 | 5 | 4 | 5 |

TABLE 6.1 MAJOR TRANSPORT MODES
(thousand million passenger-kilometres)

6.2.1 *The environmental effects of traffic on pedestrians*
Everyone entering a city has to walk for some part of the journey (approximately 37% of all journeys in urban areas are by foot). Within the urban area itself the function and purpose of existing streets have a different meaning for pedestrians and drivers. For the pedestrian the street provides access to buildings and acts as a means of communication; provides public space for play and leisure; and can be a means of orientation to the physical surroundings. For the driver a street may be seen as a space for relatively fast movement; a space for parking vehicles; it also offers access to the kerbside for loading and unloading. These conflicting requirements produce three major environmental effects which impinge on the pedestrian. First, the most obvious effect is the danger presented by motor vehicles. Second, noise, dirt and fumes may be encountered when walking near the road. Third, traffic movement may delay pedestrians wishing to cross roads and streets.

The assessment of the environmental impact of traffic on pedestrians requires the use of questionnaires, and in particular the use of opinion questions (see Appendix III). The following are examples of the type of questions which may be asked.

1. A simple question may be used to assess the danger involved in crossing the main roads in an urban area:

   How safe do you feel when crossing the main streets in ..........(name of town or city)?

   Safe | Quite safe | Not safe at all

   In addition it may be interesting (by a simple modification of the question) to ask the respondent the same question in relation to the danger posed by traffic to the young and elderly groups in the population, for example.

   Such a question provides information relating to the anxiety pedestrians experience in moving about a town or city.

2. In addition to the obvious danger caused by traffic a number of other environmental effects may produce varying degrees of annoyance.

   Noise According to the Wilson Report on noise 'Road noise ... is the predominant source of annoyance and no other single source is of comparable importance'. The degree of annoyance associated with traffic noise can be simply assessed as follows:

Which of these phrases best describes what you think of traffic noise here? (Tick one)

Bothers me, not at all
Bothers me, a little
Bothers me, a lot

Pollution Although 'pollution' in the widest sense may take many forms (subsuming noise and visual intrusion) the term usually refers to obnoxious ingredients emitted from motor vehicles (for example smoke and carbon monoxide). Again, a simple question can be used to assess the degree of annoyance associated with air pollution:

Do you find smoke and fumes from traffic (tick one)

Not annoying     Slightly annoying     Very annoying

Visual intrusion It might be useful to preface this with some general questions such as:

Would you say that traffic in this area bothers you

Very much     Quite a lot     Not very much

Which type of vehicle bothers you most?

Cars     Lorries     Vans     Two-wheeler     Buses     Others

Visual intrusion refers specifically to the impact of roads and traffic on the quality of the visual environment. For example, the building of a new by-pass may result in some houses being in the shadow of large road structures. Also a new road could effect the general view as well as resulting in the loss of character to old buildings. In a general sense traffic congestion can cause a loss of amenity to the whole community.

Although it is difficult to measure the effect of motor vehicles on the visual environment and on pedestrians, a preliminary assessment may be provided by asking the respondent how he/she feels about the visual impact of traffic:

Do you find the view of passing traffic:

Pleasant     Unpleasant     Extremely unpleasant     Don't Know

A more sophisticated version of this is to employ the semantic differential. This consists of a 7-point scale with polar adjectives located at each end of the scale (for example ugly - pretty, noisy - quiet etc.).

Neutral point

Extremely pleasant 1 2 3 4 5 6 7 Extremely unpleasant

The respondent is required to tick the appropriate box. If the respondent ticks 5, 6 or 7, it might be informative to ask what he/she finds particularly unpleasant.

Delay This refers to the degree of delay encountered by pedestrians in crossing a road. It may cause varying degrees of annoyance or result in pedestrians taking unnecessary risks. Although crossing-points are provided for pedestrians (for example, zebra and pelican crossings, underpasses and so on) a large number of pedestrians choose to ignore them. Some assessment of the difficulties encountered by pedestrians may be derived from the following question:

Do you find crossing roads in ...... (name of town/city)

Very difficult Difficult Not difficult at all

This may be followed up by asking the following question:

How often do you sue the road crossing points provided?

Always Nearly always Sometimes Never

In addition to the above questions, background data should be collected on the following: frequency with which the respondent is a pedestrian of the town or city (weekly, fortnightly etc.); reason for being in the town or city (shopping, business, for example); place of residence; mode of transport; car ownership; age; occupation. These variables could be used in the construction and testing of hypotheses. It might also be necessary to collect supplementary information on: the location and frequency of accidents; to consult local plans relating to present and future pedestrian/traffic segregation; to discover the existing provision of crossing-points for pedestrians. In other instances it might be of value to measure pedestrian flows at different locations and times. Finally, in evaluating the results of the surveys attention should focus on their implications for future pedestrian/vehicle planning.

### 6.2.2 *The impact on public transport users*

From Figure 6.1 it can be seen that higher levels of private transport ownership can also have an important impact on users of public transport. In particular more journeys by private transport leads to fewer journeys being made on public service vehicles leading to higher fares and reduced services. Congestion results in bus delays, unreliability of services and, consequently, higher bus operating costs. In sum, these factors all contribute to a lowering in quality of public transport provision. These problems could provide the basis for a relatively straightforward project focusing on the attitudes of public transport users to the service provided.

The major features which could be included in such a project are examined below.

1. The sample. The obvious location for administering a questionnaire would be at bus stops and bus stations. However, for obvious reasons students might find respondents reluctant to co-operate!

2. Background variables should be collected (see previous section).

3. At the outset it would be useful to enquire how often the respondent uses the bus service (e.g. more than once a day; more than once a week; more than once a month; less than once a month).

   If the respondent is waiting for a bus, the following questions may be asked:

   What is your destination?

   How long have you been waiting? (e.g. less than 10 minutes, less than half an hour; more than half an hour).

4. The overall satisfaction with existing services may be gauged by a question of the following form:

   How satisfied are you with existing bus services?

   | Very Satisfied | Satisfied | Neutral | Dissatisfied | Very Dissatisfied |
   |---|---|---|---|---|

5. In attempting to assess user attitudes in more detail two techniques may be employed.

   Rating scales This technique requires respondents to indicate that point on a scale they feel is the best measure of their attitude toward a particular term. A typical rating scale format is provided in Table 6.2.

| Thinking about when I use public transport for longer trips and where I might go - this feature is important to me. | Extremely important | Very important | Somewhat important | Neither important nor unimportant | Somewhat unimportant | Very unimportant | Extremely unimportant |
|---|---|---|---|---|---|---|---|
| Having a short time to wait for a vehicle | 7 | 6 | 5 | 4 | 3 | 2 | 1 |
| Having low fares | 7 | 6 | 5 | 4 | 3 | 2 | 1 |
| Having a comfortable ride in a quiet vehicle | 7 | 6 | 5 | 4 | 3 | 2 | 1 |
| Being able to get where I want to go on time | 7 | 6 | 5 | 4 | 3 | 2 | 1 |

TABLE 6.2 A TYPICAL RATING SCALE

Semantic differential This was briefly mentioned in the previous section. In essence it is a refined form of rating scale which requires each respondent to evaluate concepts with reference to a series of bi-polar statements. Table 6.3 contains a number of 5-point semantic differential scales. One end of the scale consists of a negative response (for example, uncomfortable, late) while the opposite end is positive (comfortable, on time and so on). The negative and positive ends of the scales are reversed in a random manner so as to discourage respondents from giving a stereotyped response. However, when the negative and positive responses are rearranged so that they all lie on one side (see for example Figure 6.2) a number of analytical methods can be used. The simplest method involves finding the mean score for the sample (or sub-group in the sample, for example, the old and young) for each scale. This is obtained by adding up individual scores and

ATTITUDE SURVEY

PLEASE THINK ABOUT YOUR BUS SERVICE

INDICATE YOUR OPINION OF BUS TRAVEL IN ........BY PLACING A TICK IN THE COLUMN YOU THINK MOST APPROPRIATE, FOR EACH PAIR OF STATEMENTS.

| | | | |
|---|---|---|---|
| 1. | UNCOMFORTABLE | :1 :2 :3 :4 :5 : | COMFORTABLE |
| 2. | LATE | :1 :2 :3 :4 :5 : | ON TIME |
| 3. | FARE PAYMENT COMPLICATED | :1 :2 :3 :4 :5 : | FARE PAYMENT SIMPLE |
| 4. | SAFE | :5 :4 :3 :2 :1 : | DANGEROUS |
| 5. | MODERN | :5 :4 :3 :2 :1 : | OLD FASHIONED |
| 6. | SLOW RUSH HOUR TRIP | :1 :2 :3 :4 :5 : | FAST RUSH HOUR TRIP |
| 7. | FEEL PROUD TO USE BUS | :5 :4 :3 :2 :1 : | FEEL ASHAMED TO USE BUS |
| 8. | INCONVENIENT | :1 :2 :3 :4 :5 : | CONVENIENT |
| 9. | ENJOYABLE | :5 :4 :3 :2 :1 : | UNENJOYABLE |
| 10. | FAST NON-RUSH HOUR TRIP | :5 :4 :3 :2 :1 : | SLOW NON-RUSH HOUR TRIP |
| 11. | EXPENSIVE | :1 :2 :3 :4 :5 : | INEXPENSIVE |
| 12. | RELIABLE | :5 :4 :3 :2 :1 : | UNRELIABLE |
| 13. | FEEL UNEASY USING BUS | :1 :2 :3 :4 :5 : | FEEL RELAXED USING BUS |
| 14. | UNCROWDED | :5 :4 :3 :2 :1 : | CROWDED |
| 15. | CLEAN | :5 :4 :3 :2 :1 : | DIRTY |
| 16. | LONG WALK TO BUS STOP | :1 :2 :3 :4 :5 : | SHORT WALK TO BUS STOP |
| 17. | STAFF HELPFUL | :5 :4 :3 :2 :1 : | STAFF UNHELPFUL |
| 18. | PLENTY OF LUGGAGE SPACE | :5 :4 :3 :2 :1 : | LITTLE LUGGAGE SPACE |
| 19. | QUIET | :5 :4 :3 :2 :1 : | NOISY |
| 20. | SUITABLE ROUTES | :5 :4 :3 :2 :1 : | UNSUITABLE ROUTES |
| 21. | EXPOSED TO WEATHER CONDITIONS WHILST WAITING | :1 :2 :3 :4 :5 : | NOT EXPOSED TO WEATHER CONDITIONS WHILST WAITING |
| 22. | FREQUENT | :5 :4 :3 :2 :1 : | INFREQUENT |

ANY OTHER COMMENTS ....................................

Thank you for your co-operation.

TABLE 6.3 SEMANTIC DIFFERENTIAL SCALE

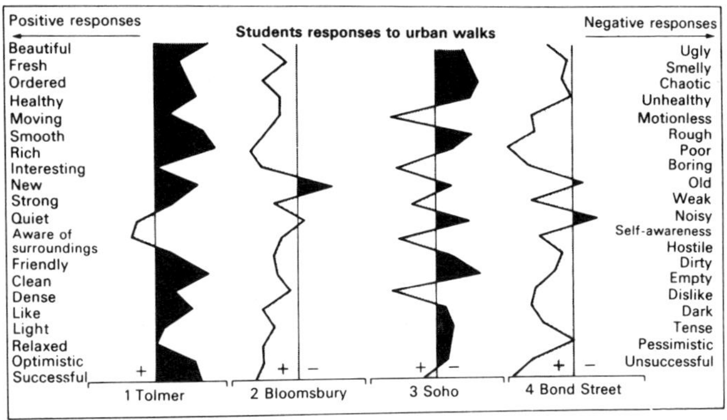


Figure 6.2 SEMANTIC DIFFERENTIAL SCALE: A GRAPHICAL EXAMPLE (from: Gold, 1980)

dividing by the sample size. Using this method a simple graphical display can be produced as illustrated in Figure 6.2. Alternatively, the scores may be used as a basis for statistical analysis. For example, to test whether attitudes towards public transport differ significantly between sub-groups (for example, age and occupational groups) in the sample.

6. Finally it might be informative to supplement the survey data with information concerning existing service provision from timetables and to discuss the problems associated with the local public transport system with the local transport manager.

### 6.2.3 *Car users*

Traffic congestion obviously creates problems for car users, particularly in terms of traffic hold-ups and difficulties in parking. These sorts of problems could easily provide the basis for a project aimed at the assessment of car owners' satisfaction with local traffic conditions and parking schemes. The type of attitude questions previously discussed could be modified without much difficulty to take account of the particular problems of car users.

## 6.3 Traffic and Pedestrian Flows: Space-time Sampling

Geographers are usually concerned with how quantities vary over space. Very often, however, it is necessary to obtain information on variation over space and time. The problem of obtaining a sample of points in space and time is the basic aim of the ever popular traffic and pedestrian censuses.

### 6.3.1 *Traffic flows*

For planners, traffic surveys represent a vital part of the transportation planning process. Data on existing travel patterns provides information on the origin and destination of journeys which, in the short term, can be used to improve the local transport system. In the long term such journeys are an important input into a transportation model which is essential in predicting future demands and network needs.

For the student, a project involving a traffic survey is a useful way of introducing the idea of variation over space and time and its implications. In more substantive terms information on traffic flows and densities is important when considering questions relating to pollution, congestion and road accidents. In other instances the results from traffic censuses may be related to local problems, for example, the need for a by-pass, the location of a public facility.

Methods of data collection

1. Planners tend to use two relatively sophisticated methods of data collection. The roadside interview involves stopping traffic at a particular point and interviewing people about their journeys. The home interview involves people at their homes, thus obtaining a detailed picture of household trip characteristics. Both these methods, however, are beyond the scope of a student project. In the first instance because police supervision is required, and in the second because fairly large samples are required for meaningful results.

2. The simplest form of traffic survey is the traffic count using a recording sheet as illustrated in Figure 6.3. Flows are usually expressed in vehicles per hour. Such a survey requires two people for the recording of information.

3. Volume and direction This type of survey is particularly relevant when studying flows at intersections so as to permit the correct phasing of traffic lights and rerouting of journeys. This type of survey requires four census takers and a more elaborate recording sheet.

4. Cordon counts This survey records all traffic entering

| Time / Vehicle | 8.00-8.15 | 8.15-8.30 | 8.30-8.45 | 8.45-9.00 | Row total by vehicle |
|---|---|---|---|---|---|
| Cars | | | | | |
| Cycles | | | | | |
| Motor cycles | | | | | |
| Vans | | | | | |
| Lorries | | | | | |
| Coaches and buses | | | | | |
| Column total (per $\frac{1}{2}$ hour) | | | | | |

Date:

Time on:
off:

Census station:

Direction of traffic:

Figure 6.3 TRAFFIC COUNT RECORDING SHEET

or leaving a particular area. A cordon is drawn around the area so that all possible roads have a checkpoint. Inward and outward traffic is recorded at each check point. From the recording sheet the total number of vehicles which left the area is subtracted from the total which entered the area - this will give an approximation of the volume of traffic in the area.

## Methods of analysis

1. Passenger Car Units (P.C.U.'s) The simple number of vehicles can give a misleading picture (for example, twenty motor cycles are not the equivalent of twenty buses). Different types of vehicles require varying amounts of road space due to individual size, performance and characteristics. Thus, only a small number of buses and lorries can seriously hamper the capacity for movement. The following P.C.U.'s have been devised for urban and rural areas (see Table 6.4). By multiplying the number in each category by the relevant P.C.U. and summating over all categories the P.C.U. per hour can be calculated. This is often used in official surveys.

| | Urban | Rural |
|---|---|---|
| Motor cycle combinations, cars, taxis, light goods vehicles | 1.00 | 1.00 |
| Motor cycles, scooters, mapeds | 0.75 | 1.00 |
| Heavy goods | 2.00 | 3.00 |
| Buses | 3.00 | 3.00 |
| Cycles | 0.33 | 0.50 |

TABLE 6.4 P.C.U.'s FOR URBAN AND RURAL AREAS

2. Occupancy Rates - Number of vehicles can be transformed into number of people travelling, by multiplying the number of vehicles in a particular class by an appropriate occupancy rate. For cars this is obtained by counting the number of people in every fourth car, for example, over a period of fifteen minutes and averaging. An occupancy rate for buses can be determined roughly in the same manner by estimating whether a bus is 'half full' or 'full' and comparing it with the known capacity. Occupancy rates for motor cycles and goods vehicles is

normally set at one person.

In a geographical context, as noted earlier, the emphasis should be on variation in both time (early and mid-morning, early and mid afternoon, early and late evening; different days of the week) and space (different locations in an urban area). The most sophisticated form of space-time sampling is based on a balanced latin-square design (see Haggett, 1975, p543). However, such an approach would be more suitable for a group project.

### 6.3.2 *Pedestrian flows*

As with the traffic census, variations over space and time should be an important feature of a pedestrian survey. Two fairly standard techniques exist for estimating the number of people in a particular street at a particular time.

1. Point surveys. This involves counting the number of people passing entry and exit points to the shopping street concerned.

2. Moving observer surveys. The observer walks along a street at a fixed speed and counts everyone that passes. The observer then returns along the same street at the same speed counting as before. The average of the two counts gives an estimate of the number of shoppers.

Although pedestrian surveys provide valuable information in their own right, it might be of more value to integrate such surveys with an overall study of a shopping centre.

## 6.4. Alternative Measures of Distance: Time and Cost

The most common form of distance metric used by geographers is spatial distance (miles and kilometres for example). In many instances, however, it is more meaningful to discuss distance in terms of time or cost. In an intuitive sense this is an every day occurrence, for example, people frequently measure journey distances in terms of temporal duration rather than kilometres or miles traversed.

When geographers examine movement, space and time coincide as space-time. Insofar as transport geographers are concerned not only with the structure of transport networks in space but also with movement over space they are intimately involved with questions of space-time. A feasible basis for student projects can be derived from two aspects of space-time studies.

### 6.4.1 *Space-time convergence*

Simply stated space-time convergence refers to the process by which travel times required to reach one place from another decreases over time (clearly, space-time divergence can also exist). As Ellsworth Huntingdon noted in 1952,

> 'the approximate travel time from Portland, Maine, to San Diego, California, may roughly be reckoned as follows: on foot ... at least two years; on horse ... at least eight months; by stage-coach and wagon ... at least four months; by rail in 1920 ... about four days; by air today (1952) ... about ten hours.'

By 1980 this had decreased to three hours. A simple measure of space-time convergence is provided by the following formula:

$$AR_{tsc} = \frac{\text{Travel time (in Y) - Travel time Y + K)}}{\text{Interval in years between Y + K}}$$

where $AR_{tsc}$ is the average rate of space-time convergence,
Y is the first year being considered and
Y + K is the last year being considered.

Thus the average rate of space-time convergence between Edinburgh and London from 1776-1966 is

$$\frac{5760 - 280}{190} = 28.84 \text{ minutes per year}$$

When rail times alone are considered for 1850-1966 the $AR_{tsc}$ is 3.4 minutes per year.

These values can then be plotted as a space-time convergence graph (on semi-log graph paper) as illustrated in Figure 6.4. Which plots the rates of space-time convergence between pairs of cities. At a smaller scale such an approach may be useful in assessing the space-time convergence (or divergence) caused by changes in bus timetables between a town or city and surrounding settlements. The information may be plotted on a space-time convergence graph and/or on a map as illustrated in Figure 6.5.

### 6.4.2 *Accessibility*

Accessibility refers to the ease with which people or goods can move from one location to another. However, accessibility is not simply a function of spatial distance but also time distance (or more generally, space-time distance). This observation suggests, therefore, that it is possible to discuss accessibility in metrics other than miles or kilometres

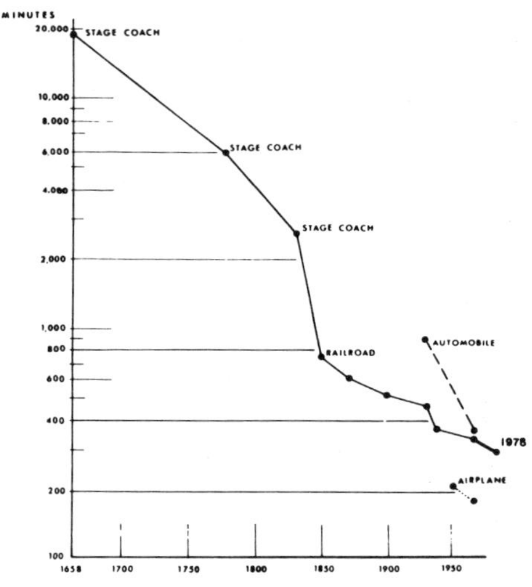


Figure 6.4 SPACE-TIME CONVERGENCE (Source: Parkes and Thrift, 1980)

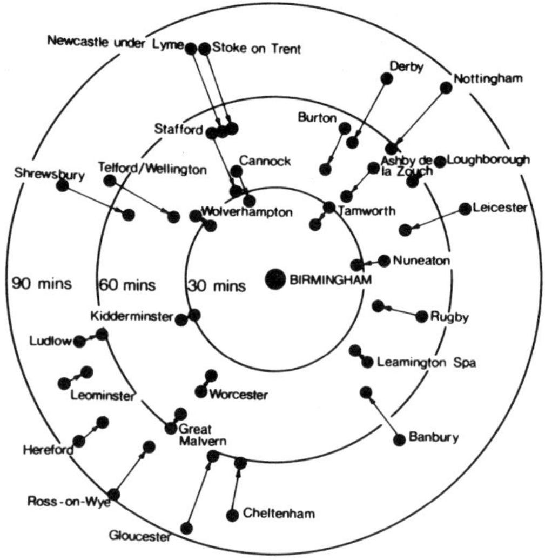


Figure 6.5 SPACE-TIME CONVERGENCE (Source: Tidswell, 1978)

etc. Accessibility is clearly dependent on the mode of transport, for example, maps depicting time distances for bus, car, rail and air journeys between London and the major cities will reveal significant variations. The general shrinkage in time distances associated with developments in transport technology has led some writers to suggest that we now live in a 'global village'. However, the extent to which this shrinkage is relevant to specific individuals and groups depends on access to particular modes of transport. This depends on a number of factors, but the most obvious is the ability to pay.

An example of the sort of analysis which might be applied to information contained in various timetables and schedules is illustrated in Figure 6.6. In the first map accessibility to the centre of London in 1972 is plotted in normal Euclidean distance. The second map expresses accessibility in terms of cost distance, while the third expresses accessibility in minutes. A more sophisticated analysis might be attempted which produces a map whose scale is expressed in pounds, minutes or hours. This type of transformation would locate places in terms of relative location associated with time or cost, (see Figure 6.7 ).

Another important idea is the concept of accessibility minutes. This is particularly relevant when examining the quality of public transport provision at the intra-urban or regional level. Although it is of interest to know how far a village is from a local centre in terms of bus travel time this is insufficient if the focus of interest is on accessibility; additional information is required on the frequency of bus services. Thus if the travel time from village A to centre B is 20 minutes but the average waiting time between buses is 90 minutes, then the distance between A and B in 'accessibility minutes' is 110 minutes. This sort of data could easily be mapped. Considerations such as these are clearly important for communities with concentrations of immobile groups (for example, mothers with children and the elderly).

## 6.5 Network Analysis

The analysis of transport networks is a relatively well developed branch of human geography. The various indices which have been developed to measure connectivity and accessibility are readily available in most standard texts (for example Bradford and Kent, 1977; Tidswell, 1978; Whynne-Hammond, 1980). In the context of advanced level geography these indices may provide the basis for a variety of projects. The aim of this section is to briefly consider the possible range of projects rather than focus on the description of the indices themselves.

At the simplest level of application the various measures of connectivity and accessibility could be used in a descrip-

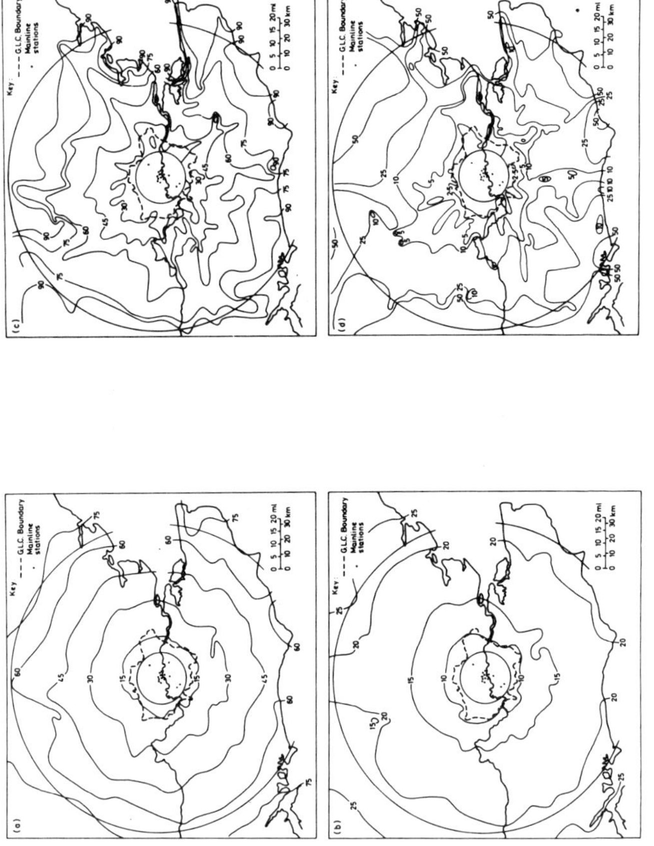


Figure 6.6 ACCESSIBILITY TO THE CENTRE OF LONDON (from Parkes and Thrift, 1980)

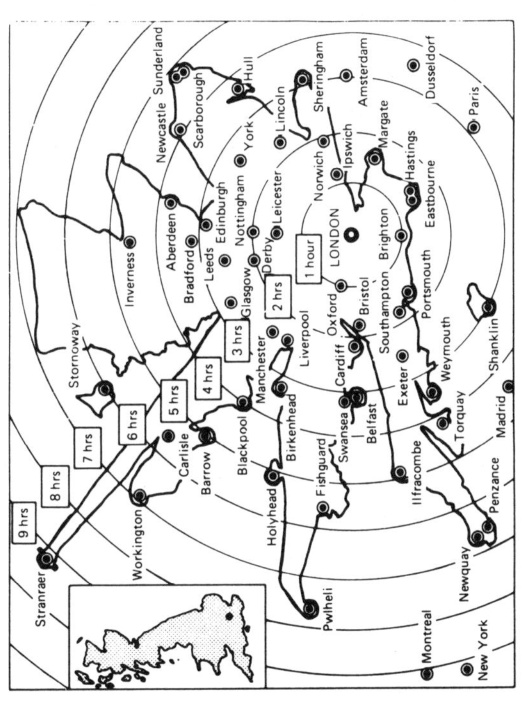


Figure 6.7 TRANSFORMATION OF SPACE : TIME FROM LONDON
(Source: Parkes and Thrift, 1980)

tive manner to examine the topographical structure of local road networks. Using the same basic rationale a comparative study could be undertaken, comparing, for example, the British motorway network with that in another country. A more advanced study (but still relatively straightforward) could focus on the evolution of the local transport network, noting the degree to which connectivity and accessibility has improved over time. In addition, in assessing the results from the analysis it may be of value to assess the suitability of the existing network for present-day needs.

Other 'problem-oriented' projects could involve the analysis of the effects of political boundaries on transport networks in an area, or an examination of the relationship between regional development and transport networks. In the latter context a suitable project might involve the analysis of a specific region. Twenty years ago, for example, North Eastern England had one of the poorest road networks in the country. However, with the Hailsham Report of 1963 the area was designated a 'growth zone' and major improvements in the road network ensured. This raises a number of questions: to what extent was connectivity and accessibility improved within the region?; how important are improvements in transport facilities in stimulating economic development as compared with other factors? Alternatively a project might be based on a comparison between transport networks of a central and peripheral region (the South East and Wales for example). An attempt could be made to examine the degree to which variations in the structure of networks contribute to regional imbalance. How does this relate to models of regional development, for example, Myrdal's idea of 'comulative causation'? (see Keeble, 1976; Bamford and Robinson, 1978). Arguably, the last two projects should only be attempted by the more able student insofar as they emphasize the ability to draw on, and synthesize, a range of ideas.

Chapter Seven

# ASPECTS OF SPATIAL INTERACTION ANALYSIS

## 7.1 Background

In the last chapter, attention was given to various possible projects involving transport studies. Analyses of transportation systems complement studies of the location of different activities, such as shopping, housing and employment; for example, the spatial distribution of residences and employment opportunities provides the structural backcloth against which journey-to-work patterns can be considered. Clearly, these topics, transport patterns *per se* or the more general link with the spatial distribution of activities, are of enormous direct interest to geographers, and in their attempt to comprehend the spatial characteristics, two types of models have been examined widely: spatial interaction models and diffusion models. The former is considered below, and, simply stated, its development stems from the Newtonian gravity model. It is noted that a distinction between communication and transport flows is made frequently. (In the former, no physical movement is involved, and, therefore, involves the exchange of ideas for example, whilst, in the former, movement of goods and people occur. Spatial interaction analysis is not restricted to describing transport flows).

Before considering various aspects of spatial interaction analysis that would provide useful components of student projects, it is necessary to comment briefly on the spatial system that is used, because it relates directly to data collection and availability and it also affects the model's parameter values. This provides the basis for the remaining sections which indicate how such work could be incorporated within student projects: first, the Newtonian gravity model is introduced, and some comments relating to its structure are presented; and, second, various applications involving extensions and reformulations of the Newtonian gravity model are examined.

## 7.2 Spatial System

In order to represent spatial interaction, it is necessary to have a suitably defined spatial system. An example of a conventional spatial referencing system based on a Euclidean coordinate geometry is the National Grid system which permits a location to be defined uniquely within the bounds of the map's accuracy. Such a description is called a continuous space representation; that is, space is portrayed as a continuous phenomenon. In contrast, a discrete space representation in which space is divided into well-defined, distinct zones or areas is often used. In empirical work, a discrete space representation is usually applied, because much data are available based on administrative areas (such as the Enumeration Districts of the Population Census). Moreover, if necessary, it is possible to transform a continuous space representation into a discrete space representation; this can be thought of as classifying spatial data.

At the outset, it should be noted that, when a discrete space representation is adopted, a variety of different definitions of zones for the same spatial system can be used. Whilst practical difficulties may arise because of great variations in size and shape, it is important to appreciate the basic influence of scale (or the number of zones). In general, the level of inter-zonal movement, for example, would be reduced if larger zones were used, because more movement would be confined to within the boundaries of a specific zone.

As suggested in the previous chapter, data on movement take two forms: flows along specific networks and flows between specified origins and destinations. Usually, in the former, no account is taken of the origins and destinations, and, in the latter, the actual route taken is not collected. Whilst network-specific data is useful, particularly for projects on congestion, pollution, and the impact of pedestrianisation schemes, the origin-destination flow counts are used more widely.

Using a discrete space representation, each possible origin and destination is a zone, and, for convenience, to summarise its location, the zone centroid is employed (see Figure 7.1). To uniquely specify a zone, each one is given a number; it is, therefore, possible to examine particular interactions, say the movement of public transport passengers from zone 5 to zone 7. Clearly, it is the spatial scale of the zones that determines whether we are dealing with intra-urban, inter-urban, inter-regional or international movement. For a general representation, if there are n zones in the spatial system of interest, there are n x n possible interactions. By defining $T_{ij}$ as the number of trips from origin i to destination j, it is straightforward to represent movement in a spatial system by a square interaction matrix. For example, for a simple four zone system, the interaction

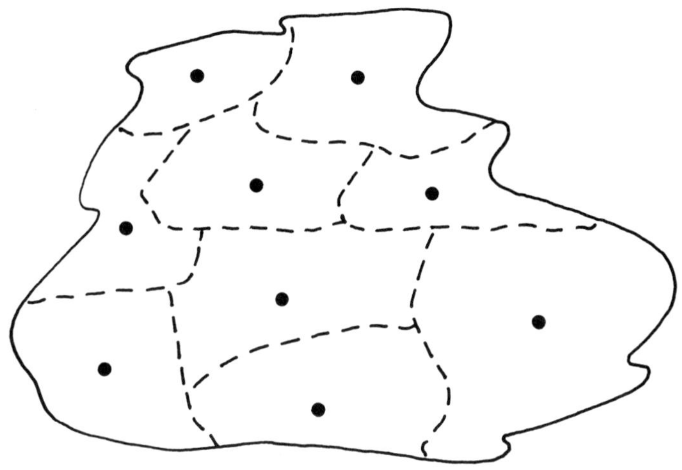

Figure 7.1 ZONES AND THEIR CENTROIDS

matrix is

| ORIGIN ZONES | DESTINATION ZONES 1 | 2 | 3 | 4 |
|---|---|---|---|---|
| 1 | $T_{11}$ | $T_{12}$ | $T_{13}$ | $T_{14}$ |
| 2 | $T_{21}$ | $T_{22}$ | $T_{23}$ | $T_{24}$ |
| 3 | $T_{31}$ | $T_{32}$ | $T_{33}$ | $T_{34}$ |
| 4 | $T_{41}$ | $T_{42}$ | $T_{43}$ | $T_{44}$ |

The elements of the principal diagonal, $T_{11}$, $T_{22}$, $T_{33}$, $T_{44}$, represent intra-zonal movements, and the remaining elements represent inter-zonal movements. It is important to note the distinction between $T_{12}$ (flows from zone 1 to zone 2) and $T_{21}$ flows from zone 2 to zone 1), because clearly they are not the same; thus, it is unlikely that an interaction matrix would be symmetrical about the principal diagonal.

## 7.3 A simple gravity model

Newton's law of gravitational attraction states that the force of attraction between two bodies i and j, $F_{ij}$, is related directly to the product of their masses, $M_i$ and $M_j$ and related inversely to the square of the distance between them, $d_{ij}$. Distance is usually defined to be straight line distance (based on Pythagoras' theorem), although intra-zonal distances must be specified exogeneously by the individual (to avoid problems of zero distance). Formally, the gravity model can be written as

$$F_{ij} = \frac{g\, M_i\, M_j}{d_{ij}^2}$$

where g is a constant, the force of gravity. Analogies of Newton's gravity model have a long heritage in geography: for example, Carey's (1858-1859) *Principles of Social Science*, Ravenstein's (1885, 1889) work on population migration, Reilly's (1931) more familiar *Law of Retail Gravitation*, Stewart's discussion of population/income potentials and Wilson's (1971) family of spatial interaction models. In the following section of this chapter, different applications of gravity-type, spatial interaction models are presented to demonstrate the broad potential of incorporating similar studies into student projects. First, however, as a general foundation to this work, it is pertinent to refer to Ullman's (1956) conceptual bases for interaction (see also Ullman, 1981).

In Ullman's general theory of spatial interaction, three bases of interaction were identified: complementarity, intervening opportunity, and transferability (or distance). As the term implies, the concept complementarity reflects that movement between two areas can be to their mutual benefit. Movement of coal, for example, in the United Kingdom occurs between regions with excess supply and those with excess demand. Clearly, it is necessary to take account of the transport costs of such movement, and, therefore, in looking at transferability, such factors as distance, type of commodity, and mode of transport are considered. The concept of intervening opportunities is important, because, although conditions of complementarity may exist between two regions, their distance of separation may be so large that they are associated with nearer regions which can satisfy their needs. Given geographers' particular interest in the impact of distance on the level of interaction, specific attention is drawn to an empirical regularity, the so-called distance-decay rule (see Taylor (1975) for a more detailed discussion). The relationship between distance and the level of interaction is indirect; the level of interaction decreases as the distance increases. Empirical evidence suggests that this relationship is not linear. That is, whilst the level of interaction decreases with increasing distance, it decreases at a decreasing rate (see Figure 7.2). Given such a non-linear relationship, it is often graphed on logarithmic paper to facilitate comprehension because it is often possible to transform it into a linear relationship.

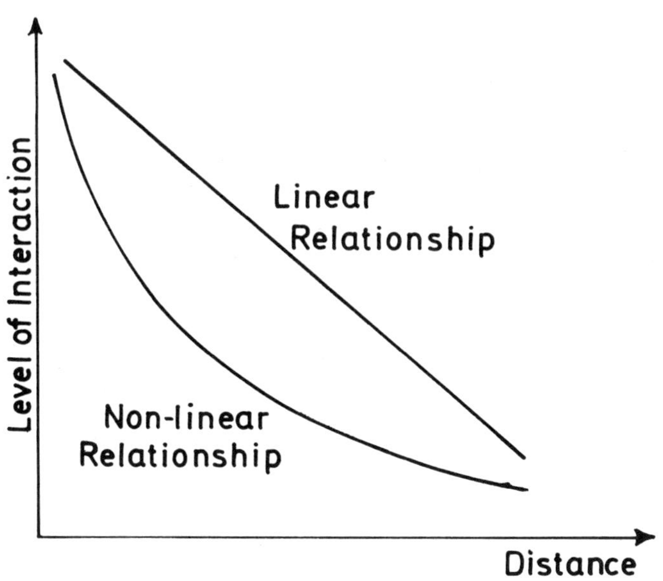


Figure 7.2 RELATIONSHIPS BETWEEN THE LEVEL OF INTERACTION AND DISTANCE

Finally, for completeness, given that it is emphasised throughout the book that good graphical presentations of data are especially informative whether it is in the form of maps or charts, attention is drawn to some of the alternative cartographic presentations of flow data that are feasible (see Figure 7.3).

## 7.4 Some applications of spatial interaction models

It is inappropriate to present the structure of a complete project based on spatial interaction analysis; the framework for projects has been well illustrated already. Instead, general discussion of a range of applications is presented. As a point of departure, some comments are made about the Newtonian formulation with regard to its geographical applicability. This provides the foundation to discuss a range of common types of applications of spatial interaction models. For example, the concept of accessibility and potential surfaces, which are well-established areas of geography, are discussed. Conventional linear regression analysis is shown to be useful in relation to linking the gravity model to classical central place concepts. The theme of market area analysis is continued by a consideration of an extension of Reilly's well-known *Law of Retail Gravitation*.

### 7.4.1 *Geographical applications of Newton's gravity model*

In geographical applications of Newton's gravity model, the mass terms ($M_i$) are often defined as the population (P) at a specific location (i), and the problem involves determining or calibrating the distance exponent ($\beta$) given a suitable value of the constant, $\alpha$, which ensures numerical consistency between the aggregate observed and predicted interaction levels. Thus, the model becomes

$$T_{ij} = \frac{\alpha P_i P_j}{d_{ij}^{\beta}}$$

In terms of 'push' and 'pull' factors behind movement, it would be possible to define $M_i$ and $M_j$, respectively, as something more appropriate than population. Obviously, the selection of variables is dependent on the specific problem of interest.

By inspecting this formula in more detail, it is possible to see that it is a more precise, analytical representation of Ullman's basic concepts. Depending on their definitions, the mass terms can be used to represent the complementarity concept (see the example on industrial movement below). For the distance concept, it is obvious that the level of interaction

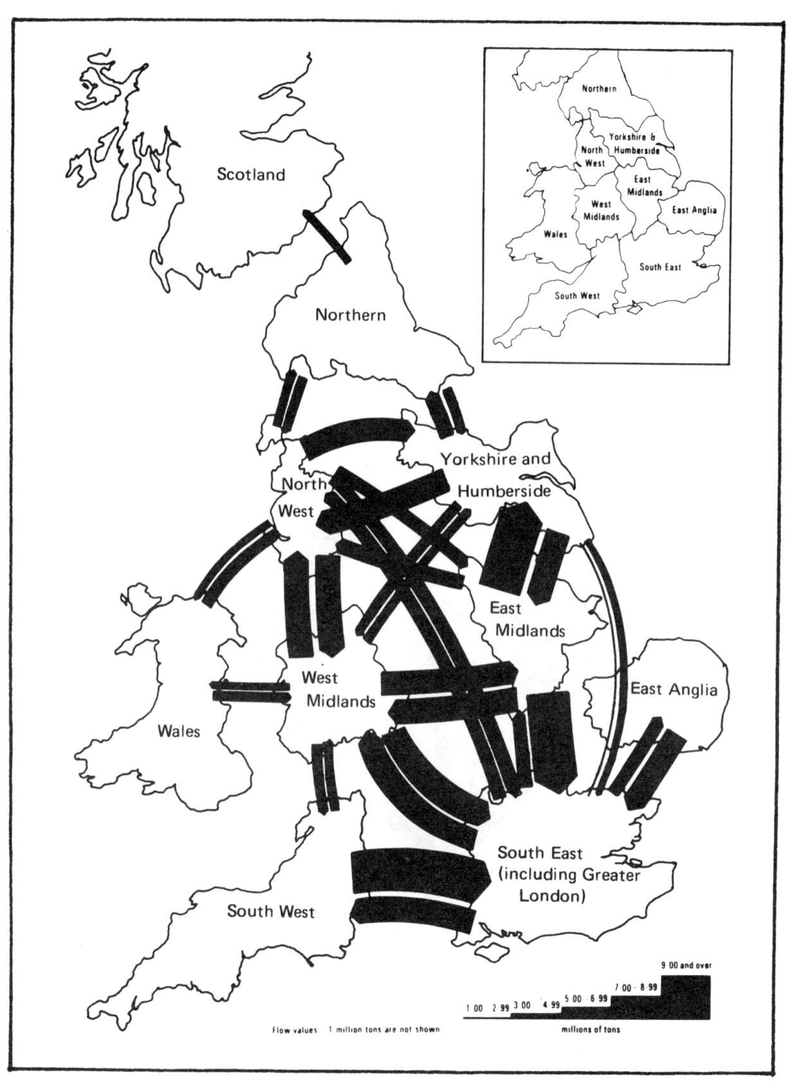


Figure 7.3a LINKAGE ANALYSIS SUMMARY OF FLOWS OF HEAVY GOODS (Source: Opėn University, 1977)

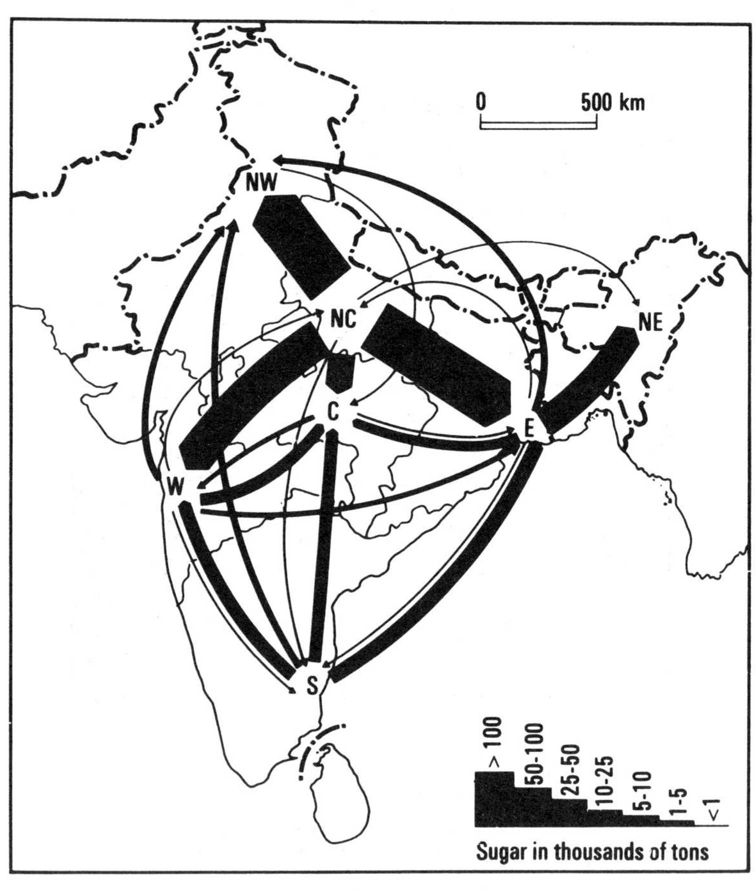


Figure 7.3b SUGAR MOVEMENT IN INDIA
(Source: Open University, 1977)

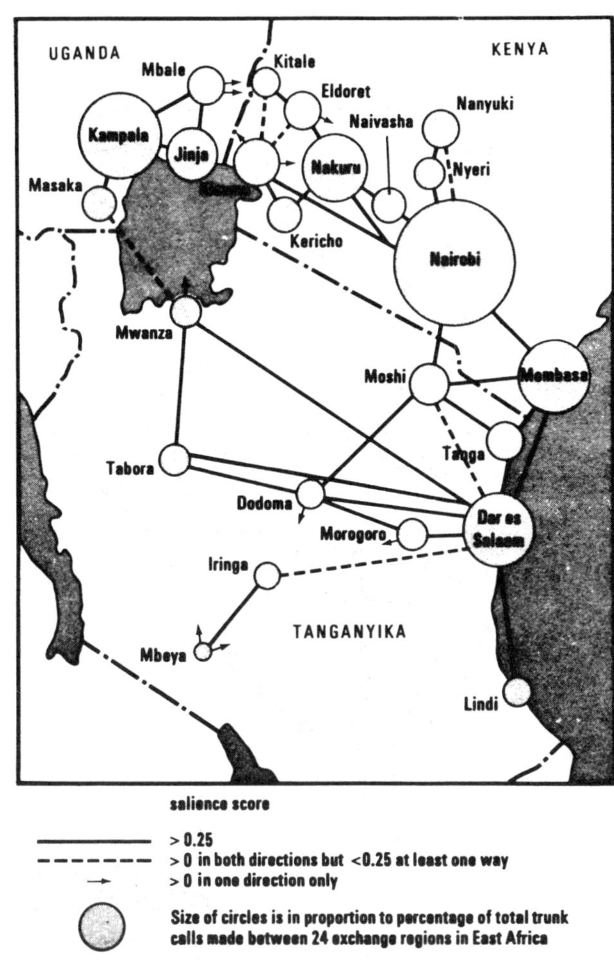


Figure 7.3c TRUNK TELEPHONE CALLS: EAST AFRICA
(Source: Open University, 1977)

decreases as distance increases; the rate of change is dependent on the value of the β parameter. That is,

$$T_{ij} \alpha \frac{1}{d_{ij}^{\beta}}$$

Figure 7.4 portrays clearly the two major influences on interaction levels: distance and a location mass. Rail dispatches from Bristol, for example, exhibit a distance-decay effect, although the effect of centre size, particularly London and Birmingham, is noticeable. Whilst the concept of intervening opportunities is linked to the previous two concepts, it is possible to represent aspects of competing destinations using the concept of accessibility which is discussed below. Alternatively, an intervening opportunities, spatial interaction model has been proposed by Stouffer (1940).
It is noted that the above formulation of the gravity model has the symmetrical property

$$T_{ij} = T_{ji}$$

Keeble (1971) studies industrial movement (which was measured by the number of jobs created) from the south-east of England to other regions in the United Kingdom between 1945 and 1965. A modified, unidirectional gravity model was applied to this data set by Huggett (1980) to attempt to describe these flows. The number of jobs created in region j (where the South East was defined as region i), $T_{ij}$, was assumed to be a function both of the 1954 unemployment level for region j, $M_j$, which was an indication of the surplus labour supply available to the potential employers, and of the straight-line distance, $d_{ij}$, between the centres of the South-East region and region j. The best-fit value for β was found to be 2.0 and the scaling parameter (α) was equated to 53,000. Table 7.1 summarises the results and provides a basis for comparing the observed and expected flows. In a project, whilst it can be reassuring to find a model is a good-fit, anomalies should be explained, because it can act as the basis for directing further investigations. For example, why was there an over prediction of movement to Wales and an under prediction of movement to the Northern region?

For completeness, it is necessary to comment briefly on the measurement of distance and the nature of the distance function. The distance between two places is, for convenience, usually measured as a straight-line distance, although it is possible to use road distances or even a composite time-distance measurement which would be different for various modes of transport. To date, all the models have used a negative power function; a negative exponential function

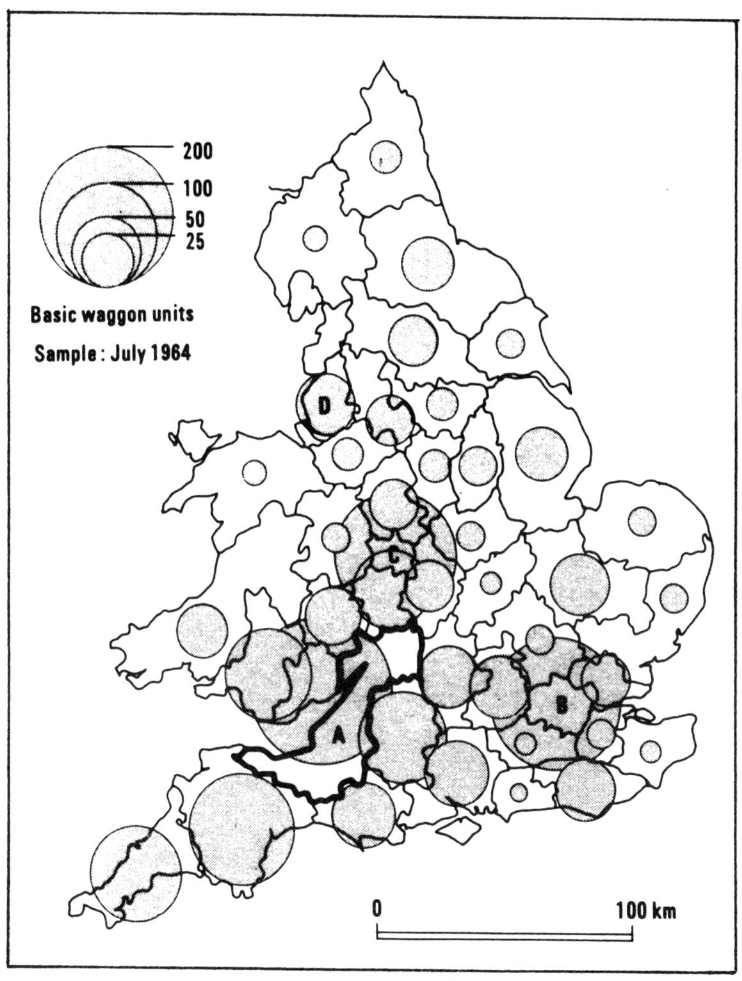


Figure 7.4 RAIL DISPATCHES FROM BRISTOL
(Source: Open University, 1977)

| REGION (j) | 1954 UNEMPLOYED WORKFORCE ($M_j$) | DISTANCE (miles) BETWEEN REGIONS ($d_{ij}$) | PREDICTED MOVEMENT OF JOBS ($T_{ij}$) | OBSERVED MOVEMENT OF JOBS |
|---|---|---|---|---|
| NORTHERN IRELAND | 33,000 | 320 | 17,079 | 16,900 |
| SCOTLAND | 59,000 | 360 | 24,332 | 24,500 |
| NORTHERN | 28,300 | 250 | 23,998 | 35,800 |
| WALES | 22,900 | 140 | 61,923 | 43,200 |
| MERSEYSIDE | 18,900 | 180 | 30,916 | 36,000 |
| DEVON AND CORNWALL | 8,600 | 190 | 12,625 | 10,600 |

TABLE 7.1

$\exp(-\beta d_{ij})$ or $e^{-\beta d_{ij}}$ has also been applied frequently, and the gravity model can be written as

$$T_{ij} = M_i M_j \exp(-\beta d_{ij})$$

This is the same as the gravity model except for the type of distance function. The selection of the function determines the value of the β parameter in the calibration process, and, whilst it is impossible to give an unambiguous indication as to when a particular function should be used, empirical evidence indicates that it is related to the scale of analysis because of the different rates of change of the functions: the negative power function is associated with inter-urban movement and the negative exponential function is associated with intra-urban movement (see Figure 7.5).

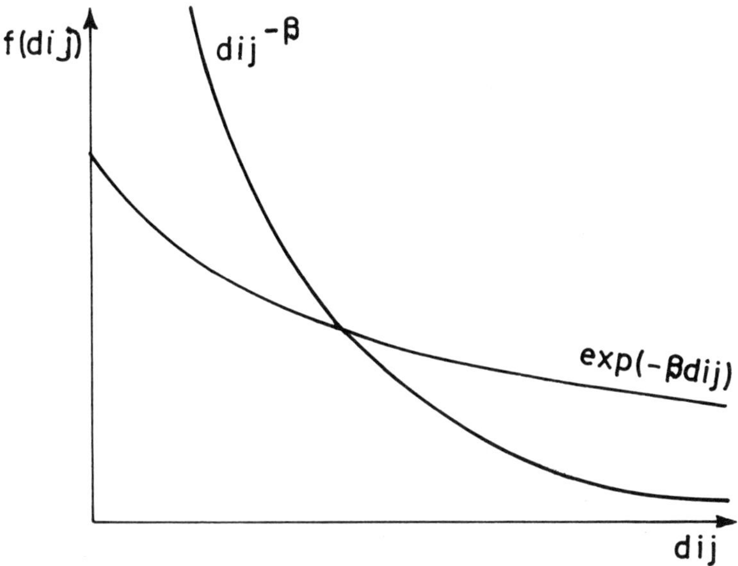


Figure 7.5 DISTANCE FUNCTIONS: NEGATIVE POWER AND NEGATIVE EXPONENTIAL

As a final comment, it is useful at this stage, to have a intuitive impression about the effect of different β values. For illustrative purposes, the negative power function is used, and Figure 7.6 graphs the effect of different β values. The influence of distance on the movement of various commodities is different, and this is represented by different values of the β parameter. Table 7.2 presents empirical evidence of this characteristic; those commodities for which distance is a relatively large impedance, such as the bulky commodities, coal and building materials, and the perishable food commodities, have relatively large β values, and the commodities with relatively low β values are the more valuable or less bulky commodities.

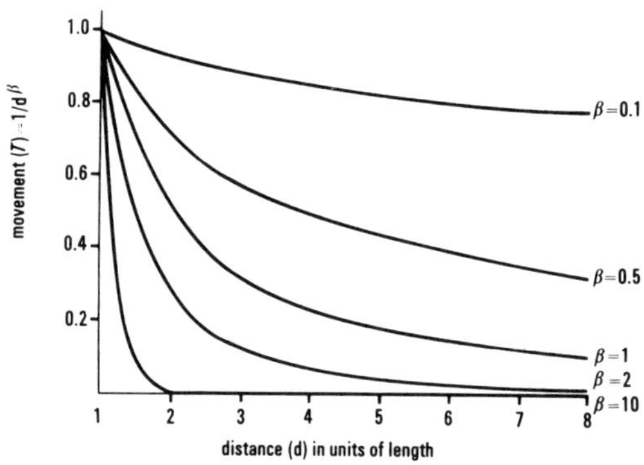


Figure 7.6 EFFECT OF DIFFERENT β VALUES
(Source: Open University, 1977)

| Commodity | β value |
|---|---|
| Building materials | 1.64 |
| Food | 1.53 |
| Coal | 1.46 |
| Miscellaneous | 1.36 |
| Oil | 1.17 |
| Other raw materials | 1.16 |
| Other manufactures | 1.09 |
| Transport equipment | 0.96 |
| Scrap | 0.82 |
| Chemicals | 0.80 |
| Steel | 0.77 |

TABLE 7.2 β VALUES FOR DIFFERENT COMMODITIES

(Source: Open University, 1977)

### 7.4.2 *Potential surfaces and accessibility*

A potential surface can be defined as a description of the opportunity of different locations for interactions over space and it is an important descriptive tool of so-called macro-geography (Warntz, 1965). Interestingly, the literature on potential surfaces (see Rich (1980) for a detailed pedagogic discussion) is related directly to gravity model formulations.

Simply stated, the potential of a particular location (i) is a relative index of aggregate possible interactions, $I_i$, which can be written formally for a unidirectional gravity model as,

$$I_i = \sum_{j=1}^{n} \frac{M_j}{d_{ij}^{\beta}} = \sum_{j=1}^{n} T_{ij}$$

The Greek capital letter, sigma, is used as the summation notation. Specifically, it represents the additions of all potential interaction with i (see Figure 7.7 ), which, in this example, can be thought to be

$$I_4 = T_{41} + T_{42} + T_{43} + T_{45} + T_{46} + T_{47} + T_{48} + T_{49} + T_{410} + T_{411}$$

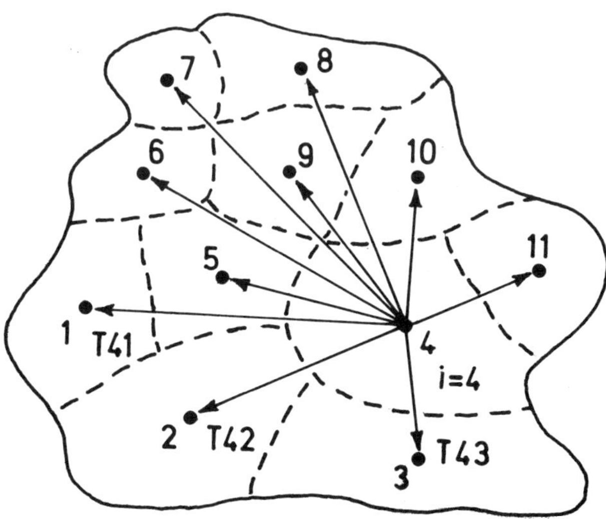


Figure 7.7 POTENTIAL FOR INTERACTION: ZONE FOUR

Thus, if the potential for each zone was calculated, it would be possible to map the relative intensities of potential interaction, a characteristic that could provide a sound foundation for projects on industrial location and the market potential for new shopping developments.

### 7.4.3 *Regression analysis and the gravity model*

In the previous discussion about the distance-decay phenomenon, attention was drawn to the desire to transform data to produce a linear relationship between interaction levels and distance. In statistical terms, this can be represented by standard bivariate, linear regression analysis (see Appendix I for more details). In this sub-section, attention is drawn to the fact that the gravity model can be rearranged into the format of a regression model which can be interpreted directly.

Restating, for convenience, the Newtonian gravity model

$$T_{ij} = \frac{\alpha P_i P_j}{d_{ij}^{\beta}}$$

it is possible to demonstrate that, using a logarithmic transformation, it can be reformulated as a standard regression model of the form,

$$y = \alpha + \beta x$$

By dividing both sides of the gravity model by the product of the masses $P_i P_j$ and then by taking the logarithm, the gravity model can be written as the following linear regression model,

$$\log \left( \frac{T_{ij}}{P_i P_j} \right) = \log \alpha - \beta \log d_{ij}$$

and graphed as a straight line which is shown on Figure 7.8 with an intercept log $\alpha$ and a slope $\beta$. This formulation is of basic geographic interest because distance is the independent or explanatory variable in the regression model and the relative level of interaction is the dependent variable. This type of description would be especially useful to compare the flows of different commodities, because the determination of the $\beta$ value for each commodity (using the least-squares criterion) would enable a direct examination of the relative influence of distance in travel behaviour associated with different commodities. In central place theory, for instance, it is suggested that consumers are willing to travel longer distances to obtain higher-order goods and services. Such a representation can be linked to Losch's (1954) demand cone

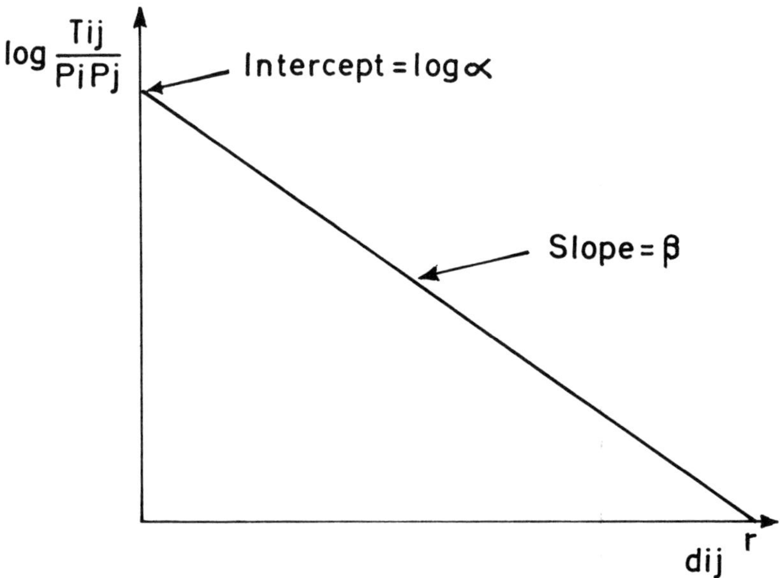


Figure 7.8 BIVARIATE LINEAR REGRESSION MODEL OF THE GRAVITY MODEL

analysis, and also to Christaller's (1966) basic concepts of the range and threshold of a good. The upper range is represented by r in Figure 7.8 and if the level of flows could be used to give an indication of expenditure, the area under the line would be a representation of aggregate revenue received.

### 7.4.4 *Market area delimitation*

Following on the above discussion about service provision, a popular topic relating to gravity models is market area delimitation, and for student projects the obvious starting place is the well-cited *Law of Retail Gravitation* by Reilly (1929). Following the underlying rationale of Newton's gravity model, it is argued that the number of visits a resident of one area makes to two competing shopping centres is inversely proportional to the distances between the residential area and the two centres and is proportional to the towns sizes (which can be thought of as an indication of their attractiveness). To determine the market boundary between two centres which are located at j and z, for example, it is noted that the levels of spatial interaction between a resident located at i and the two centres can be represented as

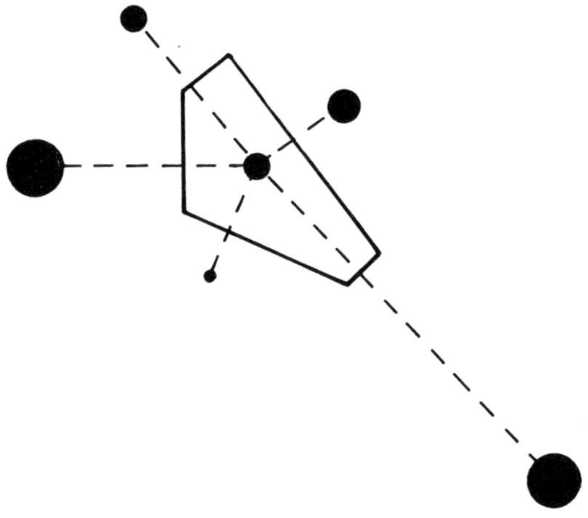

Figure 7.9 MARKET AREA DELIMITATION

$$T_{ij} = \frac{\alpha M_i M_j}{d_{ij}^{\beta}} \quad \text{and} \quad T_{iz} = \frac{\alpha M_i M_z}{d_{iz}^{\beta}}$$

Intuitively, the market boundary can be defined as the location, say i, such that the levels of interactions of residents from i to both centres are identical. That is, there is an indifference between the two competing centres. Thus, formally, if i is the market boundary,

$$T_{ij} = T_{iz}$$

which can be extended using the gravity model formulations and rearranged to give an equation for the distance of the boundary from one of the centres, $d_{iz}$

$$d_{iz} = \frac{d_{jz}}{1 + \sqrt{M_j/M_z}}$$

It should be noted that such market area delimitation is based on two centres, and, therefore, it disregards or assumes away the influence of other competitors. Indeed, as it stands, using this gravity model, it is only possible to consider movement from one zone to another zone in isolation, disregarding the important features of spatial interdependence, particularly competition. To assume that the level of movement between a particular pair of locations is independent of their spatial relationships with other locations and interactions is too restrictive and unrealistic. This can be overcome by extending the simple gravity model and by examining spatial interactions in a probablistic formulation.

Movement from a particular location to another location can be thought to depend on the relative attractiveness and relative distance of all potential destinations. If there are k possible destinations, the probability of selecting to move from location i to location j, $P_{ij}$, can be defined as

$$P_{ij} = \frac{M_j \, d_{ij}^{-\beta}}{\sum_{j=1}^{k} M_j \, d_{ij}^{-\beta}}$$

To determine the actual level of movement between i and j, it is necessary to multiply the probability by the mass term of the origin, $M_i$. Thus,

$$T_{ij} = \frac{M_i \, M_j \, d_{ij}^{-\beta}}{\sum_{j=1}^{k} M_j \, d_{ij}^{-\beta}}$$

This formulation is only a natural extension of the original Newtonian model, but it is more consistent with a geographical study of interaction between phenomena. (It should be noted that $d_{ij}^{-\beta}$ is a short-hand way of writing $1/d_{ij}^{\beta}$, and, therefore, the equation is less cumbersome). Interestingly, the denominator can be interpreted in terms of overall relative accessibility, which was discussed in sub-section 7.4.2.

## 7.5 Some concluding comments

Simply stated, spatial interaction models provide a useful description of movement patterns, and, in terms of practical applications, they could be used as a basis for impact analysis or forecasting the effect of altering the existing structure of activities. A major theoretical problem arises because of their aggregate nature; whilst they describe the total level of movement between two places, it is impossible to interpret the reasons for a particular individual's travel behaviour. Given such a situation, in a project, a questionnaire survey of individuals may be appropriate to complement the aggregate description by a spatial interaction model.

Chapter Eight

# GEOMORPHOLOGICAL AND CLIMATOLOGICAL STUDIES

## 8.1 An Introductory Comment

The preceding five chapters have included material from the field of human geography; in this chapter and the following chapter, topics from physical geography are discussed. This relative weighting, however, does reflect the potential for project work in different branches of geography. Moreover, it must be emphasised that the dichotomy 'human' and 'physical' geography is more apparent than real; already, a number of projects have been outlined that epitomise the true breadth of man-environment interrelationships. In the following chapters, because of the growing concern that physical geography is becoming increasingly irrelevant to urban children (see Naish (1983) for important, thought-provoking reflections on such developments) (or at least impractical as their potential projects), a discussion of physical geography topics *per se* (such as, drainage basin studies, slopes, and coasts) is complemented by a consideration of urban climatology, land use capability and the impact on the environment through man's activities such as recreation. First, however, some back ground comments regarding form and process.

## 8.2 Form and Process

Most of the topics covered in the present chapter are concerned with geomorphological studies, both pure and applied. Given this weighting it would be appropriate to provide a brief outline of the major developments which have occurred in the sub-discipline so as to establish a broad context for the themes introduced in the following sections.

Classical geomorphology, based on the Davisian model, was essentially historical in character. Explanation took the form of a historical description of landscapes over long time periods and wide spatial scales. The primary aim of this approach (and the attendant concern with denudation chronology) was one of classification, dominated by consideration of time and evolutionary stages of development ('youthful', 'mature', 'old', for example). Gradually, it was realised that although changes clearly occur in the landscape, of greater importance was the elucidation of those processes which induced change. This reorientation had radical implications for the study of geomorphology '...Geomorphology thus moved away from being essentially a classification procedure and adopted a more truly scientific attitude based on the use of the deductive scientific method'. (Anderson and Burt, 1981, p.4).

The 'new quantitative geomorphology', as it is sometimes called, is concerned with the nature and rate of operation of geomorphological processes. From a knowledge of processes it is possible to theorise about the landforms that may be produced. The scientific approach to landform study is, therefore, characterised by a decline in qualitative description, and an emphasis on systematic and objective data collection, statistical analysis and, increasingly, mathematical modelling. In general, these trends have involved a reduction in both the temporal and spatial scales of analysis. In addition the scale of landforms being examined has also been reduced, allowing process and form to be more easily related. Recently geomorphology has adopted the systems theory approach as an overall explanatory structure emphasising the interrelationship which exists between process and form.

Clearly, quantification is an essential part of the new geomorphology; involving an adequate definition of the variable to be measured and the existence of appropriate techniques to measure unambiguously the properties of the variable so defined. Indeed the function of any technique or measurement procedure is to allow the quantification of variables in a consistent, rigorous and precise manner. It is, therefore, important to emphasize that for many geomorphological studies the ability to measure a particular variable may well be the limiting factor.

Further developments in the field of geomorphology have been associated with its potential for application in a wide variety of contexts. Some of the issues considered by geomorphologists include the impact of geomorphology on man (natural hazard research), man as a geomorphic agent, and the use of geomorphological expertise in the solution of practical problems. This growing emphasis on applied studies, it should be noted, can also be found in many other branches of physical geography (see section 8.6 below).

## 8.3 Drainage basin studies: some quantitative analyses

### 8.3.1 *Background*

The drainage basin has been termed the fundamental geomorphic unit, and, in spite of the inherent complexity of the processes operating in such a system, it is suggested a number of manageable, field-work based projects could be undertaken successfully. Robinson et al (1978), for example, have made available data to structure students learning in their Drainage Basin Units. Gregory and Walling (1971, pp. 279-280) suggest that geomorphological studies within the drainage basin framework can be seen at a series of levels. 'Firstly, it is necessary to describe the character of the drainage basin adequately, often utilising parametric measurements of basin characteristics. Secondly, it is necessary to document the processes, particularly the output of water and sediment. The third level is that at which relationships between form and process are established, and this is often achieved, fourthly, by using two or more basins comparatively. The fifth level may be to extend the results from these four levels to a longer geomorphological time scale'. Whilst the practical difficulties, specifically time and cost constraints, probably would make detailed comparative studies impossible, students are unlikely to possess the experience and to have access to equipment necessary for studies of the level five type. For comparative studies, drainage basins formed on impermeable and on permeable rock or with different vegetation covers would be sensible choices. However, in general, they should be of the same order. Consequently, in this discussion, attention focuses on describing a drainage basin by measuring a number of its characteristics. It is noted that two scales of topographic features can be described: characteristics of the drainage basin and characteristics of the streams. Owing to practical problems of data collection for students at school, no specific consideration of the following topics is included: transportation of the sediment load; and the fluvial processes of erosion and deposition. Given their usefulness, particular attention in this section is devoted to a range of possible quantitative analyses which should be seen as providing a sound descriptive basis for later interpretation, rather than being a project in their own right.

### 8.3.2 *Measuring drainage basin characteristics*

To describe drainage basin characteristics, a number of simple, numerical measures are available. Some of the more commonly used statistics are introduced in the sub-section. Whilst such measures are likely to provide the general, background information for a fieldwork-based project, it is suggested that a comparative study of the characteristics of different, carefully selected drainage basins, which was based on map work, could be undertaken to interpret the influence of rock type, vegetation cover, soil texture and so on. For consistency, in making comparisons, it is essential that maps (or photographs) of the same scale (such as 1:25000) are used.

In such studies, information on drainage basin area and shape, and the length, number and order of streams can be obtained readily. In practice, to highlight the features of interest and to make the data collection more straightforward, it is useful to put the boundary and streams of the drainage basin onto a piece of tracing paper.

In calculating drainage basin density, the determination of drainage basin area and stream length (which, it is noted, should be measured in the same unit base) can be completed using graph paper and string, respectively. However, with the increasing availability of microcomputers (see chapter ten), such characteristics can be measured both accurately and automatically using a digitiser or a graphics tablet.

$$\text{Drainage basin density} = \frac{\text{total stream length}}{\text{drainage basin area}}$$

It is a straight forward mechanical exercise to count the total number of channels in a basin (obviously, it should be checked!). From this information, it is possible to calculate the stream frequency, which is defined as the number of channels per unit of area. Variations in such measures, for instance, could be related directly to rock permeability.

Horton (1945) proposed a scheme for ordering streams, because it was desirable to be able to classify drainage basin according to size. As figure 8. indicates, each finger-tip

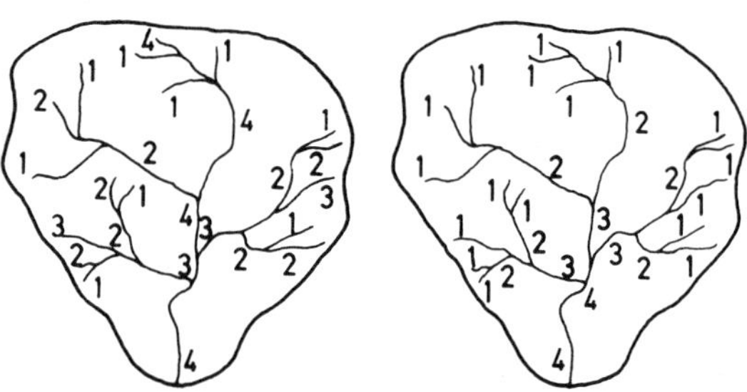

HORTON'S STREAM-ORDERING STRAHLER'S STREAM-ORDERING

Figure 8.1 STREAM ORDERING: HORTON AND STRAHLER

tributary is described as a first-order stream, two first-order streams combine to give a second-order stream, two third-order streams combine to give a fourth-order stream, and so on. Once this initial ordering is determined the highest order streams are taken back to their headwaters. Consequently, all the smallest tributaries are not the same order. To overcome criticisms of Horton's system, Strahler (1952) proposed to follow Horton's initial ordering stage because it is consistent with combinatorial analysis. Each segment of a stream, rather than the whole stream, is assigned an order (see Figure 8.1 ). The drainage basin is described usually by the highest order stream that flows in it. It is noted that one difficulty associated with Strahler's system arises, because the order of the main stream (and, hence, the drainage basin) is not changed necessarily when the number of low-order streams alters.

In terms of stream ordering, Horton proposed a number of laws of morphometry, which reflect empirical regularities in the geometrical characteristics of the organisation of drainage networks. Whilst it would be usual to describe a single drainage network in a project, comparative analyses, if possible, would be illuminating. Moreover, given the linearity of the relationships, the 'best' fit lines can be determined using simple linear regression analysis (which is described in Appendix I).

For a single basin, Horton's first law, the so-called law of stream numbers, is portrayed as a straight line if the number of stream segments of specific orders are plotted against stream order on semi-logarithmic graph paper (one axis, the vertical axis, is measured on a logarithmic scale) (see Figure 8.2). This relationship describes the fact that the number of stream segments of successively lower orders form a geometric series increasing at a constant ratio. This ratio is termed the bifurcation ration, $R_b$, which is defined formally as the mean of the individual ratios between the aggregate number of streams of each order and the total for the next higher order. This is an important index, because it provides an indication of the rate of discharge that would occur after a sudden rainfall. More specifically, in general, the higher the bifurcation ratio the lower the concentration of discharge in the main stream from its tributaries. For example, Table 8.1 provides the necessary information for the river Yealm in Devon.

| Order | Number of streams |
|---|---|
| first | 29 |
| second | 7 |
| third | 1 |

TABLE 8.1 River Yealm: numbers of streams of different orders

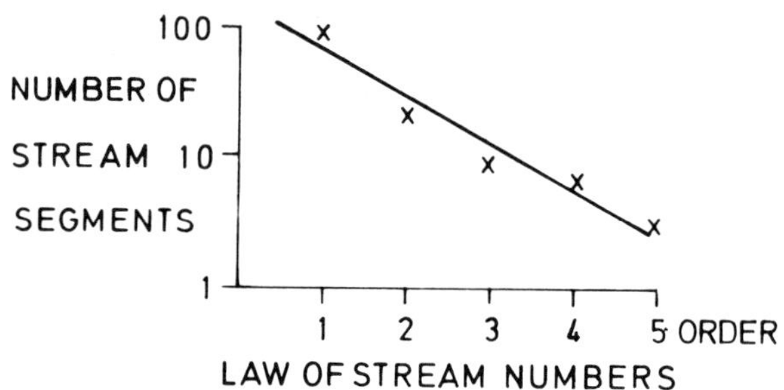


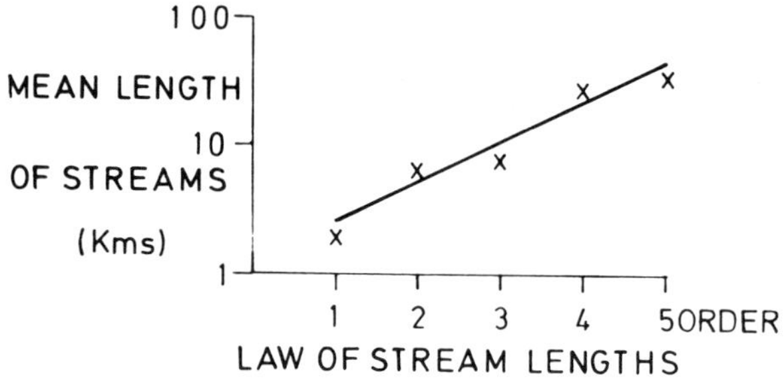


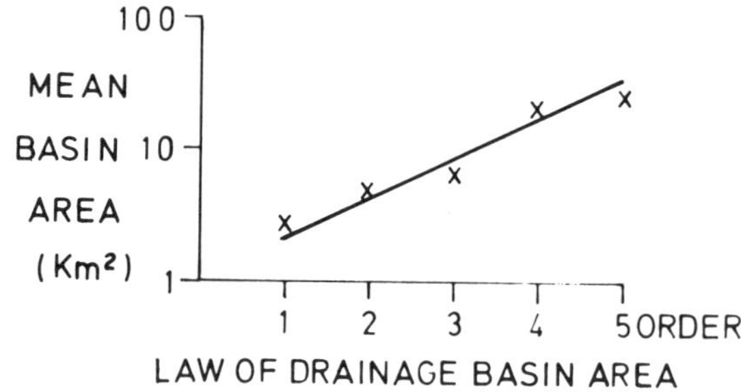


Figure 8.2 STREAM ORDERING: LAWS OF MORPHOMETRY

The associated bifurcation ratio is

$$R_b = \frac{\frac{29}{7} + \frac{7}{1}}{2} = 5.57$$

It is noted that the slope of the best fit line of a regression model also measures the logarithm of the bifurcation ratio.

A second law, the so-called law of stream lengths considers a similar relationship between the mean length of streams of each of the different orders against stream order (see Figure 8. 2). The mean stream length increases geometrically with order according to a constant ratio, the length ratio. A final law, the so-called law of drainage basin area, illustrates the relationship between the mean basin areas of streams of different orders against stream order (see Figure 8.2 ). The mean basin area increases geometrically with order according to a constant ratio, the area ratio.

In comparative studies, attention would focus on variations in the value of each ratio as indicated by the different gradients of the lines representing the different regions. Explanations could involve such factors as tectonic history, geology and climate. Interestingly, whilst Horton developed a logical argument to explain such geometrical regularities, it is noted that random simulations of drainage networks have produced similar characteristics.

Finally, for completeness, it is appropriate to mention other examples of quantitative analyses that provide a description of drainage basins. Simply stated, it is useful to have numerical descriptions of their shape and relief. For instance, if a basin is very long, the throughflow of water after a rainstorm will take a relatively long time.

It is possible to recognise two components of shape: circularity and elongation. They can be calculated using the following formulae:

$$\text{Circularity} = \frac{\text{Drainage basin area}}{\text{Area of the circle with the same perimeter as the drainage basin}}$$

and

$$\text{Elongation} = \frac{\text{Diameter of a circle with the same area as the drainage basin}}{\text{Basin length}}$$

To date, the three-dimensional nature of a drainage basin has not been described adequately. Morphometric maps describing slope direction and angle do give some indication of of these features, say the asymmetry of valleys. An alternative approach is to use area-altitude, or hypsometric, curves, specifically the integral, as a numerical index of a drainage basin's morphology. Simply stated, the range in altitude is determined by finding both the highest point on the watershed and the local base level. To facilitate direct comparisons, the altitudes are converted so that they vary between zero and one. To achieve this result, each location's observed altitude is made a proportion of the altitude range. For example, if the range for a drainage basin is between 350 metres and 650 metres, a location which is 450 metres would have a value of

$$\frac{450 - 350}{300} = 0.3$$

A similar method is used for area, because it is necessary to have an indication of the proportion of the basin's area above (or below) specified altitudes. If a drainage basin of 10 square kilometres has four square kilometres below an altitude of 450 metres, the proportion of the basin lying above this altitude is

$$\frac{10 - 4}{10} = 0.6$$

(It is noted that both these proportions are often used in the form of percentages). The hypsometric curve is a plot of altitude proportions against area proportions (see Figure 8.3). Clearly if hypsometric curves are plotted for different drainage basins, this representation provides a clear graphical indication of variations in drainage basin morphology. Moreover, it is also possible to calculate the hypsometric integral which is the area under the curve expressed as a proportion of the total area of the square formed by the two axes. If it is assumed that, at one point in time, the drainage basin was a flat surface at the existing maximum height, the hypsometric integral provides an indication of the proportion of material that has not been eroded away. Strahler, for example, recognised two distinct phases based on this statistic: an early, equilibrium phase when the drainage system is still developing; and a later, equilibrium state when the form of the basin was time-independent. The former has hypsometric integrals of greater than 0.6, and the latter has values in the range 0.35 to 0.6. It was suggested that the only situation when hypsometric integral would be less than 0.35 would occur when isolated residual hills were present in a drainage basin; anyway, it was argued that this is a transitionary stage before equilibrium.

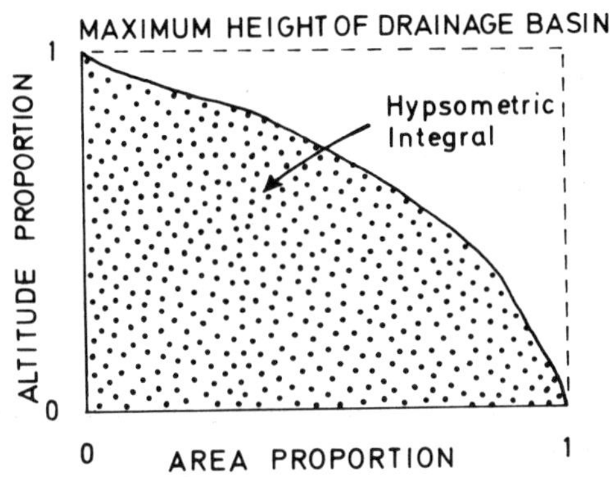


Figure 8.3 THE HYPSOMETRIC CURVE

### 8.3.3 *Some drainage basin input-output interrelationships*

#### 8.3.3.1 *Introductory comments*

To understand many of the interrelationships found in drainage basin networks, it is important to view a stream as part of the hydrologic cycle. In essence, there is a fixed quantity of water, which is transferred in an orderly sequence. A stream, therefore, is only a single component within this closed system. The hydrologic cycle can be represented as

$$P = RO + E \pm \Delta S$$

where P is the total precipitation input, RO is the runoff, E is the total evapotranspiration loss, and $\Delta S$ is the change in storage. That is, an indication of the proportion of precipitation input that is runoff, evapotranspiration loss and storage is given. For convenience of presentation, aspects of each component are considered separately in this subsection.

#### 8.3.3.2 *Total precipitation*

The term precipitation has been selected carefully. In such studies, it would be necessary to measure, if possible, more than rainfall. Precipitation also includes snowfall, and it is also possible to measure dew, fog and mist. Simple instruments, such as a rain gauge, can be used to measure precipitation at different locations over periods of time. In practice, for a specific investigation, a student would probably use a non-recording, storage device, and, therefore, regular visits must be practical. Whilst particular measurement problems associated with specific devices should be discussed explicitly in a study, in an estimation of precipitation over a drainage basin, special care must be given to the sampling distribution of data points, specifically their distribution over time and space. Given a set of sample point precipitation values, interpolation (using either isohyets or thiessen polygons) is applied to find a drainage basin's aggregate precipitation. However, it should be appreciated that variations in precipitation levels are likely to occur because of variations in geology, relief and vegetation. For instance, when there is a dense vegetation cover, it is often appropriate to measure net precipitation, because, to consider either run-off or the transportation of sediment load, gross precipitation values could introduce bias. Gauges can be placed under trees to measure throughfall (although it is probably impractical under other types of vegetation) and it is possible to ascertain stemflow by fixing devices to tree trunks. Net precipitation is the total throughflow and stemflow. Given the variations in the density of such vegetation, if it is decided to measure net precipitation, it is advisable to have a large number of sampling points. Obviously, the level of interception by vegetation is dependent not only on the type of vegetation, but also on the season of the year. Consequently, often the location of sampling points should be stratified to reflect such characteristics. Finally, it is noted that such methods would not provide satisfactory information on the duration and intensity of storms.

In trying to determine what happens to the total precipitation in a drainage basin, it is possible to measure runoff, evapotranspiration losses and storage alterations.

#### 8.3.3.3 *Infiltration capacity*

As precipitation reaches the ground, streams are not formed immediately because of the permeability of the rocks and soils. Clearly, the infiltration capacity or the rate at which water enters the ground varies with the types of rocks and soils. Moreover, this capacity starts with a relatively large initial value, and decreases quickly to a constant value for a specific type of rock or soil. For example, if the soil is dry at the onset of rain, the infiltration capacity would be relatively large at the beginning, but soon the rate would decrease as water replaces the air between soil particles. When rain falls with an intensity greater than the infiltration capacity, surface runoff is produced.

Thus, the infiltration capacity is a basic determinant of the runoff in a drainage basin. This capacity is dependent on rock type, more particularly its permeability, soil texture and structure, more particularly its antecedent soil moisture, and the vegetation cover. Any examination of a drainage basin, therefore, should include such background details to provide a firm foundation for more specific, descriptions and explanations.

#### 8.3.3.4 *Runoff*

Total streamflow can be measured relatively easily, and it is recorded at many stations throughout the country. Generally, it is reduced if the basin is elongated, the relief is gently sloping, and if the vegetation cover is thick. Discharge in cubic metres per second, Q, is

$$Q = AV$$

where A is the cross-sectional area of the channel and V is the mean velocity at the cross section. Stream velocity is measured usually by a current meter, but if this device is unavailable, floats provide a suitable, simple alternative. However, floats can only be used to measure mean velocity over a stream's length; it cannot measure velocities at specific cross-sectional points. It is noted that stream velocity varies with depth and with distance from the channel sides, and this feature should be examined explicitly.

A description of variations in the rate of flow at different cross-sections over the length of a stream would be interesting, particularly if it was linked to the location's altitude. Similarly, although it would be less practical for an individual project, for specific cross-sections it would be possible to consider how the rate of runoff varies over time. This can then be summarised graphically on a stream hydrograph. More specifically, after a storm over a basin, the rate of runoff increases rapidly and this can be portrayed clearly on a hydrograph. From such descriptions, it is possible to consider the passage of flood waves.

#### 8.3.3.5 *Evapotranspiration*

Evapotranspiration covers losses from both the evaporation of water from (soil and rock) surfaces and the transpiration of vegetation. The measurement of these characteristics is difficult, although useful approximations can be obtained from evaporation pans. Vessels of water are used, and it is calculated how much water is required to maintain a constant level.

#### 8.3.3.6 *Changes in storage*

As with evapotranspiration, changes in water storage are very difficult to calculate. Given this fact, it is appropriate to undertake and complete any field work during the period of minimum storage. In the United Kingdom, for example, this is at the end of summer, because precipitation has been relatively low and evapotranspiration has been relatively high. Consequently, the 'water year' is from 1st October to 30th September.

## 8.4 Slopes

### 8.4.1 *Introduction*

Arguably, hill slopes are the characteristic feature of the earth's land surface. Apart from areas such as the alluvial plains of the Indus and Ganges, most of the land surface of the earth is formed by valley slopes.

For the geomorphologist, hillslopes are basic landscape units and are, therefore, of crucial significance in any explanations of landscape development. As with other branches of geomorphology, the main areas of study can be identified. First, the study of hillslope form focuses on the shape of the ground surface utilising the techniques of slope profiling, contour survey, morphometry and morphological mapping. In addition, form also encompasses the thickness and composition of the regolith and environmental conditions (geology, climate and vegetation cover). Process studies, in contrast, focus on the agents acting on slopes and which cuase the form to change (for example, soil creep, surface wash, land slides and the processes of weathering). The study of process and form combine to elucidate the nature of slope evolution: the change in form over time as brought about by the action of processes.

In this section, attention is focused specifically on the measurement and analysis of hillslope form for two major reasons. First, from a practical perspective, studies of hillslope form, can be carried out with a minimum of equipment and in a reasonable period of time. In contrast, meaningful process studies in general require specialist equipment and a time-span generally unavailable in the context of student projects. Second, it is important to stress that the study of hillslope form is an important research area in its own right. As Young (1972, p. 136) has noted:

> ...There is a need ... for a branch of slope geomorphology concerned wholly with the description and comparison of form, without reference to their explanation; this would complement the existing independent subject of process study, the two branches together providing the basis for consideration of slope evolution.

### 8.4.2 *The terminology of hillslope profiles*

A slope profile may be defined as a line along the ground surface measured down the line of the steepest slope (that is, at right-angles to the contours). The highest point on a slope profile is termed the crest and the lowest point the base. Slope profiles are obtained usually by a series of pairs of measurements of angle and distance. Profile stations are the points between which measurements are taken; the interval between profile stations is a measured length. The inclined distance measured along the ground surface of a specified length is the ground surface distance. When the ground surface distance is projected onto the horizontal and vertical planes the horizontal and vertical equivalent lengths can be obtained (see Figure 8.4).

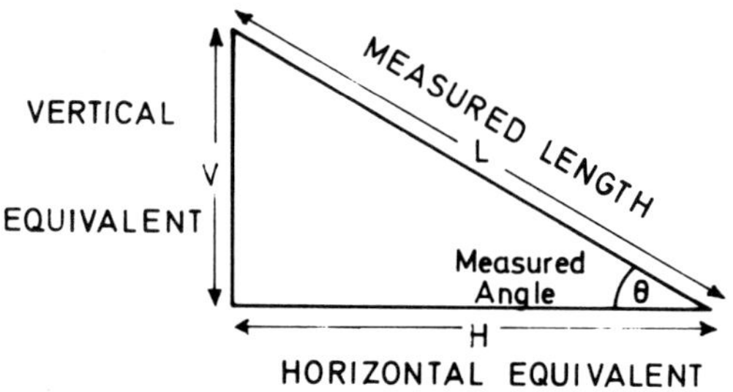


Figure 8.4 A HILLSLOPE PROFILE: SOME BASIC FEATURES

### 8.4.3 *Profile survey*

#### 8.4.3.1 *Profile sampling and location*

In determining the location of profiles two major methods are available. The first involves the use of purposive sampling procedures whereby profile lines are deliberately selected. The second method, in contrast, employs a controlled sampling procedure (random or systematic) for the selection of profile lines. Purposive sampling is subjective and can lead to the selection of profiles which will demonstrate some preconceived hypothesis. In addition results obtained from purposive sampling are not representative of a particular area as a whole, and refer only to the surveyed profiles. If the profiles are to be analysed statistically as a sample drawn from the total ground surface of an area, then some form of controlled sampling is necessary (for more detail see Young, 1972, 1974). For descriptive purposes, however, and for the analysis of particular subclasses of slopes purposive sampling is permissible (for example scarp slopes; slopes on particular rock types; profiles along the sides of a river valley to show the progressive change in form down valley).

A slope profile line may be defined as a '...straight line in an area where the major portion of the slope is made up of rectilinear contours (that is, contours which are nearly straight and run parallel to each other). This includes valley-heads and spur-ends where contours are strongly curved' (Finlayson and Statham, 1980, p. 147) (see Figure 8.5). It is important that the profile extends from the ridge crest to the local base level.

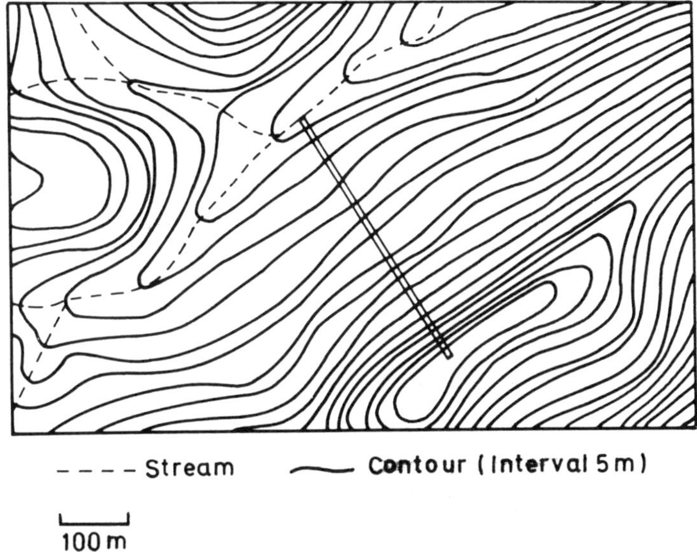


Figure 8.5 LOCATION OF A SLOPE PROFILE
(Source: Finlayson and Statham, 1980)

8.4.3.2 *Methods of profile survey*
Having located the slope profile line(s) the next stage is to survey the slope profile by angle and distance measurements. Two principal methods are suggested for profile surveying:

1. Tape and Abney level. This procedure requires a measuring tape or similar measure (for example, a rope or piano wire), ranging poles (or broom handles!) and an Abney level. The tape and ranging poles are used to divide the profile in a series of measured lengths. Where possible standard measured lengths of 5 m. should be used, if this is unsuitable then either 2 m, 10 m, or 20 m lengths should be used.

    Moving downslope from the crest the slope angle of a particular measured length can now be determined (see Figure 8.6 ). The ranging poles are located vertically, such that a line drawn from the top of one pole to the top of the other is parallel to the ground. The angle between this line and the horizontal is equivalent to the

(a)

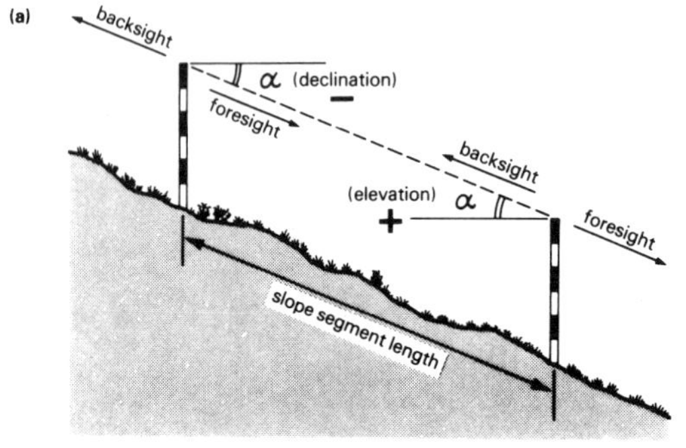


(b)

Slope location PIKE HILL
GR top of slope 764183
Orientation of profile 174° (GRID)
Geology Permian Sandstones and Marls.

| Segment no. | Length (m) | Foresight | Backsight | Corrected slope angle (sign as foresight) | Comments |
|---|---|---|---|---|---|
| 1 | 9.4 | + 2.5 | − 2.5 | + 2.5 | CRESTAL SEGMENT |
| 2 | 10 | −1 | + 0.5 | − 0.75 | |
| 3 | 8 | −4 | + 4 | − 4 | |
| 4 | 11 | −11 | + 11 | − 11 | Bare rock exposed. |

Figure 8.6 SLOPE PROFILING: THE ABNEY LEVEL (Source: Clowes and Comfort, 1982)

(b)

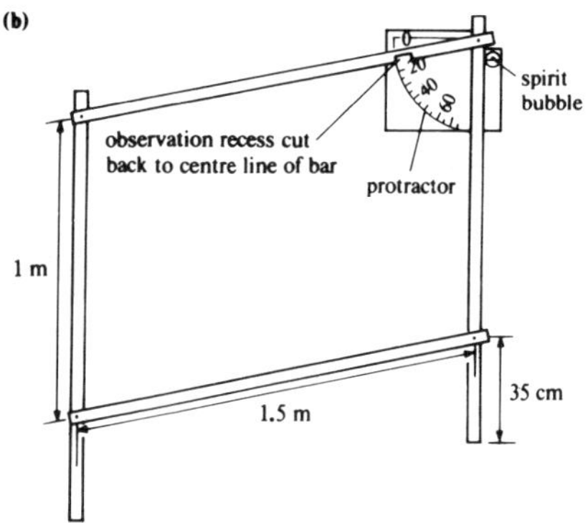


Figure 8.7 SLOPE PROFILING: THE PANTOMETER (Source: Goudie, 1981)

slope angle of the measured length. The Abney level rests on top of or against one of the ranging poles and the corresponding height on the other pole is sighted through the eye piece. It is preferable to take two readings, upslope and downslope. The downslope reading is referred to as declinations and are recorded as negative values (see Figure 8.6 ), elevations as positive values. This procedure continues downslope until the base is reached.

2. Pantometer. A cheaper method is to use a pantometer (Pitty, 1968). This consists of four pieces of wood (or aluminium), a protractor and a spirit level. This equipment is accurate to 0.5° and can easily be used by one person (see Figure 8.7 ). The measured length is the distance between the two uprights; the recommended length is 2 m. The pantometer is 'leap frogged' down the slope profile until the base is reached.

TABLE 8.2

Profile survey - suggested observations

<u>Suggested observations for each measured length</u>

1. Vegetation or landuse
2. Any visible features of the ground surface e.g. micro-relief, mass-movements, rock outcrops, stones, nature of regolith.
3. Man made features (e.g. roads, hedges) which may be useful in locating the profile.
4. Disturbed ground. Where a measured length passes in whole or part across artificially disturbed ground e.g. road, mark the length as "D" or, if uncertain whether there is a disturbance "D?".

<u>Supplementary observations of profile form</u>

1. Identification - profile code number, date of survey
2. Location - approximate latitude and longitude, grid reference of crest and base.
3. Climate - where possible mean annual rainfall.
4. Geology - description of rocks (including superficial deposits) forming slope on which profile is sited. Source of data (maps, field observations).
5. Vegetation and landuse - general features of area, slope on which profile is sited, and profile line.
6. Regolith, soil - general features of area, slope on which profile is sited, and profile line.
7. Micro- relief - general features of area, slope on which profile is sited, and profile line.
8. Landforms - general features of area; relation of

profiled slope to nearby landforms and slopes; relation of form along profile line to form on other parts of slope on which profile is sited.

9. River channel (where possible) - estimated width, depth, speed of flow and volume of flow. Note whether the slope is or is not undercut by the river.
10. Aspect - at the steepest point on the profile take a compass bearing of the aspect.

Whichever technique is used it is suggested that the observations noted in Table 8.2 should be recorded.

### 8.4.4 *Plotting the profile*

Having obtained a set of angle and distance measurements for a particular slope profile the next step is to plot the profile graphically by converting the angle and distance measurements into rectangular coordinates. Using simple trigonometry the vertical and horizontal equivalents are determined as follows:

$$\text{Vertical (V)} = D \sin \theta$$

$$\text{Horizontal (H)} = D \cos \theta$$

The coordinates are then summed from end and plotted without vertical exaggeration in the first instance; a suitable scale for an initial plot is 1:500. An example of the conversion of angle and distance measurements into rectangular coordinates is provided in Table 8.3 . The points are joined by straight lines and the angle inserted above each length.

### 8.4.5 *Profile analysis*

Profile analysis involves the division of a profile into a number of parts which possess certain properties of form. From the preceding discussion it is clear that most slopes are not completely straight from crest to base; usually the angle of slope varies along the profile. Parts of a slope profile in which the angle remains approximately constant are termed segments. Portions of the profile which are smoothly curved are called convex or concave elements. Curved slopes are described in terms of their rate of curvature in degrees per 100 metres ($^{o}$/100 m). By convention this value is negative for convex elements and positive for concave elements. To calculate curvature over a measured length the following formula is used:

$$C_b = 100 \frac{\theta a - \theta c}{2L} \quad {}^{o}/100 \text{ m}$$

| Measured Length | D distance (metres) | θ Angle (degrees) | Cos θ | Sin θ | Horizontal difference = D.Cos θ | Vertical difference = D.Sin θ | Co-ordinates (Cumulative) Horizontal metres | Vertical metres |
|---|---|---|---|---|---|---|---|---|
| 1 | 10 | - 2.5 | .999 | .044 | 10.0 | - 0.4 | 10.0 | - 0.4 |
| 2 | 10 | - 2.0 | .999 | .035 | 10.0 | - 0.4 | 20.0 | - 0.8 |
| 3 | 10 | - 0.5 | 1.000 | .009 | 10.0 | - 0.1 | 30.0 | - 0.9 |
| 4 | 10 | 0.0 | 1.000 | .000 | 10.0 | 0.0 | 40.0 | - 0.9 |
| 5 | 10 | + 1.0 | 1.000 | .018 | 10.0 | + 0.2 | 50.0 | - 0.7 |
| 6 | 10 | + 2.0 | .999 | .035 | 10.0 | + 0.4 | 60.0 | - 0.3 |
| 7 | 10 | + 4.0 | .998 | .070 | 10.0 | + 0.7 | 70.0 | + 0.4 |
| 8 | 20 | + 4.0 | .998 | .078 | 20.0 | + 1.4 | 90.0 | + 1.8 |
| 9 | 20 | + 6.0 | .995 | .105 | 19.9 | + 2.1 | 109.9 | + 3.9 |
| 10 | 20 | + 8.0 | .990 | .139 | 19.8 | + 2.8 | 139.7 | + 6.7 |
| . | . | . | . | . | . | . | . | . |
| . | . | . | . | . | . | . | . | . |
| . | . | . | . | . | . | . | . | . |
| . | . | . | . | . | . | . | . | . |
| 31 | 10 | +24.0 | .914 | .407 | 9.1 | + 4.1 | 420.1 | +80.2 |
| 32 | 10 | +26.0 | .899 | .438 | 9.0 | + 4.4 | 429.1 | +84.6 |
| 33 | 10 | +27.5 | .887 | .462 | 8.9 | + 4.6 | 438.0 | +89.2 |
| 34 | 10 | +29.5 | .870 | .492 | 8.7 | + 4.9 | 446.7 | +94.1 |
| 35 | 10 | +30.0 | .866 | .500 | 8.7 | + 5.0 | 455.5 | +99.1 |

TABLE 8.3 Conversion of angle and distance measurement into rectangular co-ordinates (Source: Young, 1972)

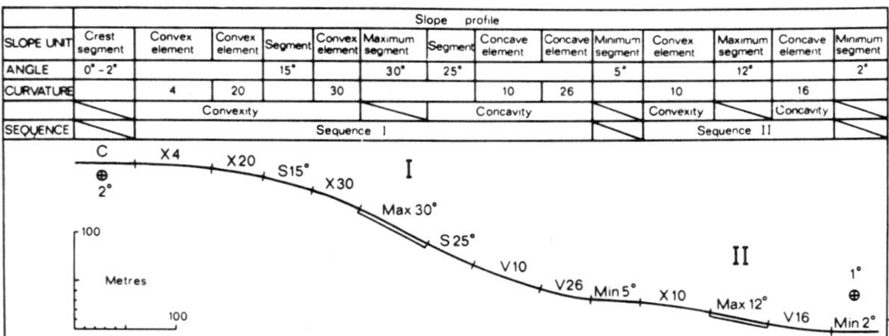

Slope profile
SLOPE UNIT
Crest segment
Convex element
Convex element
Segment
Convex element
Maximum segment
Segment
Concave element
Concave element
Minimum segment
Convex element
Maximum segment
Concave element
Minimum segment
ANGLE
0°-2°
15°
30°
25°
5°
12°
2°
CURVATURE
4
20
30
10
26
10
16
Convexity
Concavity
Convexity
Concavity
SEQUENCE
Sequence I
Sequence II
C
2°
X4
X20
S15°
X30
I
Max 30°
S 25°
V10
V26
Min 5°
X 10
II
Max 12°
V16
1°
Min 2°
100
Metres
100


*Slope unit* — a segment or an element.

*Segment* — a portion of a slope profile on which the angle remains approximately constant.

*Element* — a portion of a slope profile on which the curvature remains approximately constant.

*Convex element* — an element with a downslope increase in angle (i.e. with positive curvature).

*Concave element* — an element with a downslope decrease in angle (i.e. with negative curvature).

*Maximum segment* — a segment which is steeper in angle than the slope units above and below it; it may also form the lowest unit on a profile, having a gentler unit above it.

*Minimum segment* — a segment which is gentler in angle than the slope units above and below it.

*Crest segment* — a segment bounded by downward slopes in opposite directions.

*Basal segment* — a segment bounded by upward slopes in opposite directions.

*Irregular unit* — a portion of a slope profile within which there are frequent changes of both angle and curvature.

*Convexity* — all parts of a slope profile on which there is no decrease in angle downslope, but excluding maximum, minimum, and crest segments.

*Concavity* — all parts of a slope profile on which there is no increase in angle downslope, but excluding maximum, minimum, and crest segments.

*Profile sequence* — a portion of a slope profile consisting successively of a convexity, a maximum segment, and a concavity.

Figure 8.8 THE TERMINOLOGY OF HILLSLOPE ANALYSIS (Source: Young, 1972)

where a, b and c are consecutive slope measurements having fixed length L and angles θa, θb and θc. $C_b$ is the rate of curvature of section b. The curvature at a point on the profile is calculated by:

$$L = 100 \frac{\theta a - \theta b}{L} \quad ^{o}/100 \text{ m}$$

where L is the curvature at the point at which measured lengths a and b meet. A graphical portrayal of the terminology of hill slope analysis is contained in Figure 8.8

The distribution of angles over a single profile can be presented in the form of a graph (see Figure 8.9 ) where angle is plotted against the cumulated distance from the top of the slope. In conjunction with the cross section this procedure should help distinguish between major units and local irregularities.

Another approach to the analysis of slope angles is to produce a graph of angle frequency distribution (Figure 8.10). On such a graph the mode of the distribution is termed the characteristic angle. The characteristic angle may apply to all slopes or may be typical of one rock type, one location or one type of landform (for example, glacial). Slopes of man-made waste (for example, slag heaps) demonstrate typical angles.

### 8.4.6 *Slopes in plan*

A useful supplementary method of analysis in certain types of slope studies is the construction of a morphological map. This technique attempts to divide the landscape into uniform morphological units, bounded by morphological discontinuities (see Figure 8.11 and Table 8.4 ) However, morphological maps are time-consuming to construct and difficult to interpret, but are useful for reconnaissance purposes. Normally, morphological maps are compiled at a scale of 1:10000.

### 8.4.7 *Projects*

Using the techniques described above a number of hypotheses could be devised (see, for example, Hilton (1979) and Clowes and Comfort (1982).

1. that lithology influences slope - slope histograms and/or slope frequency graphs could be used to compare slopes. The significance of the difference between the means of slope angles on the different rock types could be assessed using the Students' 't' test.

2. that basal removal produces a steeper slope

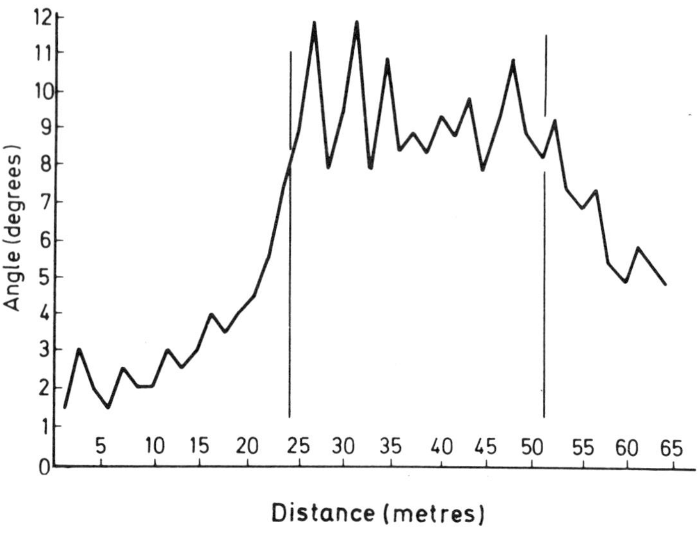


Figure 8.9 DISTRIBUTION OF ANGLES ALONG A SLOPE PROFILE
(Source: Finlayson and Statham, 1980)

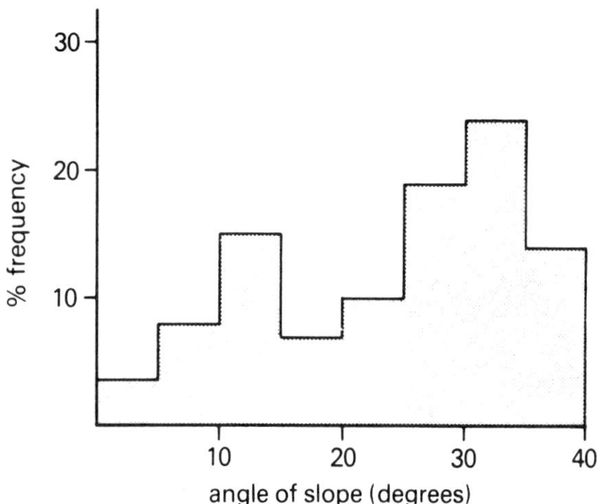


Figure 8.10 ANGLE FREQUENCY DISTRIBUTION
(Source: Clowes and Comfort, 1982)

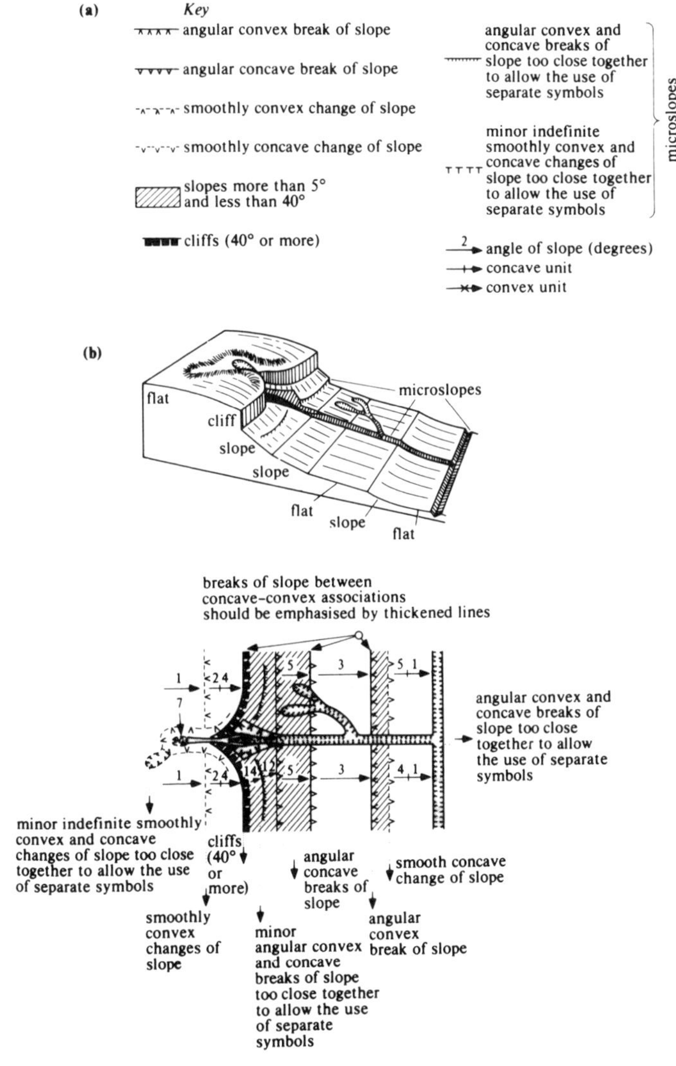


Figure 8.11 SLOPES IN PLAN
(Source: Goudie, 1981)

Slope facet (or plane morphological unit) - sloping plane area of the ground surface.

Flat - level area of the ground surface.

Curved morphological unit - area of the ground surface which has approximately constant profile curvature.

Convex morphological unit - curved morphological unit with positive profile curvature.

Concave morphological unit - curved morphological unit with negative profile curvature.

Cliff - slope of more than $40^{\circ}$ formed of base rock.

Irregular morphological unit - a unit which possesses surface irregularities too small to be represented on the scale of the map.

Morphological discontinuities

Break of slope - angular discontinuity of the ground surface seperating morphological units.

Change of slope - gradual change of angle of the ground surface seperating morphological units.

Inflection - line or zone seperating adjacent convex and concave morphological units.

TABLE 8.4 Morphological units and discontinuities (after Young, 1972)

3. that profiles on the same rock type are influenced by aspect - reflecting differences in dampness, temperature and so on, which may affect weathering and mass wasting.

4. specific hypotheses relating to scree slopes could also be tested. For example, that screes are sorted with the smallest particles lying highest on the slope, or that the curvature of the scree profile varies with the height of the free fall which feeds the scree.

5. Other hypothesis could be derived from the consideration of specific aspects of geomorphology (for example, fluvial and coastal geomorphology). While man-made slopes could also form the basis of a manageable project. Finally, an interesting integrated study could be based on the relationships between slope, vegetation, soil and landuse in a selected area.

## 8.5 Coastal Studies

The coastal environment provides the student of physical geography with several potential projects utilizing simple and straightforward methods of data collection. Following a brief outline of the major developments which have occurred in coastal geomorphology, the remainder of this sub-section focuses on a number of themes which could form the basis for student project work (an excellent general text is King, 1972).

In many respects the development of coastal geomorphology reflects the development of geomorphology in general. The early preoccupation of coastal geomorphologists was with the description and classification of coastal forms. Explanations were essentially historical in character with the emphasis placed on long-term events such as the emergence and submergence of coasts. During the 1950's attention focused on changes occurring over a much shorter time period. In practice this reorientation was characterised by an increasing emphasis on process rather than form. By the 1960's the principal concern of coastal geomorphologists was with the explanation of form in terms of process which was closely associated with the introduction of quantitative methods. Quantification was both inevitable and invaluable, as Clark (1982, p. 238) notes:

> Simple observation and mapping proved quite inadequate and the only workable approach was found to lie in the actual measurement of form and process, and then the use of statistical methods to generalise the measured information to establish its reliability and to investigate relationships between different sets of information.

Initially studies focused on relatively straightforward process-response relationships such as an increase in wave height produces energy concentration, leading to sediment transport and hence a steepening of the seabed slope. A more recent development is associated with the introduction of systems analysis which stresses the mutual relationship which exists between process and form. In terms of the above example this would mean that as the seabed slope becomes steeper, water offshore becomes deeper, reducing wave height and energy concentration. This effectively controls the tendency for the seabed to steepen producing a balance or equilibrium between form and process. The most recent phase of development in coastal geomorphology has been an emphasis on applied studies focusing on problem-solving and environmental management and represents an important interface between physical and human geography. (For an introduction see Cooke and Doornkamp, 1978, particularly Chapter 8; Clark, 1978).

Over the last decade, the above developments have been paralleled by an increasing sophistication in instrumentation which has enabled coastal geomorphologists to measure accurately many of the process variables operative in the coastal zone (see for example, Dugdale, 1981). Clearly, these instruments are beyond the reach of students in schools, however, a number of feasible projects could be devised combining observation with relatively simple instrumentation.

1. Wave height and period can both be determined by visual observation. Wave height can be measured by comparison with a graduated pole in the breazer zone. Wave period may be determined by noting the number of wave crests passing a fixed point in a given time interval. Wave length is the distance between two successive crests. The deep water length (L) is given by the formula

   $$L \text{ (metres)} = 1.56T^2 \text{ (seconds)}$$

   where T is wave period.

   The ratio of height and length (H/L) is the wave steepness which in general cannot exceed 1 in 7 (0.14) or the wave breaks. (Significant positive correlations, for example, have been found between beach slope and steepness and between wave period and beach slope). This information should also be supplemented by meteorological observations, particularly wind speed and direction. These can be obtained from the Meteorological Office in their Monthly Weather Reports.

2. Beach profile survey. Using the methods described in section 8.4 it is possible to examine how a beach

changes under particular conditions. Surveys could be carried out at regular time periods, after one tidal cycle, or after storm conditions. Under the same conditions differences in exposure on beach materials may have a significant impact on the beach's response. It might be possible to compare beaches with similar exposures but consisting of different materials, alternatively a comparison could be made between sheltered and exposed beaches with similar deposits. Using the profile approach a transect survey could also be carried out at several points to examine, for example, the form and nature of a shingle ridge. Relevant information would include the angle and length of different slopes, beach material and different types of plants (see for example, Sampson, 1981).

3. Longshore drift. In appropriate locations the rate of longshore drift may be assessed by using pebbles marked with waterproof paint. These pebbles are placed against pebbles of a similar size at low tide. At the next low tide their positions can be mapped and movement measured. Under storm conditions only 1-2 per cent of pebbles may be recovered. In conjunction with a study of longshore drift an analysis of pebble size, shape and roundness may be undertaken. Roundness could be measured using Powers' Roundness Index, while Zingg's classification is a useful measure of three-dimensional shape. Size may be determined by measuring a pebble's long axis using a pair of calipers or 'pebble-ometer' (see Hanwell and Newson, 1974, p. 166). Details of the various techniques are contained in Briggs (1977). Using these techniques variations in the character of beach material in the direction of transport can be examined. Variation in beach material also occurs from the low water mark to the highest point reached by storm tides. The data collected can be used as the basis for a number of statistical analysis. These may range from straightforward descriptive measures such as the mean and standard deviation to hypothesis testing. (for example, see Gill, 1979).

Problems of applied and practical geomorphology in the coastal environment (see for example Clark, 1978) demands a thorough understanding of coastal processes and environmental management, and would be beyond the experience and knowledge of sixth-form students. However, an acknowledgment and limited analysis of man's interference would provide an added perspective to a coastal project. For example, coastal protection and beach improvement will undoubtedly influence the form of the beach. While intensive recreational use may encourage wind erosion in the form of blow-outs. This is a result of the trampling of vegetation by visitors in dunes thus weakening the vegetation cover. Such action may eventually result in dunes migrating inland.

## 8.6 Applied climatology: some aspect of urban climates

### 8.6.1 *Background*

In terms of the availability of quality data over long time periods, meteorological data are especially attractive. However, such data sets are unlikely to act as much more than a background for more detailed, local studies. For this reason, in this sub-section, attention focuses on some aspects of urban climates because this topic would provide a practical project area for students (see also Unwin (1978)). Moreover, given it was stated at the outset of this book that it would be desirable for projects to be applied in nature, interest focuses on air pollution (note the broad impacts of the 1974 Control of Pollution Act and the interest in the effect of air pollution on soils and vegetation (Moss, 1975)). However, in general, the impact of climate has wide-ranging implications for man's activities (see Thornes (1977) interesting discussion of the effect of weather on sport), and this would appear to be a fruitful avenue to pursue in terms of projects capturing the full essence of man-environment relationships. In connection with the discussion in Chapter three, perception studies of the implications of natural hazards on man's behaviour, which are often related to extremes of weather, offer many opportunities for studies. (For reference, attention is drawn to Doornkamp et al.'s (1980) *Atlas of drought in Britain 1975-76*).

### 8.6.2 *Urban Climate*

In geography, a recurrent area of interest has been the effect of climate on man's activities. However, research is being undertaken to explore the possibilities of man controlling the weather, specifically ameliorating adverse conditions - rain-making to enhance agricultural yields. In fact, it is man's built environment, especially urban areas, which most affects climate.

This modification of a local climate has a variety of dimensions, and, as many students have the experience of taking some meteorological readings for local (school) stations, a study of urban climates would be a natural and interesting extension of their education. If it is decided to link a study to aspects of air pollution, it is noted that, in the majority of cities and towns, numerous air pollution survey sites collect data on smoke and sulphur dioxide concentrations on a daily basis. In analysing such information and collecting local climatic data, very careful attention must be paid to the selection of the number and location of sampling sites. (For any chosen sampling frame, it is essential that it is described clearly and the reasons for its selection are stated explicitly).

As a general description, in contrast to the climate of surrounding areas, the urban climate exhibits higher temperatures, greater cloud cover, more rainfall and increases in air pollution. Barry and Chorley (1971) recognise three general effects of urban form (the architecture, density and function of this built environment): heat production; modification of atmospheric composition; and alteration of surface configuration and roughness. Urban heat production occurs directly by combustion and there is also the radiant heat from buildings. Given the nature of heating, diurnal and seasonal variations are present. In winter, such sources are a significant proportion of the incoming solar radiant energy. Differences between the temperatures of an urban area and of its surrounding region, which is referred to as an 'heat island' effect, are especially marked during the night, although this is modified by other factors: For instance, still-air conditions do not facilitate heat transfer; and high density developments mean that, in general, less energy is lost through evapotranspiration and more heat is radiated from buildings. The overriding importance of radiant heat from buildings is demonstrated by the fact that London's heat island effect is greater in summer, when direct heat combustion is at a minimum and solar radiation is at a maximum. This fact is reinforced because atmospheric pollution is at a minimum.

Whilst aspects of air pollution are considered in the following sub-section, the presence of pollution does modify the atmospheric composition of the urban environment which affects the local climate by reducing incoming radiation and sunshine, providing condensation nuclei and by altering the atmosphere's thermal properties. Moreover, as stated above, wind speeds are reduced because of a sheltering effect of buildings. Other impacts of the changes to surface characteristics are related to the lack of extensive vegetational cover and large bodies of standing water. The reduced evapotranspiration results in cities exhibiting lower absolute humidities.

Thus, sufficient details have been given to indicate that a detailed monitoring of a local climate would provide a stimulating project topic (see also Meyer and Huggett (1977)). For example, to what extent do air temperature and relative humidity alter with proximity to dense, urban areas? Thus, in keeping with the philosophy of project work being an integral component of the curriculum, it would provide a direct extension of course work with useful, practical training experience.

### 8.6.3 *Urban air pollution*

In general, discussions of air pollution focus attention on the concentrations of smoke and sulphur dioxide. Owing to the cost and lack of widespread availability of the necessary monitoring equipment, student projects involving primary data collection are not feasible. However, in the United Kingdom, there is a suitable, detailed, daily data base for many towns and cities, dating from the late 1950s or early 1960s, which is coordinated by the Department of Industry's Warren Spring Laboratory at Stevenage. It is suggested that it would be possible to examine how weather conditions, particularly wind speed and direction, affect spatial variations of pollution. Alternative issues to consider could be the effect of locating noxious facilities or spatial associations between air pollution and human health.

Following Elsom (1980), one suitable project could involve describing and explaining relationships between weather and pollution in a study area. Simply stated, two basic factors, the locations of emission sources and the varied atmospheric conditions associated with different air masses, are significant, and, from wind direction and pollution roses, it is possible to infer about the operating processes. For example, in a study of pollution in Oxford, for comparison , two pollution survey sites were selected: a city centre site; and a rural site $\frac{1}{2}$ kilometre southeast of Oxford. To construct a pollution rose diagram (see Figure 8.12 ), at the outset, it is necessary to specify a set of wind directions

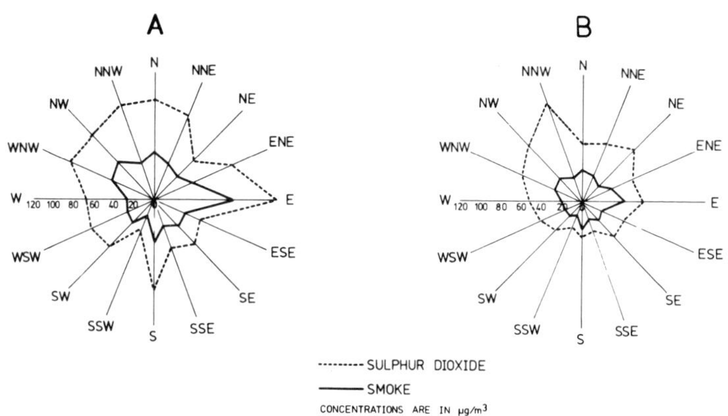


Figure 8.12 WIND DIRECTION-POLLUTION RELATIONSHIPS AT (A) A CITY CENTRE SITE AND (B) A RURAL SITE $\frac{1}{2}$ km SOUTHEAST OF OXFORD (Source: Elsom, 1980)

and a time period for analysis. (One possibility would be to take account of potential seasonal variations using local meteorological information. Mean (smoke and sulphur dioxide) pollution concentrations are plotted easily, and the result begs the question why winds from specific directions are more polluted than others. Clearly, this variation would be accounted by the pattern of possible emission sources. In addition, the fact that winds from different directions exhibit different dispersive properties could be important. 'For example, westerlies are associated with generally higher wind speeds while easterlies are associated with more frequent inversions. Furthermore, northerlies with generally cooler temperatures although not affecting dispersion of pollutants, create the need for more heating of homes and therefore increase pollutant emissions indirectly' (Elsom, 1980, p. 57). As Figure 8.13 portrays, this kind of weather-pollution study can examine other relationships, such as wind speed.

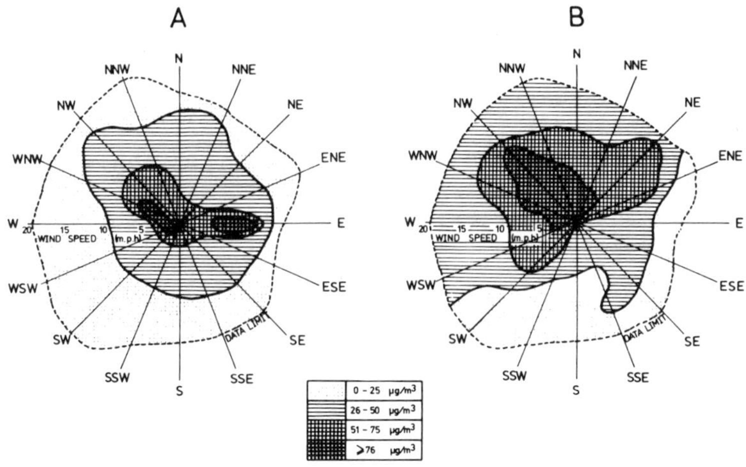


Figure 8.13 WIND-SPEED AND WEATHER POLLUTION

Finally, if a city has a number of pollution survey sites, it would be possible to draw isoline maps based on the sample sites to describe the spatial distribution of pollution over the city for different weather conditions.

# Chapter Nine

# BIOGEOGRAPHICAL STUDIES

## 9.1 Introductory overview

In the past, biogeography was a cinderella subject area, the poor relative of more traditional topics of interest such as geomorphology, climatology and pedology. On the other hand, as indicated in the previous chapter, important relationships with geomorphology and hydrology are well established. In the past decade, enormous developments have taken place, and it is suggested that biogeographical studies per se offer a range of potential, manageable student projects. At the outset, it should be stressed that there has been a move away from a preoccupation with descriptions of the spatial distribution of soils and vegetation (which, incidently, is not a topic that is confined to geographers' interest); indeed, Stoddart (1977) argues for a change from the traditional emphasis on spatial distributions and classifications. However, it should be noted that, in the last few years, much progress has been made to establish a National Vegetation Classification for the United Kingdom, which supercedes Tansley's standard reference from the early part of this century, *The British Isles and their Vegetation*. Using quadrat sampling based on 10 kilometres by 10 kilometres grid squares, detailed information on habitat, species lists and on any zonation and succession will be available. While this data set will prove to be a valuable secondary source for information, specific fieldwork-based projects involving a detailed mapping of vegetation and/or soils using quadrat or transect sampling would be appropriate (see Watts (1978) for a discussion of opportunities in this direction). Such studies could include a simple, statistical data analysis, such as the much-used 'association analysis' (see Johnston (1976) or Williams and Lambert (1959)), which is based on the $\chi^2$-statistic that is presented in Appendix I. From this type of investigation, it would be possible to describe different communities founded on the presence or absence of specified species, and then to map the communities in association with other characteristics, such as soils.

In terms of suggesting project areas for students, as Pears (1977) stresses, it is really necessary for students to have first-hand experience of their selected subject matter, and, consequently, a number of practical issues should be given careful consideration when assessing the feasibility of specific projects: inaccessibility to an area for fieldwork; the lack of a detailed background in the concepts and methodology (because of the present content of syllabi); and the unavailability of equipment (although it may be possible to borrow it from a local university). Such reasons mean that, at least at present, much of the diversity exhibited by contemporary biogeography is beyond the scope of individual projects. However, a brief overview of some of the research areas could help to make both teachers and students aware of potential areas (and may preceed minor alterations in syllabi).

In terms of British research, recent attention has focused on historical biogeography and palaeoecology involving a reconstruction of past environments (see, for instance Simmonds (1979)). In practice, however, school geography students cannot be expected to have the practical experience of pollen analysis (for details, see Moore and Webb, 1978) to be able to undertake individual projects, although a sound, general knowledge of alterations in vegetation would be useful for geomorphological studies. An alternative area of biota-environment relationships involves a comparative contemporary investigation of different areas. Whilst such studies are interesting, clearly, they would never be really practical for student projects.

Current constraints in relation to equipment availability means the increasing use of satellite data for soil and vegetation mapping is unlikely to provide many opportunities for student projects in schools, although, again, a background knowledge of remote sensing would be apposite (see Allan (1980) and Barratt (1975) for more details). However, as Hilton (1981, p. 162) demonstrates, 'The coincidence of availability and syllabi changes ... makes Landsat an innovation at a turning point' (see for example Smith, 1980). Indeed, the use of such satellite imagery can be seen as a natural extension to studies based on aerial photographs. (Such resources can be obtained directly from either Focal Point Audiovisual Limited, Portsmouth or Thomas Nelson and Sons, Walton-on-Thames). In other chapters, projects founded on map analyses have been outlined, and, given the minimal edge distortion of the images, it is possible to interpret the information using a range of different maps. For example, in an examination of drainage basin characteristics, the information on relief and vegetation would be useful. However, it should be appreciated that the images are unlikely to possess the necessary detail for small-scale, local studies.

At present, the depth of coverage of biogeography in schools is insufficient to permit or to expect students to be able to examine different functional interrelationships of ecosystems (in a way similar to the descriptions and explanations of drainage basins presented in chapter eight). To establish significant relationships even between a small number of components would be very difficult. For instance, school geography students do not have the knowledge to consider soil and vegetation systems in terms of the movement of minerals (see Trudgill (1977)). Mottershead (1980), however, suggests that energy flows could be studied, specifically correlating productivity with data on solar radiation. Whilst the practical difficulties of such productivity studies, for example the accurate weighing of plants, should not be underestimated, if this direction is pursued, a natural extension would be to link it to aspects of man's activities through agricultural geography (see, for example, Jones (1979)).

One wide-ranging area of interest that does offer potential for interesting projects concerns man's relationship with biota, usually the impacts of his activities. (For completeness, the interest in 'disturbance ecology', examining the effect on natural ecosystems of disturbances not caused by man such as flooding, fires and mass movements, should be noted (Vale and Parker, 1980)). Reference to the influence of pollution on soils and vegetation was made in the preceding chapter, and reference to aspects of conservation and ecosystem management was made in the chapter on rural studies. In many cases, such issues are multidimensional, and, therefore, students should select their particular emphasis with regard to their chosen project. For example, in discussing recreational and leisure activities, concern could focus on either the difficulties of ecosystem management because of the pressure placed on the natural resources or the range of attitudes and perceptions to the natural environment. In chapter three, a detailed discussion of urban images was presented, but it should be appreciated that man's image of his natural environment directly affects both his conservation and exploitation of this resource.

In the following subsections, soil studies and community biogeography are examined as areas where project work may be sensibly pursued. The coverage of topics is not intended to be exhaustive, however, the discussion does provide a reasonable indication of the types of projects that may be undertaken with fairly limited resources.

## 9.2. Soil Studies

### 9.2.1 *Some background comments*

Arguably, soils and water constitute man's most important resources, and, whilst soils are one component in the complex environment, it is possible to outline a range of feasible project areas within the field of soil studies per se. Specifically, following a recurrent desire for applied work, it is suggested that descriptions of soil characteristics and pedological processes could often be extended sensibly to incorporate discussions of the implications for agricultural activities, environmental conservation, land reclamation, recreation, urban planning (particularly expansion) and so on. Obviously, the emphasis of specific studies must be considered very carefully.

Following the theme of soil as a resource, reference is made to the Soil Survey of England and Wales, which is available readily in maps at scales of 1:63 360 or 1:25 000. Whilst no single soil classification system exists (see Clayden's (1982) recent review), this secondary data source could provide the foundation for detailed, local study of the spatial variations in soil type (remembering the Soil Survey presents the 'modal' soil class) or an examination of land capability in terms of possible future developments. In the following subsection, these ideas are developed further.

### 9.2.2 *Local spatial variations in soil type: aspects of a mapping project*

By their scale of analysis, national soil surveys cannot be expected to describe the local spatial variability of soils. Consequently, it is suggested detailed soil mapping projects would be appropriate topics of study, particularly if differences would be related to the landscape. For instance, the concept of a soil catena, which is a series of topographically associated soils on the same parent material, would provide a direct framework for investigation. Simply stated in an analogous way to the translocation process in the formation of a soil profile with different horizons, a selective transportation of soil materials by throughflow produces a concatenation of soils down-slope. Given that a fundamental component of soil mapping is a description of a soil profile, some general comments are made regarding soil characteristics (and, for more details, readers should consult introductory tests by Fitzpatrick (1974) and Knapp (1979)). First, however, it is necessary to highlight certain aspects of sampling methodology.

In previous chapters, ideas on systematic and random sampling and on the possibility of stratification have been presented. In soil mapping, there are two distinct components to the sampling design: sampling a set of locations in the study area; and sampling at different depths at a selected site. (Hodgson (1978) provides a sound discussion of *Soil sampling and soil description*). With regard to the former, systematic sampling using an arbitrary grid, spatial referencing system would permit an examination of whether spatial autocorrelation in specified characteristics exists (see Appendix II for a brief discussion of spatial auto-correlation and Campbell (1979) for recent examples of research in this direction). For random sampling, a set of random number tables would be required to determine the co-ordinates of sample sites. If attention, however, was to be concerned with studying a soil catena, traverse sampling would be apposite, and care must be given to the selection of the transect's orientation. In addition, in systematic and transect sampling, it is necessary to decide on an appropriate sampling interval (which is obviously dependent on the scale of analysis). If a 1:25000 map is used, for example, an interval of between 20 metres to 25 metres would probably be satisfactory. Moreover, as specific soil types are related to characteristic parent materials, spatial stratification taking account of variations in such associated features should be considered; studies of aerial photographs, satellite imagery or other kinds of maps could assist in this direction. (It is noted that in Appendix I, the appropriate statistic to examine such associations using a cross-tabulation of soil types against parent material based on the sample sites is presented). For sampling at particular sites, stratification in relation to the different horizons or simply at different depths (say 20 cms.-30 cms. depth, 40 cms.-50 cms. depth, and 60 cms. - 70 cms. depth) would be sensible.

In soil text books, a range of soil characteristics, such as the colour, texture, structure, porosity, consistence, acidity and the chemical and mineralogical composition, are defined and described. A variety of combinations of these features could be studied depending on the aims of a particular project, a student's experience and knowledge, and the availability of equipment for laboratory analyses. For example, in a mapping study that was concerned primarily with the origin of the parent material, at the outset, soil samples from different sites and from different depths could be sieved using say a 2 mm. sieve. This would produce three types of material: unsieved material; material coarser than 2 mm.; and material finer than 2 mm. The unsieved samples can be used to describe the soil colours for each horizon at all the sample sites. Colours are obtained by reference to an available colour system, such as the Munsell soil colour chart. To ensure a correct matching, the comparison should be made when the soil is in both the air-dry and moist states. It is noted that the occurrence of any mottling of colours is significant, and, if it is found, the proportion of each colour should be estimated. Colour is an important soil characteristic, because it is determined by the parent material's minerals and the presence of organic matter, and from it it is possible to make inferences about the stage of weathering.

Another important soil characteristic is texture, and this could be examined using the samples of material finer than 2 mm. Texture describes the distribution and size of particles. It is a feature that is very difficult to change (although the addition of organic matter or lime can improve its agricultural potential), and, therefore, it is the most permanent attribute. In practice, an analysis is concerned with the percentage of particles that are sand (2.0 mm - 60 mm. in diameter), silt (60 mm. - 2 mm. in diameter) and clay (less than 2 mm. in diameter), and this information can be summarised diagrammatically using a so-called 'textural triangle'. A simple method to describe soil texture is to examine the effects on a small handful of soil when it is moistened gradually and kneaded into a ball and then pressed out between one's thumb and forefinger.

Other soil features that would be described easily by students are the soil's pH value (using the sample material that is finer than 2 mm.), the soil's structure (by considering the shape, size and lithology of stones in the sample material that is coarser than 2 mm.), and the content of organic matter from plant and animal sources (particularly in the surface horizon). In addition, brief comments on the type of boundaries between horizons, such as abrupt, gradual or diffuse, would be appropriate.

With their background knowledge and the availability of useful reference books such as Curtis _et.al._'s (1976) introduction to British soils, students should be able to produce good quality, detailed maps showing the distribution of different soil types. A natural extension to this description of the physical nature of the soils is to consider the land's capability (note, for instance, Stamp's much-cited Land Utilization Survey in the 1940s). In Britain, a Land Use Capability Classification has been developed by Bibby and Mackney (1969) (see Table 9.1 ), to translate the results of soil surveys into a format that is more appropriate for farmers, planners and so on. It is noted that sub-classes are defined to indicate other types of constraints affecting land use, such as wetness limitations, soil limitations, gradient and soil pattern limitations, climatic limitations and a liability to erosion. Moreover, it should be appreciated that the classification is founded on the land's potential, and, consequently, it assumes, at least implicitly, the existence of sound management practices. It is suggested that a description and explanation of any differences between the land's expected and observed potential would probably be very worthwhile, particularly if prescriptive suggestions were forthcoming.

| | |
|---|---|
| CLASS 1 | Land with very minor or no physical limitations to use. (slopes usually less than $3^{o}$; depth less than 75 cm.; altitutde less than 150 m.) |
| CLASS 2 | Land with minor limitations that reduce the choice of crop and interfere with cultivation. (moderate or imperfect drainage; less than ideal rooting depth; slightly unfavourable soil texture and structure; moderate slopes usually less than $7^{o}$; depth 50 cm - 70 cm.; altitude less than 230 m.) |
| CLASS 3 | Land with moderate limitations that restrict the choice of crops and/or demand careful management. (imperfect or poor drainage; restricted rooting depth; unfavourable structure and texture; strongly sloping ground usually less than $11^{o}$; depth 25 cm-50 cm.; altitude less than 380 m.) |
| CLASS 4 | Land with moderately sever limitations that restrict the choice of crops and/or require very careful management practices (poor drainage; occasional flooding; shallow or stony soils; moderately steep gradients usually less than $15^{o}$; depth less than 25 cm.; altitude less than 460 m.) |
| CLASS 5 | Land with severe limitations that restrict use to rough grazing, forestry and recreation (poor to very poor drainage with improvement not possible; frequent flooding; steep slopes less than $25^{o}$; altitude less than 530 m.) |
| CLASS 6 | Land with very severe limitations that restrict use to rough grazing, forestry and recreation (very poor drainage, frequent floods, stones or boulders; slopes may be greater than $25^{o}$; altitude less than 610 m.). |
| CLASS 7 | Land with extremely severe limitations that cannot be rectified (such as bogs, sands and screes). |

TABLE 9.1 Land Use Capability Classification (source: Bibby and Mackney (1969)).

## 9.3 Vegetation description and analysis

Biogeography has witnessed substantial advances in the development and application of numerical methods in recent years, this, however, does not preclude interesting and valuable projects being undertaken by students. Indeed vegetation data collected from quadrat and transect field surveys can be analysed readily on a hand calculator and is 'admirably suited for use in schools, equally as well as at research level' (Watts, 1978).

In sub-sections 9.3.2 and 9.3.3 particular techniques of vegetation description and analysis are discussed. However, logically prior to these considerations is the problem of sampling design; this is examined in the following sub-section.

### 9.3.1 *Sampling procedures*

Quadrats are usually (but not always) rectangular wooden frames subdivided into a grid measuring 5 units by 5 units. The size and number of quadrats should be adequate to sample the degree of variation present. The size and number can be determined by constructing a species-area curve (based on the relationship that as area increases the number of species present also increases). In terms of quadrat size a species-curve can be produced by using quadrats of increasing size (x-axis) and recording the number of species recorded in each quadrat (y-axis). The point at which the curve flattens out indicates the optimal quadrat size (for herbaceious vegetation the most suitable size is 1-2 square metres). A similar procedure can be used to determine the optimal number of quadrats required; in this case, however, the x-axis would be the number of quadrats (running from 0 to 50, for example). In general 25 quadrats are desirable and for more precise studies 50 to a 100 are necessary.

The application of statistical techniques to the analysis and description of vegetation has led to the adoption of random sampling methods in many studies. This involves laying out a grid across the area to be examined and to sample points using a random number table and the co-ordinates of the grid. This method is very often time-consuming and, in some instances, impracticable. An alternative involves a blind-folded person throwing a quadrat in any direction and sampling where it lands. This method, however, is not truly random and results tend to be biased. A more realistic, and, in certain instances, more satisfactory alternative to random sampling is the systematic or regularly spaced sampling procedure. A special use of systematic sampling is the transect. This is usually located along a topographic or environmental gradient so as to sample change along the transect line. The sample quadrats can be regularly spaced (for example, at 10 metre intervals) or contiguous, and as long as the transect line is randomly located statistical techniques can be employed (Randall, 1978).

### 9.3.2 *Vegetation description and analysis: Physiognomic methods*

Physiognomic methods are essentially non-floristic in character, insofar as they do not require a detailed knowledge of community floristics. According to the physiognomic approach the plants present are divided into various life-forms. The essential idea is that plant communities possess a structure 'the set of life-forms foundin a plant community and their spatial arrangement constitute the primary morphological feature of that vegetation and bestow upon it a more or less definite structure.' (Pears, 1977, p. 14). The vertical arrangement of plants may be examined using the scheme illustrated in Figure 9.1in conjunction with a transect through a forest (normally measuring 60 m. x 8 m.) to produce a profile diagram (Figure 9.2). The idea of life-form has been used in a more specialized sense to define life-form classes, whereby plants are classified according to their regenerating parts (see Pears, 1977, p. 20-21 and Tivy, 1982, pp. 174-188). The significance of life-form for the geographer is the association which may exist between the form of plants and particular environmental conditions. Although the description and classification of plants using the physiognomic approach is difficult, it may be used in a quick reconnaissance survey of a large area, or as part of a geographical synthesis.

### 9.3.3 *Vegetation description and analysis: floristic methods*

The floristic approach, in contrast to physiognomic methods, is based on a detailed analysis of the species composition of a vegetation stand. Both qualitative and quantitative methods can be employed. Qualitative assessments usually involve species being placed in classes in a series of scales usually running from 1 to 5. These methods tend to be subjective, requiring both careful observation and field experience. Estimates can easily be affected by colour, form, ease of identification and so on. Examples of qualitative measures are contained in Table 9.2 and are based on the early work of Braun-Blanquet.

Quantitative measures require the observer to make accurate counts of particular properties such as number of individuals per unit area.

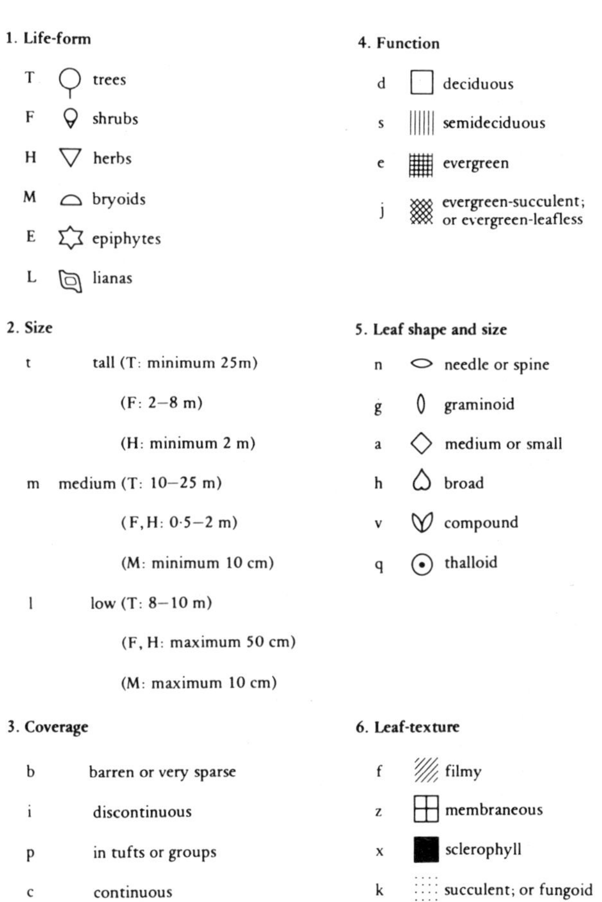


Figure 9.1 SIX CATEGORIES OF CRITERIA TO BE APPLIED TO A STRUCTURAL DESCRIPTION OF VEGETATION TYPES (Source: Pears, 1977)

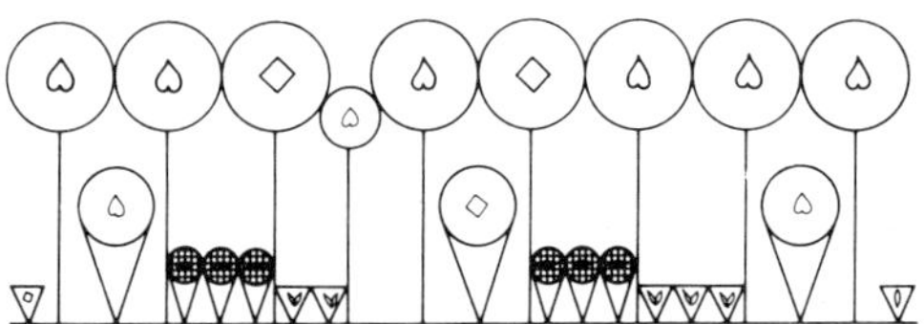

Aceretum Saccharophori Betulosum

Ttdh(a)zc.    Tldhzbi.    Ftdh(a)zi.    Flenxp.    Hldvzp(g,azb).

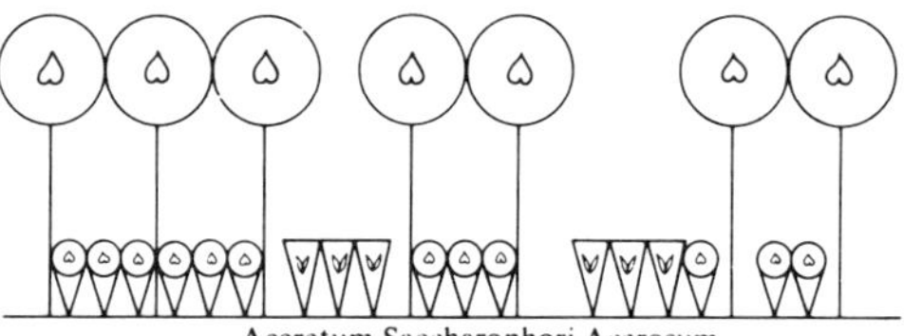

Aceretum Saccharophori Acerosum

Ttdhzc.    Fldhzp.    Hmdvzp.

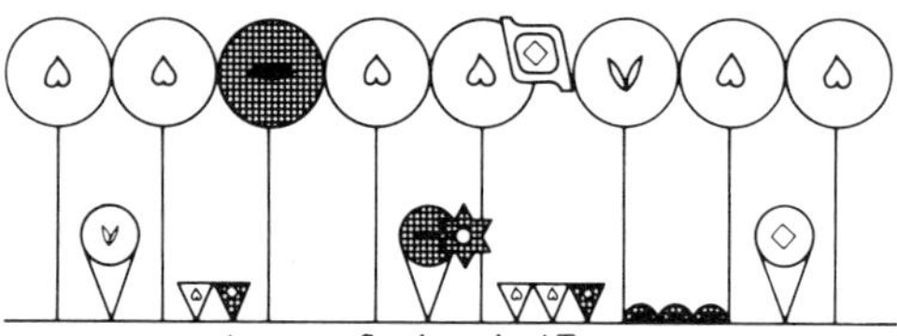

Aceretum Saccharophori Tsugosum

Ttdh(v) zc(enxb). Ltdazb. Fmda(v)z(enx)i.
Emeqzb. Hldhz(eax)p. Mlenfp.

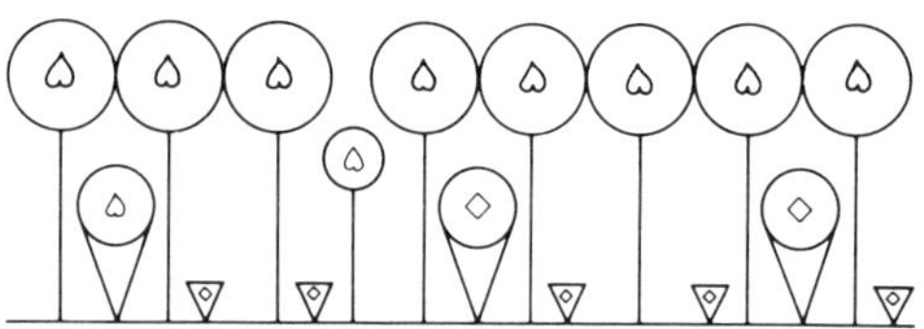

Aceretum Saccharophori

Ttdhzc.    Tldhzb.    Ftda(h)zi.    Hldazi.

Figure 9.2 EXAMPLES OF MAPLE-DOMINATED WOODLAND STANDS IN THE ST. LAWRENCE VALLEY, CANADA (Source: Pears, 1977)

Sociability Classes

| | |
|---|---|
| Class 1 | Plants growing in one place, singly |
| Class 2 | Plants grouped or tufted |
| Class 3 | Plants in troops, small patches or cushions |
| Class 4 | Plants in small colonies, in extensive patches or forming carpets |
| Class 5 | Plants occurring 'in great crowds', or pure populations |

Vitality classes

| | |
|---|---|
| Class 1 | Ephemeral adventives, germinate occasionally but cannot increase their area |
| Class 2 | Plants maintaining themselves by vetetative re-production but not completing a full life cycle |
| Class 3 | Well developed plants, regularly completing a full life cycle. |

(use of this scale requires familiarity with the behaviour of the plant throughout the year(s))

Abundance classes

| | |
|---|---|
| Class 1 | Very sparse |
| Class 2 | Sparse |
| Class 3 | Not numerous |
| Class 4 | Numerous |
| Class 5 | Very numerous |

TABLE 9.2 FLORISTIC METHODS: QUALITATIVE MEASURES (Source: Pears, 1977)

#### 9.3.3.1 *Density*

Plant density measures the number of species per unit area. The number of the individuals in each quadrat is recorded and the average number per quadrat determined. It should be noted, however, that density estimation is a time-consuming process and accuracy may be affected by factors such as suckering and layering.

#### 9.3.3.2 *Frequency*

Density values fail to give any indication of spatial variability in species abundance due to their pooled nature. Frequency attempts to remedy this by providing a measure of the regularity with which a species is distributed throughout a community. The frequency is the percentage of quadrats in which a species is present. Thus, if out of fifty quadrats ten contained *Erica tetralia*, the frequency would be 20%. the measurement of frequency is easily and rapidly made and is widely used in surveys. However, it is necessary to decide beforehand whether to record rooted presence or shoot presence. The former refers to a plant rooting in the quadrat, while the latter refers to the plant's shoot occurring vertically above a quadrat. It is necessary to note that frequency may be influenced by quadrat size and shape and also by the size of the plant under study (see Pears, 1977, p. 24). Frequency is often expressed as a limited number of frequency classes:

| | |
|---|---|
| Class 1 | 1 - 20% |
| Class 2 | 21 - 40% |
| Class 3 | 41 - 60% |
| Class 4 | 61 - 80% |
| Class 5 | 81 - 100% |

#### 9.3.3.3 *Cover*

Cover refers to the proportion of ground occupied by a species when the aerial parts of each plant are projected perpendicularly down on to it. Because of the existence of overlap amongst species the combined percentage cover for all species usually exceeds 100%. Cover is determined by estimating the approximate cover of the various species as they occur in each quadrat. Cover values are frequently arranged into a limited number of cover classes:

| Class | Observed range of count (%) | mean cover (%) |
|---|---|---|
| + | < 1.0 | 0.1 |
| 1 | 1.0- <10 | 5.0 |
| 2 | 10 - <25 | 17.5 |
| 3 | 25 - <50 | 37.5 |
| 4 | 50 - <75 | 62.5 |
| 5 | 75 - 100 | 87.5 |

In practical terms each species is recorded in each quadrat, in the field, in terms of its cover class. At a later date the mean percentage cover can be calculated for each species using the values in the above classification.

Individual scales may be combined to form an integrated scale, as in the case of Domin's cover-abundance scale:

| | |
|---|---|
| Class + | Occurring as a single individual with reduced vigour; no measurable cover. |
| Class 1 | Occurring as one or two individuals with normal vigour; no measurable cover. |
| Class 2 | Occurring as several individuals; no measurable cover. |
| Class 3 | Occurring as numerous individuals but cover less than 4% of total area |
| Class 4 | Cover up to 1/10th (4-10%) of total area |
| Class 5 | Cover about 1/5th (11-25%) of total area |
| Class 6 | Cover 1/4 to 1/3 (26-33%) of total area |
| Class 7 | Cover 1/3 to 1/2 (34-50%) of total area |
| Class 8 | Cover 1/2 to 3/4 (51-75%) of total area |
| Class 9 | Cover 3/4 to 9/10 (76-90%) of total area |
| Class 10 | Cover 9/10 to complete (91-100%) of total area |

As was noted in section 9.1 many areas of biogeographical study are beyond the scope of individual student projects. However, studies of community biogeography, involving the description and analysis of plant communities, could provide the basis for a number of manageable projects. Using transect surveys and relatively simple techniques of data collection a number of interesting projects may be developed. Williamson (1982), for example, reports the use of these methods in an 'A' level context to provide the basis for the study of a hydrosere, a lithosphere and a structural survey of wood. Clearly these methods may also be applied to other topographic or environmental gradients.

A more ambitious project would involve an examination of the relationships which might exist between various environmental factors (for example, vegetation, climate, soils and slope). Following Hanwell and Newson (1973), the first step could involve the identification of two contrasting slopes (one north-facing and one south-facing, for example) and making a quadrat survey of each slop. For both slopes a species list should be prepared, together with an appropriate measure of frequency. At this point an assessment should be made of the influence of soils and climate in explaining the differences between the two slopes. Large scale plans of the slopes should be prepared locating quadrat sites, these can then be used to plot the frequency of occurrence of individual plants at each site. On tracing paper isonome maps of the density and cover of each species can also be drawn. Finally, the isonome maps can be overlaid providing the basis for an analysis of vegetation distribution on both slopes.

Further analysis of plant communities can be made using the $\chi^2$ statistic. The objective would be to examine how various species in the same community relate to each other. The method involves comparing the joint occurrence of pairs of species in quadrats, and allows groups of positively associated species to be determined. Results can be presented in both a statistical and cartographic manner. Details of this method can be found in Willis (1977), Hanwell and Newson (1973) and Kershaw (1973). A more sophisticated method involving the $\chi^2$ statistic is association-analysis, and details of this method can be found in the sources referred to in section 9.1 and in many standard biogeographical texts.

As noted in the introduction to this chapter project work is also possible in the examination of man's impact on biota. Clearly, relevant topics will depend to a large extent on local conditions. However, the influence of pollution on vegetation (Moss, 1975), the impact of recreational activities (Liddle, 1975; Rees and Tivy, 1978) and the pattern and role of planted and preserved vegetation in urban ecosystems (Schmid, 1975) are themes which are applicable to many areas.

Chapter Ten

# COMPUTING AND PROJECTS IN GEOGRAPHY: SOME GENERAL CONSIDERATIONS

## 10.1 Introduction

The impact of the computer revolution has been wide-ranging, and, indeed, today, it can be argued that it is now a vital necessity to the continuation of normal everyday life. In the so-called 'post-industrial society', information processing has become an end in itself, rather than a means to an end, and, as Weizenbaum (1976) suggests, the commitment to computers is nearly irreversible. The present Government has introduced a policy to try to ensure that all secondary schools possess their own micro computer in the near future. One general problem is that the choice of micro computers (such as Apples, B.B.C. machines, Commodore Pets, Research Machines 380-Zs, Sinclair ZX81s or Spectrums, Tandberg EC10s, Tandy TRS-80s) is large, and there is, obviously, no single machine which is the most appropriate for all a school's needs. Fortunately, assistance is available to help teachers in this direction (see, for example, the Council for Education Technology (1980), Eldridge and Shaldrick (1980), the Micro Information Database Advisory Service (1980), and Webster (1979)), and in the last few years, a range of training courses and seminars have been run to keep teachers informed about recent developments.

Although there are important political, social, economic, legal and ethical issues relating to the role of computers, attention here is focused on certain educational issues (specifically in conjunction with a geographic curriculum (see also Watson, 1979)). Given that computers have arrived to stay, arguably, it can be suggested that a basic comprehension of computers should be an essential component of education, and its ultimate effectiveness as an integral part of any curriculum must be assessed in relation to its educational usefulness. At the outset, it is important to appreciate that this additional 'intellectual resource' (Gould, 1981) has to be learned by the students rather than taught to them in a conventional way by teachers. Clearly, whilst financial constraints will restrict the number of machines that are available in each school, students should be given every opportunity to practise. As Bruner suggests, "Mastery of the fundamental ideas of a field involves not only the grasping of general principles, but also the development of an attitude toward the possibility of solving problems on one's own". There are, therefore, general implications for the method of teaching geography, and, perhaps more fundamentally, for the nature of geography that is taught. (See also Unwin, 1980). These two features complement each other and some of the important issues are addressed in sections two and three. Obviously, as with any such proposed changes, there will be difficulties in implementation and dangers to be avoided, and, in the final section some of the envisaged problems are highlighted. First, however, it should be appreciated that additional more practical problems, associated with teaching arrangements and assessments, must also be given careful consideration.

Finally, given the increasing availability of computer programs (or software) that have been designed for teaching geography in schools (see, for example, Geographical Association Programme Exchange (GAPE)), it is unlikely that students should be involved with computer programming *per se* (at least in the geography sessions). It is, however, realised that to enable one to make modifications to purchased software and to permit the writing of specific programs for particular projects, a large number of teachers desire to have a broad comprehension of writing computer programs. Accordingly, an introduction to BASIC, the computer language that is used most frequently on micro-computers, is presented in Appendix IV.

## 10.2 The computer as an educational tool

In their excellent discussion of computer assisted learning in geography, Shepherd et al. (1980) highlight five reasons why the computer is a useful educational tool. For instance, in relation to its most obvious application in conjunction with statistics, a computer's speed removes the need for repetitive and boring calculations and its reliability overcomes the problem of the inevitable calculation errors. Thus, if the more tedious component of their work can be reduced, students can spend more time on gaining insights into important geographical concepts. (It is, however, important to appreciate that the availability of statistical packages for the computers must not remove the basic need that the students comprehend the rationale of the statistics and their underlying assumptions). Consequently, the computer can be employed as a useful aid in statistical exercises. However, it is its impact on the form of teaching that is likely to be more wide-ranging, and it should not be confined to methodological aspects of courses. It is these aspects that are considered briefly in this section.

From our experience, it is a basic tenet that students' comprehension of rather abstract concepts is enhanced by active involvement in the learning process (and, consequently, it easily outweighs the disadvantage of being a relatively slow procedure). Let the students actually participate and think for themselves! Do we give them sufficient opportunity for self-motivated, semi-independent and creative thinking? Obviously, these questions relate directly to the underlying rationale of project work.

First, however, is it desirable for geography teachers and students to be able to write their own computer programs? Without even discussing the advantages of different languages - say, BASIC, FORTRAN or PASCAL - because most programs for micros are written in BASIC, it can be argued that, in general, the existence of a range of pre-prepared and well-tested programs removes this requirement. If this argument is accepted, there remains the need (especially for geography specific programs) for the development of a formal network to exchange individuals' programs; dissemination of information and experience is essential, particularly in the formative early stages. Attention is drawn to the fact that geography is one of four subjects that is most likely to be affected by micro-computers according to the present D.E.S. Micro-electronics Education Programme. More specifically, valuable funding has been obtained by the Geographical Association from the National Development Programme in Computer-Assisted Learning to encourage the development and dissemination of good quality software and to provide teachers with up-to-date information about the potential of computer-assisted learning in geography. Clearly, this process of education and software exchange is essential for computer-assisted learning to develop successfully, and the Geographical Association Programme Exchange (GAPE), which is based at Loughborough University, provides a useful service in this direction.

The exclusive use of pre-prepared programs could eventually constrain a student's analysis. One of the potential educational advantages is the scope for individual exploration and development, and this would be hindered by this 'black box' type approach. There would be an inherent tendency to fit the problems and projects to available programs, rather than vice-versa. Anyway, a short programming course would provide additional training in logical and systematic reasoning and also, increasingly important, vocational knowledge.

It would be a misconception to believe the role of the teacher is being substituted. Indeed, it is important to appreciate that, at least in the first instance, the required commitment would be relatively large. If required, in the writing of computer programs for student use, great care must be taken to accommodate every possible eventuality, foreseeing students' behaviour and difficulties (as they are directed through a fairly rigid sequence of actions). Briefing and follow-up sessions are essential for success. At the outset, students' background knowledge must be detailed and the objectives of the exercise must be specified unambiguously, otherwise students would probably proceed aimlessly without direction and the venture would be a dismal failure. Obviously, it would be essential to go beyond a collection of computer print-outs - the computer says ... One of the advantages of the computer is that more time can be spent on the analysis of results and on the obtaining of a greater understanding of the geographic concepts and techniques involved. Thus, although the use of a computer as an educational tool may prove more demanding for the teacher (particularly as the various projects would probably have different requirements), in terms of student learning, it is likely to prove beneficial. In fact, given the time constraints for completing the course, it can be suggested that the use of computers by geography students is probably most appropriate in conjunction with their project work. Clearly, it must remain a geographical project, and not be simply a computer exercise!

## 10.3 What can the computer be used for?

The use of the computer in teaching affects not only the method of teaching but also what is actually taught. Many of the conceptual developments in geography since the mid-1950's, for example, have been facilitated by the increased availability of computer power. In this section, a brief overview of the potential areas of computer applications is presented, and, whilst the discussion is not restricted to project applications, it is believed that insights into their usefulness in project work will be forthcoming.

It would be incorrect to suggest that the educational usefulness of a computer is restricted to a (statistical) analysis of data. Clearly, it is this role as a fast calculator that is likely to be the most significant. However, it must be appreciated that, in addition to performing calculations, the computer can be used to present data graphically in a non-verbal format. This graphics capability is significant from a general educational perspective, because, if a computer can draw shapes and patterns, it can often communicate information more effectively. For a geographer, the opportunity to draw maps on a computer is especially appropriate, and it is possible to link a digitiser or graphics tablet to most micro computers.

Given the continued empirical foundation of much of geography, as stated above, computers will be particularly useful for data analyses in student projects. (It should be noted that the operating conditions of micro-computers are sufficiently flexible to enable them to be taken into the field for both data collection and analysis). Indeed, at least in the short term, it is envisaged that it is this type of application that will be most important. Attention is drawn to the fact that to make things more straightforward for students, interactive computing facilities (rather than a batch system), which involve a continual question and answer service, would be attractive; this is another advantage of micro-computers. In this type of application, if students have a knowledge of computer programming, they should be encouraged to write their own programs. Whilst it is highly likely, of course, that such programs would be readily available everywhere, this kind of exercise would serve two important functions. It would not only provide practice in structuring and writing computer programs, but also, simultaneously, ensure students possess a working knowledge of the mechanics of the techniques. (In fact, it is being suggested increasingly that introductory statistics and computer programming complement each other and can be taught together).

From an educational perspective, simulation modelling and gaming are two areas which the computer has made available for teachers (Walford, 1969). For instance, simulation is already a popular way of introducing concepts, such as diffusion, which have complicated effects that are difficult to appreciate using traditional, more formal, teaching methods. Moreover, it allows important aspects to be studied that were impossible previously. The implications of different transport management policies, for example, can be seen from simulation results without the events actually occurring. The effects can be relatively easily observed (particularly if graphical facilities are applied). One attraction of simulation, therefore, is that it offers an operational form in which to readily analyse complexity (interrelated phenomena). In the present context, it is important to realise that some simulation programs will not be especially useful in teaching; a large number of university research programs exist, and, in general, the incentive to directly transfer them for student teaching should be resisted.

Instructional simulation programs should enable students to obtain an intuitive feel for concepts and problems involved and to learn through progressive and varied analyses. In practice, this type of teaching proves to be enjoyable for the students, and, if the problem is sufficiently complicated, this kind of problem solving could form part of a student's project. Instructional simulation programs are frequently of the 'what if ...' variety; a student would examine the results for computer runs with different starting values for the variables and different values for the parameters. This approach has been found to not only enhance students' comprehension but also to improve their cognitive skills. For example, after students have gained some insights into the nature of a problem by simulation, real data sets should be examined. Why are there differences between reality and the simulation results? How would we extend our simulation program to be more realistic?

In general, students can only alter the input of a simulation model, and they must follow its structure directly. However, if students were encouraged to develop their own simple simulation programs as part of their project (to which credit must be given), this would be an excellent way of training logical and systematic thought and of really ensuring students fully comprehend their projects. (Clearly, this sort of project must be restricted to a small set of students who have had the necessary training in other disciplines). However, the restrictions of this 'black box' approach will be particularly significant until more sophisticated instructional simulation programs are developed; at present, the students' learning potential is constrained because the programs are too simplistic, although this is a direction that is being developed actively.

The use of the computer is seen in terms of its assistance it can provide in different situations. There is no suggestion that it removes the need for teachers, but their role must change if the benefits are to be enjoyed. Moreover, of course, computers should not be applied in a geographic curriculum for their own sake, although project work is one area in which they will be used increasingly. Teaching partly founded on the use of a computer is an exciting and an imposing recent development. However, before its potential can be realised fully, a number of obstacles must be overcome. Some of these issues are considered briefly in the next section.

## 10.4 Some concluding thoughts

For the majority of teachers, the development of coherent and integrated courses, incorporating many of the characteristics that have been considered already, is a daunting task filled with uncertainty. Given this situation, there is a need for greater liason between colleagues; formal networks of communication to disseminate information should be developed. A sufficiently firm foundation, based on the experiences of others, is a prerequisite for successful implementation. This would help stop one possible problem occuring; a proliferation of very similar computer programs. Fundamentally, it would avoid unnecessarily wasting time on writing computer programs, and, anyway, no enormous degree of standardisation would be forthcoming. It should be appreciated that it is this question of software quality which will determine ultimately the effectiveness of the computer in teaching. Such issues relating to a teacher's deployment of effort is likely to be especially important in the formative stages. In addition to the distribution of software, advice must be given to enhance teachers' appreciation of the computer's potential role in teaching. In the excitement, there would be a natural tendency to try to progress too quickly by concentrating on the relatively more advanced forms of analysis. Indeed, there are some obvious dangers of indiscriminate applications. Students should not become dependent solely on the computer (like many of the present generation of school children brought up on calculators who have difficulty doing basic arithmetic operations themselves). Clearly, the ultimate analysis of the relevance of computers in a school geographic curriculum is dependent on the students' geographical interpretation of the results.

If insufficient attention is given to debriefing and analysis sessions, there is a danger that students are only being introduced to an additional, rather novel form of abstraction. This situation would be both unfortunate and unnecessary. Location theory, for example, could be introduced from a prescriptive, problem-solving perspective as well as from a descriptive perspective (see Beaumont (1983) for more details). This would enable a geography student's training to become more vocational, a characteristic that will continue to become of increasing importance in the future. Moreover, projects that are directed towards an examination of local issues would be an appropriate component of such redirection.

Arguably, in conclusion, it would appear that, in fact, a computer is a necessary, extra educational tool to add to more traditional ways of teaching and instruction. Indeed, at present, its full potential is not fully appreciated and the challenge should be tackled immediately. It would appear that an obvious starting place is in association with student projects (although, clearly, there is no suggestion that all projects should be computer-based).

The danger is that '... we could find ourselves with a generation of children sharply divided between those who have amplified their own brain power with that of the computer and those who remain wedded to the ignorance of the past' (Evans, 1979, p.128). Is this prospect perhaps more evident among teachers themselves? Such duality must not be permitted to develop.

# Appendix I

# AN INTRODUCTION TO STATISTICAL METHODS FOR ADVANCED LEVEL GEOGRAPHY

## I.1 Introduction

Simple statistical methods are now an important tool in geographical investigations, and are an integral part of new curricula at Advanced Level Geography. At the outset, it should be appreciated that the so-called 'new' geography is often associated directly with quantification; however, many geographical concepts and principles are not dependent on quantification.

Whilst quantity is clearly no substitute for quality, it is suggested that, in the 1980s, all geography students should have an adequate working knowledge of elementary statistical techniques. It should be an integrated part of their training so that they are confident to apply appropriate methods to add rigour and precision to their geographical description and analysis. It is recognised that problems are likely to arise because of the limited mathematical backgrounds of students and teachers, and therefore, in this introduction, nothing is assumed beyond a basic comprehension of the simple arithmetic operations - addition, subtraction, multiplication and division. Moreover, obfuscating jargon is avoided, and the necessary notation is introduced from first principles. Techniques are introduced by examples.

A wide range of descriptive statistics are included in Advanced Level Geography curricula, and it is often difficult to determine which particular method is pertinent to a specific problem. In this introduction, the techniques are classified according to two criteria: the level of measurement of the data and the type of statistic. This structure provides the framework for the entire discussion and should facilitate selection of appropriate techniques. First however, some basic definitions and a discussion of visual displays of data are presented.

## I.2 Some Basic Definitions

### I.2.1 *Statistics*

In everyday use, the term, 'statistics', refers to facts and figures, such as official Government statistics, sports results, women's so-called 'vital statistics' and so on. This, however, is not the definition that is employed throughout this discussion. To avoid possible confusion over terminology, such information in the form of numerical facts will be called 'data'. 'Statistics' on the other hand are seen to involve the body of knowledge with four recognisable phases '.....dealing with the collection, organisation, analysis and interpretation of numerical data (Adler and Roessler, 1972, p.1).

For the application of any statistical techniques to be possible, a representative data set or sample, must be collected. The need for a carefully designed and executed survey is a fundamental prerequisite. An examination of sampling frames, interview techniques, questionnaire structuring and wording, fieldwork mapping and so on is outside the scope of this discussion (see Appendix III); it is assumed that an adequate data set exists.

The second stage is organisation, presenting the data in the most appropriate form. Graphical and tabular methods of presenting data remain of fundamental significance. A clear presentation aids comprehension and interpretation.

The stage of analysis involves essentially numerically summarising the data. A variety of descriptive statistics are available, and they form the bulk of the material examined below. More specifically, three kinds of statistics are considered; measures of central tendency, measures of dispersion and measures of association.

The other major area of interest is the fourth stage, interpretation, which involves making inferences about a population from a sample data set. Whilst a comprehension of the use of statistical tests and statistical inference, such as the chi-squared test and student t-test, is expected, it is argued that the majority of 'A' level geography students would have insufficient understanding of statistical inference. Indeed, it is believed that the actual returns or benefits of teaching 'A' level geography students would be small, and therefore, they are not considered here.

### I.2.2 *Data Measurement Levels*

The range of statistics considered below are differentiated on the basis of the data's measurement level. Four levels are recognised: nominal, ordinal, interval and ratio. It is of fundamental importance to appreciate that particular statistics are applied to specific levels of data. For example, the only appropriate statistic of central tendency for nominal data is the mode. Table I.1 classifies the statistics required for 'A' level geography.

| TYPE OF STATISTICS | DATA MEASUREMENT LEVEL: Nominal | Ordinal | Interval/ Ratio |
|---|---|---|---|
| Measures of Central Tendency | Mode | Median | Mean |
| Measures of Dispersion | Variation Ratio | Range, interquartile range, Quartile Deviation | Standard Deviation |
| Measures of Association | Phi Coefficient, Contingency Coefficient | Spearman's Rank Correlation Coefficient | Pearson's Product Correlation Coefficient |

TABLE I.1: STATISTICS: CLASSIFIED BY DATA'S MEASUREMENT LEVEL

The lowest level of measurement, the nominal scale, involves a simple classification of information into specific categories. (The categories are 'mutually exclusive' and 'exhaustive'; simply stated it is impossible to belong to more than one category and all the information must fit into one of the categories). For example, land use (industrial, residential.......) and a general classification of settlement, such as city, town, village and hamlet, are both nominal data sets. A special kind of nominal data is a binary (or two category) form, such as urban/rural, above average/below average, present/absent, and so on.

An important characteristic of nominal data is that there is no knowledge about the relationships between different categories. In contrast, ordinal level data involves ranking different categories in some order of preference; no indication, however, is given to the numerical magnitude of the difference between categories. For example, a sample of foreign tourists could rank the following places in order of preference: York, Chester, London, and Stratford.

Both the nominal and ordinal scales are low order measurement levels; interval and ratio scales are high order measurement levels. When numerical values for variates are given, exact differences between them are known. For example, using a standard measurement unit, such as degrees centigrade, it is possible to say how much warmer January temperatures are in Harare in contrast to in London. Temperature data are interval level. However, if a non-arbitrary, fixed zero level exists, the measurement scale is called ratio. Distance measurements between places in say kilometres, for instance, are ratio data, because zero distance is meaningful; it is the same location! To highlight this distinction between interval and ratio scales, compare the following contrived examples. Two places A and B are 100 kilometres and 200 kilometres away from the capital city respectively, and therefore it is possible to say that B is twice as far away as A from the capital (because of a non-arbitrary, fixed zero). On the other hand, if January temperatures at A and B are 10°C and 5°C, respectively, it is not possible to say that A is twice as hot as B (because 0°C is an arbitrary level). It is noted that, in general, for the presentation of the statistics, the distinction between interval and ratio is not required.

From the preceding, systematic description of the measurement levels, it is clear that it is possible to convert data from one level to a lower level. That is, ordinal data can be converted to nominal data, and interval and ratio data can be converted to both ordinal and nominal data. It is important to appreciate that such conversions involve an associated information loss. Such a change is undertaken, because particular statistical techniques are applied to specific data measurement levels. Thus, as a measure of dispersion for ratio data, it would be possible to calculate not only the standard deviation, but also the range, interquartile range, and variation ratio. However, unless there are specific reasons why the entire set of statistics are calculated, it is suggested that one statistic is chosen on the basis of Table I.1.

Repeating the basic argument, it is of fundamental importance to ensure that the statistics are appropriate for a specific data set's measurement level; this framework should actually facilitate the selection procedure. For completeness, it is noted that much of the data analysed by geographers are of a low measurement level (and many of the common statistical techniques assume a high level of data measurement).

## I.3 Data Presentation

As indicated already, data presentation is a basic component of statistical analysis, which should be emphasised much more (particularly in 'A' level geography). Numerous types of

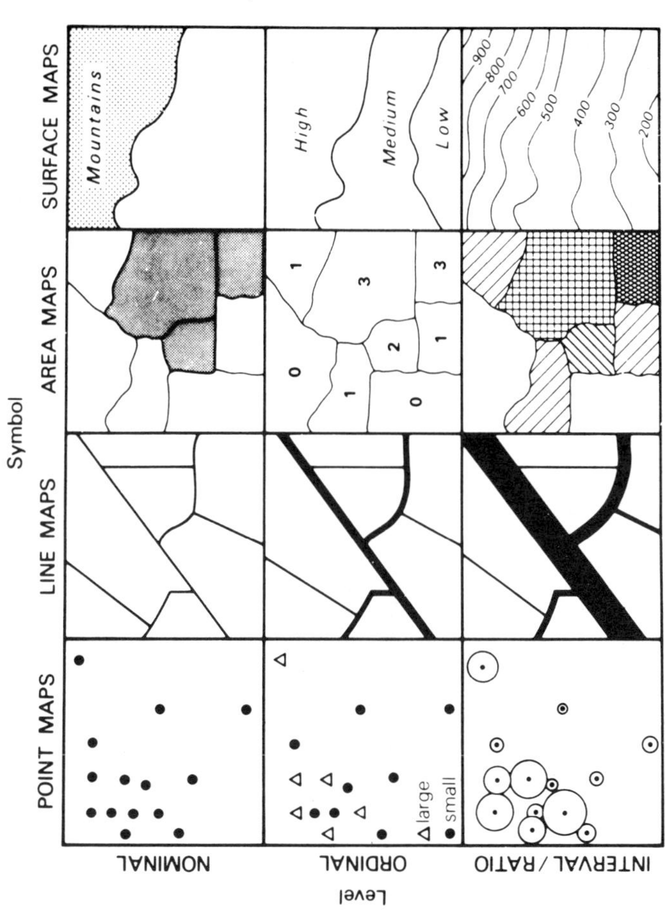


Figure I.1 A TYPOLOGY OF MAPS
(Source: Unwin, 1981)

presentation are possible, although it is common to use either a tabular or graphical form. These are exemplified briefly below, but it should be remembered that a geographer's traditional tool for presenting information is the map. Moreover, in Figure I.1 Unwin (1981) has presented a useful classification of all types of maps on the basis of two criteria: the level of measurement of the data (nominal, ordinal or interval/ratio) and the dimensionality of the spatial analysis (points, lines, areas or surfaces).

Visual presentations of data sets help transform them into useful information. They have the advantages of conciseness and clarity over a verbal description. It is important to remember that their effectivity is dependent, to a large extent, on their simplicity. For example, Figure I.2 is a pie chart indicating the relative importance of different energy sources to the United Kingdom. Figure I.3 is a scatter graph which portrays a large amount of information simply, specifically the changing composition of the U.K.'s fuel economy in the post-war period. Scatter graphs which are plots of the values of one variable against those of another (not necessarily including time) are widely applied, and they often give insights into the association between the two variables (see the discussion about correlation and linear regression below).

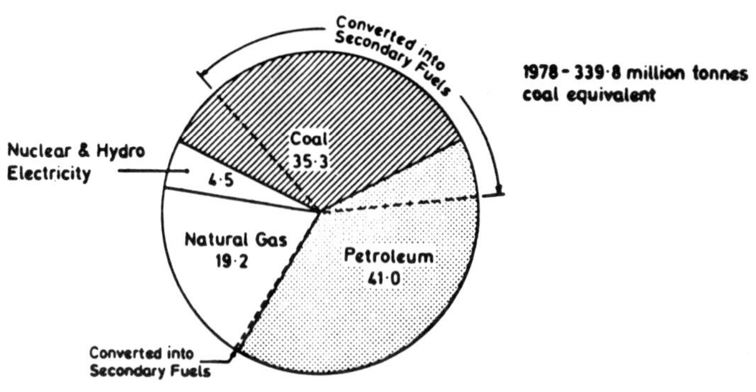


Figure I.2 TOTAL CONSUMPTION OF FUELS IN THE UK - 1978

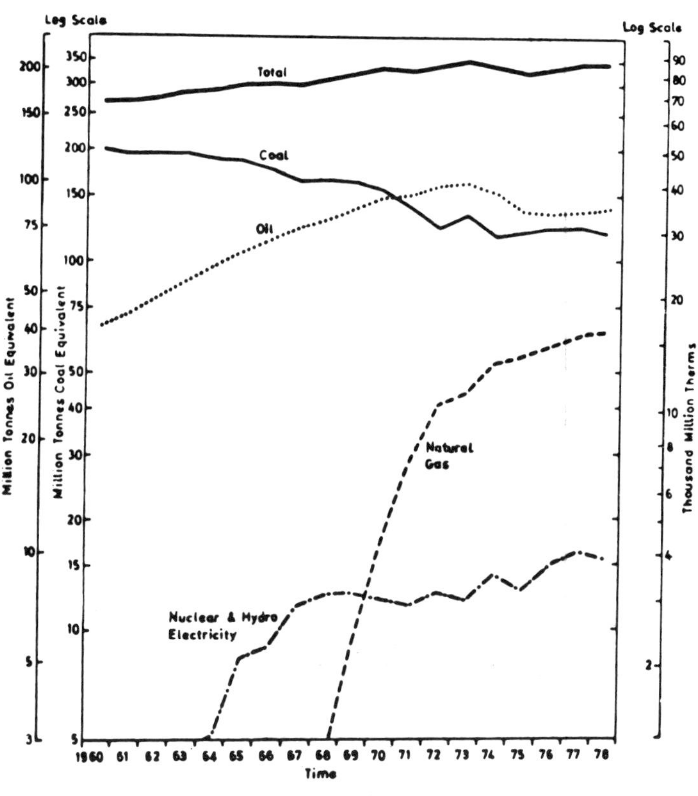


Figure I.3 ENERGY CONSUMPTION IN THE UNITED KINGDOM

| Country | LOCALITY SIZE | | | | |
|---|---|---|---|---|---|
| | 1m & over | 100,000-999,999 | 50,000 - 99,999 | 20,000 - 49,000 | 10,000 - 19,999 |
| North Africa | | | | | |
| Morocco | - | 8 | 3 | 12 | 22 |
| Algeria | - | 4 | 13 | 51 | 24 |
| Tunisia | - | 1 | 3 | 14 | 22 |
| Libya | - | 2 | - | 1 | 8 |
| Egypt | 2 | 14 | 9 | 57 | ? |
| Middle East | | | | | |
| Turkey | 1 | 13 | 16 | 74 | 85 |
| Iran | 1 | 13 | 15 | 41 | 48 |
| Syria | - | 3 | 2 | 4 | 14 |
| Jordan | - | 1 | 2 | 6 | 6 |
| Israel | - | 3 | 2 | 15 | 14 |
| Cyprus | - | - | 1 | 2 | 1 |
| Saudi Arabia | - | 3 | 3 | 5 | 12 |
| Kuwait | - | 1 | - | 3 | 4 |
| Trucial States | - | - | 1 | 2 | - |
| Bahrain | - | - | 1 | 1 | - |

(Source: Clarke and Fisher, 1972)

TABLE I.2: FREQUENCY OF LOCALITIES WITH 10,000 INHABITANTS OR MORE: MIDDLE EAST AND NORTH AFRICA

| SIZE | FREQUENCY OF CENTRES |
|---|---|
| Over 1 million | 9 |
| 500,000 - 1,000,000 | 24 |
| 250,000 - 500,000 | 41 |
| 100,000 - 250,000 | 130 |
| 50,000 - 100,000 | 185 |
| 10,000 - 50,000 | 1581 |
| Below 10,000 | 3418 |

(Source: Dewdney, 1971)

TABLE I.3: SOVIET CENTRES BY SIZE GROUPS: 1968

TABLE I.4 KEELE UNIVERSITY MONTHLY SUNSHINE TOTALS

| | JAN. | FEB. | MARCH | APRIL | MAY | JUNE | JULY | AUG. | SEPT. | OCT. | NOV. | DEC. |
|---|---|---|---|---|---|---|---|---|---|---|---|---|
| 1952 | 70.2 | 75.4 | 82.3 | 137.9 | 173.3 | 140.9 | 140.3 | 132.6 | 100.9 | 107.7 | 70.1 | 41.2 |
| 1953 | 33.4 | 68.7 | 138.9 | 161.0 | 208.6 | 153.2 | 174.0 | 177.8 | 131.7 | 81.1 | 48.7 | 28.6 |
| 1954 | 46.2 | 58.6 | 99.4 | 172.3 | 119.0 | 127.2 | 108.4 | 115.0 | 155.7 | 71.2 | 50.9 | 50.6 |
| 1955 | 32.8 | 70.6 | 153.4 | 136.9 | 208.9 | 132.0 | 271.1 | 197.9 | 149.7 | 125.8 | 58.9 | 55.5 |
| 1956 | 54.2 | 82.1 | 131.6 | 137.2 | 238.0 | 135.4 | 153.1 | 135.4 | 94.8 | 104.2 | 53.4 | 18.8 |
| 1957 | 47.1 | 81.0 | 97.6 | 138.3 | 177.3 | 283.8 | 114.3 | 113.1 | 91.7 | 73.4 | 59.1 | 57.8 |
| 1958 | 50.5 | 54.4 | 99.7 | 132.2 | 149.5 | 127.2 | 154.2 | 114.9 | 94.6 | 66.6 | 39.5 | 26.7 |
| 1959 | 68.9 | 45.5 | 72.0 | 137.7 | 227.0 | 230.9 | 213.9 | 181.1 | 188.7 | 137.2 | 52.7 | 34.0 |
| 1960 | 22.7 | 65.2 | 61.3 | 136.9 | 173.7 | 271.1 | 142.5 | 114.2 | 124.4 | 45.5 | 50.9 | 60.2 |
| 1961 | 26.5 | 56.4 | 130.7 | 77.2 | 182.4 | 188.3 | 146.0 | 169.5 | 122.5 | 105.7 | 76.7 | 51.3 |
| 1962 | 67.0 | 61.4 | 130.9 | 163.7 | 143.2 | 188.7 | 129.6 | 148.3 | 90.0 | 88.6 | 38.6 | 51.5 |
| 1963 | 67.0 | 79.9 | 90.0 | 107.5 | 165.7 | 198.0 | 165.2 | 106.2 | 124.7 | 79.6 | 53.4 | 42.5 |
| 1964 | 35.4 | 51.5 | 48.5 | 106.2 | 189.5 | 108.3 | 159.8 | 169.7 | 157.2 | 106.6 | 52.2 | 61.6 |
| 1965 | 62.5 | 25.0 | 128.8 | 131.4 | 124.5 | 174.0 | 105.7 | 164.8 | 83.0 | 97.6 | 66.2 | 53.0 |
| 1966 | 29.2 | 33.8 | 89.5 | 89.3 | 226.2 | 146.2 | 125.2 | 153.8 | 130.5 | 80.1 | 34.2 | 30.0 |
| 1967 | 52.0 | 56.2 | 134.5 | 92.0 | 138.5 | 216.6 | 174.5 | 152.4 | 95.0 | 83.3 | 57.2 | 50.5 |
| 1968 | 26.3 | 56.7 | 78.0 | 148.3 | 117.0 | 152.7 | 107.9 | 124.4 | 80.8 | 57.2 | 34.7 | 29.1 |
| 1969 | 26.0 | 57.9 | 64.5 | 166.9 | 116.3 | 239.6 | 213.5 | 126.8 | 84.5 | 75.2 | 81.0 | 36.8 |
| 1970 | 19.6 | 95.3 | 110.8 | 127.7 | 213.7 | 259.7 | 140.9 | 164.5 | 115.6 | 99.1 | 50.3 | 45.7 |
| 1971 | 33.6 | 52.3 | 79.8 | 90.7 | 213.9 | 134.8 | 215.6 | 135.0 | 164.9 | 132.6 | 75.6 | 28.9 |
| 1972 | 41.5 | 40.9 | 136.2 | 125.7 | 142.5 | 142.2 | 166.6 | 165.9 | 94.0 | 96.4 | 80.5 | 25.7 |
| 1973 | 33.2 | 67.7 | 128.6 | 142.6 | 172.6 | 200.3 | 153.9 | 191.7 | 124.5 | 86.0 | 73.1 | 53.5 |
| 1974 | 53.4 | 45.9 | 107.4 | 161.8 | 184.9 | 175.3 | 161.3 | 17718 | 117.8 | 82.9 | 50.9 | 53.9 |
| 1975 | 33.5 | 52.1 | 86.0 | 111.4 | 208.0 | 247.8 | 190.6 | 221.9 | 147.6 | 96.1 | 79.4 | 29.9 |
| 1976 | 39.6 | 51.2 | 88.4 | 116.9 | 133.5 | 219.1 | 230.0 | 226.7 | 73.9 | 50.5 | 60.4 | 31.9 |

TABLE I.4 KEELE UNIVERSITY MONTHLY SUNSHINE TOTALS

Frequently, in practice, geographers would be faced with a data set containing a large number of observations. In such circumstances, it is often convenient, in a presentation of the distribution of the values, to classify the data into pre-specified groups. The selection of class intervals is of fundamental importance, and in general, it is dependent on the range and number of values of interest. (For this reason, the statistics package listed in Appendix V has a range of options for defining classes). It should be noted that if only a few classes are defined much of the original information would be lost. On the other hand if a large number of classes are used, many of the classes would be empty. No precise rules exist for class selection in this trade-off between generalisation and detail; commonsense must prevail. In defining class limits, it must be remembered that classes must be mutually exclusive (that is no individual can be a member of more than one group). For example, information on the frequency of localities with 10,000 inhabitants or more in the Middle East and North Africa, presented in Table I.2, illustrates the importance of primate cities. However, care must be taken to avoid ambiguities in the class limits (see Table I.3 on the distribution of the urban population in the Soviet Union.) For example, in which class would a settlement of exactly 100,000 people be included?

As a basis for the exemplification of many of the statistics, meteorological data collected from Keele University are used. Table 1.4 presents monthly sunshine totals (hours) for Keele University during the period 1952 to 1976. By defining the following classes, it is possible to construct a frequency table: 0.0 - 49.9; 50.0 - 99.9; 100.0 - 149.9; 200.0 - 249.9; 250.0 - 299.9. It is noted that the accuracy of the class intervals (that is, the number of significant numbers) is the same as the original data.

| Class Interval | Tally Marks | Frequency | Cumulative Frequency |
|---|---|---|---|
| 0.0 - 49.9 | | 42 | 42 |
| 50.0 - 99.9 | | 102 | 144 |
| 100.0 - 149.9 | | 84 | 228 |
| 150.0 - 199.9 | | 48 | 276 |
| 200.0 - 249.9 | | 20 | 296 |
| 250.0 - 299.9 | | 4 | 300 |
| | | 300 | |

TABLE 1.5: FREQUENCY TABLE

The tally marks are used as an aid to adding up the numbers in each category (see Table I.5). Given that the class intervals are a constant size, the visual presentation of this frequency table as a histogram is relatively straightforward. At the outset, it should be appreciated that it is the area of each bar (not its height) that reflects the relative frequency of each class interval. Figure I.4 is a histogram for this information and it should be compared with the histogram that is constructed from the same data set but which has unequal class intervals (see Figure I.5). Clearly, the selection of class intervals affects the presentation and interpretation of data! (Note, it is important to label all axes correctly and give a title for each diagram).

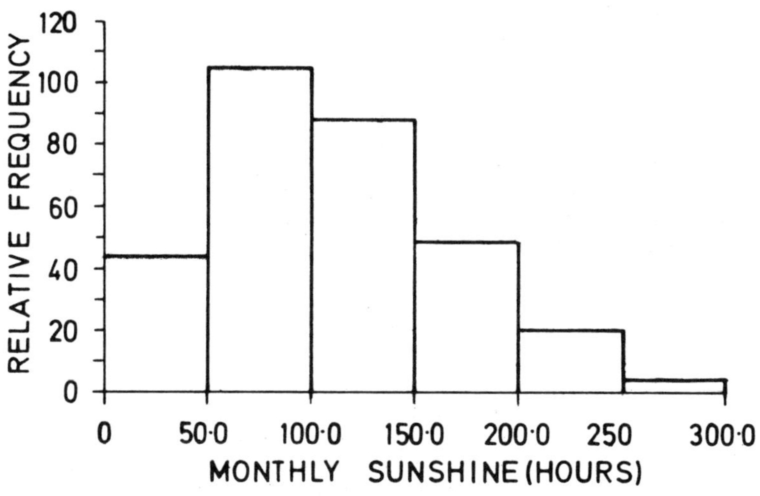


Figure I.4 HISTOGRAM FOR EQUAL CLASSES

| Class Interval | Tally Marks | Frequency | Cumulative Frequency |
|---|---|---|---|
| 0.0 - 49.9 | | 42 | 42 |
| 50.0 - 74.9 | | 57 | 99 |
| 75.0 - 99.9 | | 45 | 144 |
| 100.0 - 124.9 | | 29 | 173 |
| 125.0 - 149.9 | | 55 | 228 |
| 150.0 - 174.9 | | 30 | 258 |
| 175.0 - 199.9 | | 18 | 276 |
| 200.0 - 299.9 | | 24 | 300 |

TABLE I.6: FREQUENCY TABLE

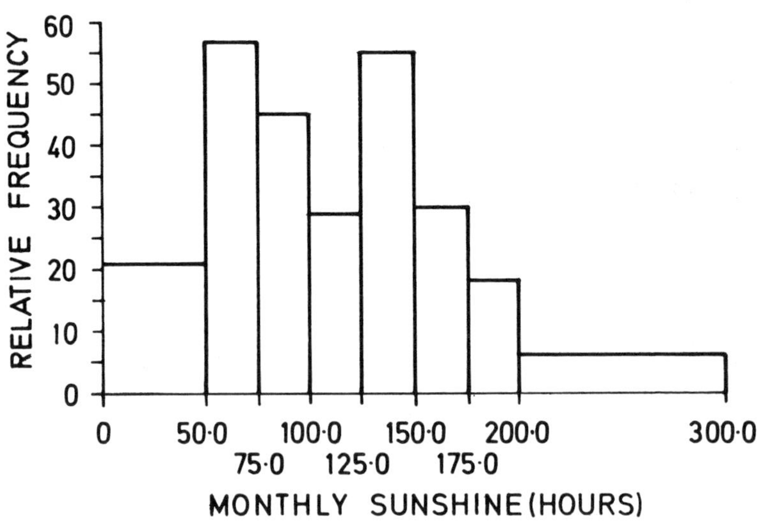


Figure I.5 HISTOGRAM FOR UNEQUAL CLASSES

In figure I.5, attention is drawn to the fact that the height of the bars compensates for the different class intervals to ensure each unit of frequency is represented by the same unit of area. For example, assuming the relative frequency equals the frequency for the class intervals of 25 hours, for class intervals of 50 hours, the relative frequency is half the frequency, and for class intervals of 100 hours, the relative frequency is a quarter of the frequency.

From the frequency tables, such as Table 1.6, it is possible to plot the associated commulative frequency diagram (or ogive). This can be used as a basis to determining the so-called median and inter quartile range of a data set, which are examined below.

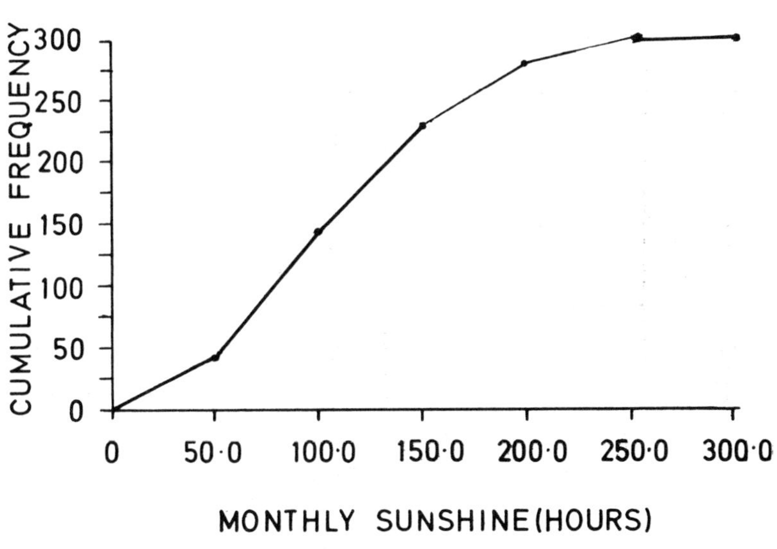


Figure I.6 CUMULATIVE FREQUENCY DIAGRAM FOR EQUAL CLASS INTERVALS

## I.4 Description /Summarising Data: Measures of central Tendency, Dispersion and Association

### I.4.1 *Introduction*

In the preceding discussion about frequency distributions and histograms, the aim was to present simple, visual summaries of large data sets in order to facilitate a comprehension of the original information. It is, however, possible to make further generalisations by summarising data in relation to two, numerical properties: central tendency, and dispersion. In addition, it is useful to be able to describe the characteristics of more than one variable, specifically the relationship between variables. Consequently, a third type of statistic, a measure of association, is discussed. The specific characteristics of the statistics are considered in detail below, and are illustrated through the presentation of a number of examples. It is noted paranthetically that two additional statistics which summarise the characteristics of frequency distribution, the so-called 'skewness' and 'kurtosis', are not included, because they are not in any 'A' level geography course.

Following on from the discussion on data measurement levels, particular emphasis is given to indicating which statistics are appropriate for different measurement levels. It should be remembered that data which are of a high level, such as interval, may, for some reason, be converted to a lower level, and therefore a number of different statistics can be applied to one data set. In general, the statistics that are apposite for high level data are more informative.

### I.4.2 *Measures of Central Tendency*

In this sub-section, a number of measures of central tendency are presented, such as the mean, median and mode which relate to the various levels of data measurement, such as ratio/interval, ordinal and nominal scales, respectively.

#### I.4.2.1 *Mean*

In everyday terminology, the statistic termed the mean is called the "average" - a cricketer's batting average! Although it requires high level data, it is the most common measure of central tendency; it cannot be used with ordinal or nominal data. By convention, the mean of a population (that is, the entire possible data set) is represented by the Greek letter, $\mu$, and the mean of a sample is represented by $\bar{x}$ (In the vast majority of studies, analysis is concerned with samples).

Before formally presenting the mean, it is necessary to introduce a conventional (short-hand) notation which is used throughout this discussion and it is found in introductory geography/statistics text books. By way of an illustration, assume a data set of one variable, angles (in degrees) on a particular scree slope, comprised of seven values.

23; 21; 27; 15; 40; 36; 23.

By convention, the variable, angle of slope, could be denoted by the letter x, and each individual value or variate is represented by a lower case number or subscript which is generally denoted by i. For example,

$x_1 = 23$; $x_2 = 21$; $x_3 = 27$; $x_4 = 15$; $x_5 = 40$; $x_6 = 36$; $x_7 = 23$.

and, it is noted that the value 23 is both $x_1$ and $x_7$. A shorthand representation of this data is,

$$x_i \ (i = 1, 2, \ldots, 7)$$

That is, i is a subscript which has seven values. More generally, if the sample size of the data is n, interest is centred on

$$x_i \ (i = 1, 2, \ldots, n)$$

A second piece of notation that should be comprehended is the sigma sign, $\Sigma$. Simply stated, it means add up all the variates. For example, if the variates to be added are $x_i$, the term

$$\sum_{i=1}^{n} x_i$$

means take the sum of all the subscripted x's, from subscript 1 up to and including subscript n. That is,

$$\sum_{i=1}^{n} x_i = x_1 + x_2 + \ldots\ldots + x_n$$

and

$$\sum_{i=2}^{4} x_i = x_2 + x_3 + x_4$$

Formally, the mean, $\bar{x}$, of a sample set of n variates, $x_i$ $(i = 1, 2, \ldots, n)$, is given as

| Year | Jan. | Feb. | Mar. | Apr. | May | Jun. | Jul. | Aug. | Sept | Oct. | Nov. | Dec. | Yearly Total |
|---|---|---|---|---|---|---|---|---|---|---|---|---|---|
| | ins. | ins. | ins. | ins. | ins. | ins. | ins. | ins. | ins. | ins. | ins. | ins. | ins. |
| 1954 | 1.66 | 2.93 | 2.16 | 0.78 | 2.62 | 4.53 | 3.13 | 4.87 | 3.40 | 5.70 | 5.49 | 3.41 | 40.68 |
| 1955 | 2.63 | 1.67 | 2.64 | 1.57 | 3.14 | 3.85 | .51 | 1.20 | 2.89 | 1.56 | 1.89 | 2.61 | 26.16 |
| 1956 | 3.92 | 0.67 | 1.72 | 1.39 | 0.80 | 2.41 | 5.29 | 5.62 | 2.17 | 2.27 | 1.23 | 2.98 | 30.47 |
| 1957 | 1.66 | 2.36 | 2.78 | 0.40 | 1.13 | 2.07 | 6.00 | 6.62 | 5.69 | 2.51 | 1.65 | 1.66 | 34.55 |
| 1958 | 2.72 | 3.95 | 1.39 | 1.04 | 3.04 | 4.77 | 3.68 | 3.70 | 4.04 | 3.61 | 1.81 | 2.77 | 36.72 |
| 1959 | 3.03 | 0.44 | 1.59 | 3.44 | 1.43 | 1.22 | 1.50 | 1.65 | 0.18 | 1.86 | 3.27 | 5.18 | 24.79 |
| 1960 | 4.94 | 2.23 | 1.35 | 1.33 | 1.87 | 3.71 | 3.83 | 4.57 | 4.66 | 4.31 | 3.96 | 2.60 | 39.36 |
| 1961 | 3.50 | 2.47 | 0.77 | 3.09 | 1.39 | 1.23 | 3.91 | 3.04 | 3.15 | 3.71 | 1.50 | 2.17 | 29.93 |
| 1962 | 2.55 | 1.62 | 1.58 | 2.77 | 3.32 | 1.02 | 2.61 | 3.69 | 3.07 | 1.43 | 1.90 | 2.47 | 28.03 |
| 1963 | 0.50 | 0.27 | 2.08 | 1.58 | 1.52 | 3.36 | 1.35 | 2.61 | 3.06 | 2.44 | 4.02 | 0.47 | 23.26 |
| 1964 | 0.66 | 0.91 | 3.17 | 3.00 | 1.80 | 2.50 | 2.29 | 2.43 | 0.83 | 2.70 | 1.59 | 2.90 | 24.66 |
| 1965 | 3.70 | 0.46 | 2.75 | 2.27 | 2.88 | 2.87 | 3.27 | 3.36 | 5.76 | 0.88 | 3.17 | 6.23 | 37.6 |
| 1966 | 1.94 | 3.35 | 1.80 | 2.95 | 2.87 | 3.55 | 3.88 | 3.82 | 1.99 | 3.93 | 2.20 | 4.96 | 37.24 |
| 1967 | 2.01 | 2.35 | 1.78 | 1.44 | 5.29 | 1.09 | 2.23 | 1.62 | 3.69 | 5.85 | 2.56 | 2.88 | 32.79 |
| 1968 | 3.14 | 1.50 | 1.66 | 2.22 | 3.23 | 2.77 | 3.39 | 2.31 | 3.78 | 2.50 | 2.29 | 1.81 | 30.60 |
| 1969 | 2.38 | 2.57 | 1.39 | 2.64 | 6.82 | 1.63 | 1.88 | 3.85 | 2.72 | 0.45 | 5.08 | 3.05 | 34.46 |
| 1970 | 3.17 | 3.23 | 2.64 | 3.15 | 0.78 | 1.85 | 2.42 | 3.63 | 2.26 | 2.97 | 6.21 | 1.37 | 33.68 |
| 1971 | 2.91 | 1.10 | 2.24 | 3.10 | 2.13 | 4.03 | 1.01 | 5.39 | 0.95 | 2.21 | 3.35 | 0.66 | 29.08 |
| 1972 | 3.62 | 2.34 | 3.32 | 2.63 | 2.95 | 3.79 | 2.10 | 1.76 | 1.92 | 1.41 | 3.56 | 2.91 | 32.31 |
| 1973 | 1.52 | 1.80 | 0.99 | 2.33 | 3.74 | 1.41 | 4.47 | 3.31 | 2.75 | 3.07 | 2.77 | 2.87 | 31.03 |
| 1974 | 3.50 | 2.67 | 1.66 | 0.38 | 1.82 | 2.33 | 3.50 | 3.10 | 5.15 | 3.90 | 4.14 | 1.97 | 34.12 |
| 1975 | 3.49 | 1.07 | 1.93 | 3.29 | 1.42 | 0.61 | 4.31 | 2.22 | 2.09 | 1.15 | 2.13 | 2.73 | 26.42 |
| 1976 | 2.03 | 2.02 | 2.18 | 0.93 | 3.39 | 0.73 | 0.49 | 0.20 | 4.87 | 5.23 | 1.78 | 2.17 | 26.01 |

TABLE I.7 UNIVERSITY OF KEELE: RAINFALL 1954-1976

$$\bar{x} = \frac{\sum_{i=1}^{n} x_i}{n}$$

That is, add up all the variates and divide the total by the number of variates. Correspondingly, the mean of a population is,

$$\mu = \frac{\sum_{i=1}^{n} x_i}{n}$$

For example, from table I.7 which indicates monthly rainfall (inches for Keele (1954-1976), it is easily demonstrated that the mean January level is,

$$\bar{x} = \frac{\sum_{i=1}^{23} x_i}{23} = \frac{61.18}{23} = 2.66 \text{ inches}$$

It has already been indicated that data are generally available in one of two forms: either a list of individual variates or classified into groups. It is therefore, not a surprise to know that it is a simple procedure to calculate the mean of a classified data set. It should be noted that its exact value is dependent on the specific definition of the classes; different class definitions for the same data set would produce slightly different answers. A crude estimate of the mean is likely to be derived if there is a very small number of classes.

Formally, if $x_k$ is the value of the mid-point of the $k^{th}$ class (the mid-point is acceptable intuitively as the representative value of a particular class), $f_k$ is the frequency of occurences in class k and there are c classes, the sample mean is

$$\bar{x} = \frac{\sum_{k=1}^{c} x_k f_k}{\sum_{k=1}^{c} f_k}$$

where, by definition,

$$\sum_{k=1}^{c} f_k = n$$

Formally,

$$M_d = a + b\left(\frac{\frac{n}{2} - c}{f}\right)$$

where a is the lower class limit of the median class; b is the size or class interval of the median class; c is the cumulative frequency for the class immediately lower than the median class (that is, the number of variates less than a; f is the frequency of the median class; and n is the sample size.

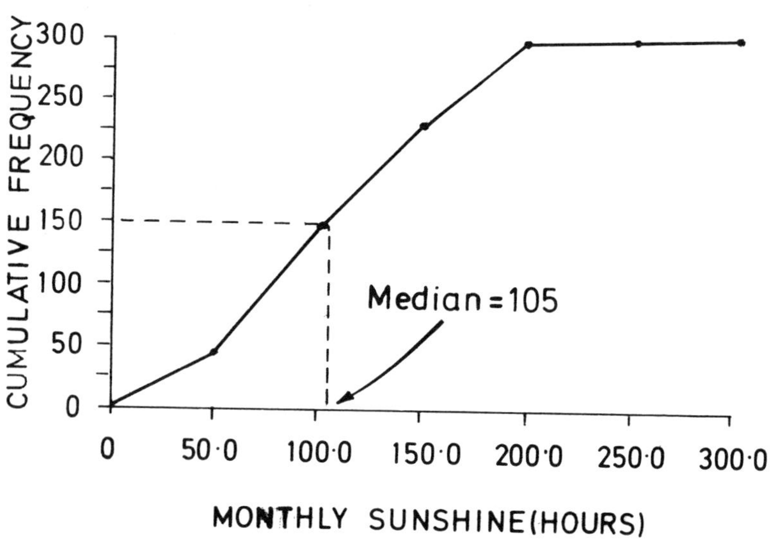


Figure I.7 CUMULATIVE FREQUENCY DIAGRAM TO CALCULATE THE MEDIAN

I.4.2.2 *Median*
If a data set is measured at a lower scale than the interval scale, the mean is an inappropriate measure of central tendency. If ordinal scale data are available, it is the median that should be calculated. For a set of data arranged in ascending (or descending) order of magnitude, the calculation of the median, $m_d$, depends on whether the sample size (n variates) is an odd or even number. If n is an even number, the median is

$$m_d = \frac{x_{(\frac{n}{2})} + x_{(\frac{n}{2}+1)}}{2}$$

and, if n is an odd number it is

$$m_d = x_{\frac{(n+1)}{2}}$$

It is important to reemphasise that the variates are ordered, and, therefore, the subscripts indicate relative values.

As an example, for the following set of data,

12; 13; 15; 19; 27; 29; 30;

the median is 19. The data can be rewritten as,

$x_1 = 12$; $x_2 = 13$; $x_3 = 15$; $x_4 = 19$; $x_5 = 27$; $x_6 = 29$; $x_7 = 30$.

and as n is odd,

$$m_d = x_{(\frac{n+1}{2})} = x_{(\frac{7+1}{2})} = x_4 = 19.$$

For completeness, it is noted that the median can be ascertained by graphical methods as well as by arithmetic calculations. It is based on the cumulative frequency distribution or ogive, and it is determined by inspection. Using the example that illustrated the construction of the cumulative frequency distribution (see Figure I.7), the median can be readily calculated.

This measure of central tendency can be calculated for classified data. Whilst the actual equation is rather daunting, at least at first sight, it is based simply on the idea of ascertaining the median class and then of determining the specific value of the median by linear interpolation (by assuming the variates' values are distributed uniformly across the median class).

I.4.2.3 *Mode*

The mode is a rarely used measure of central tendency, although, if only nominal level data exist, it is the only suitable statistic for measuring central tendency. Simply stated, it is the most frequently occurring variate in a data set. For classified data, the class which has the largest frequency is called the modal class. If there are more than one mode in a data set, the terms bimodal, trimodal and so on are used.

I.4.2.4 *Concluding comments: Measures of Central Tendency*

In addition to reemphasising the fact that the measures of central tendency are measurement level - specific, it is possible to make a number of concluding remarks of a comparative nature. The mean is the most frequently used statistic. It is easy to calculate, and gives equal weight to all variates. Moreover, the mean has been found to be a more stable statistic than the median- that is, if a number of samples are taken from the same population data set, in general, there would be more congruency between the values of the sample means than the values of the sample medians.

However, unlike the mean, the median is unaffected by extreme values. Due to this property, the median is often used as a measure of central tendency if the frequency distribution of a data set is asymmetrical (that is, 'skewed' - see below). For example, the 'average' personal income in a country is often presented as the median personal income, because the income frequency distribution is positively skewed (with a large number of people earning relatively low incomes and a small number of people earning relatively high incomes).

Only when a frequency distribution is perfectly symmetrical do the mean, median and mode posses identical values (see Figure I.8). Frequency distributions which are assymetrical are called skewed. It is usual to differentiate between positively skewed and negatively skewed distributions. Figure I.9 portrays the relationships between the various measures of central tendency for both positively and negatively skewed frequency distributions.

An incorrect interpretation can be given to the values of these statistics if a skewed frequency distribution is examined, and, consequently, it is important to ensure that the form of the frequency distribution is examined before using the descriptive statistics. Calculation of one of the statistics should not replace the graphical presentation of a frequency distribution; they complement each other and both components are necessary for sound description.

Whilst it is unlikely, that any geographical data set would be perfectly symmetrical and unimodal, the application of the measures of central tendency is tolerated if a

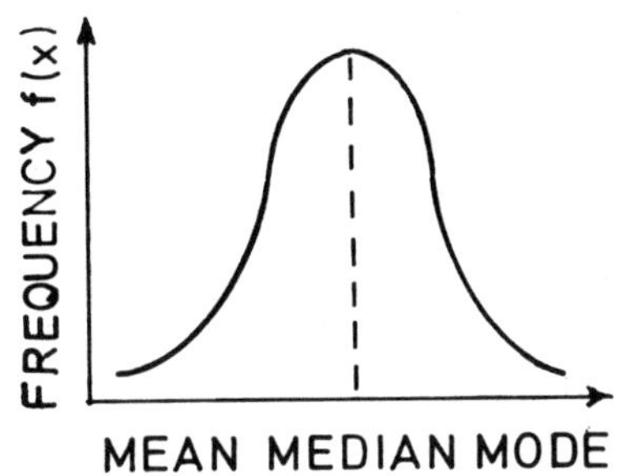


Figure I.8 A SYMMETRICAL FREQUENCY DISTRIBUTION

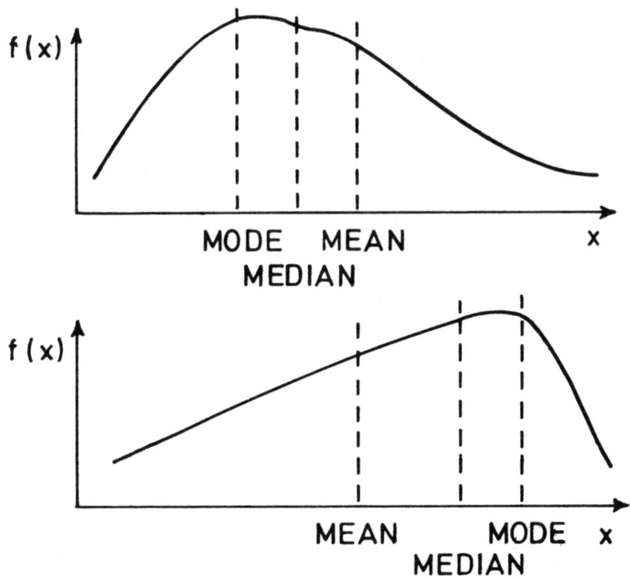


Figure I.9 RELATIONSHIPS BETWEEN MEASURES OF CENTRAL TENDENCY FOR ASSYMMETRICAL FREQUENCY DISTRIBUTIONS

frequency distribution is not too skewed. Figure I.10, for example, presents a bimodal frequency distribution in which the calculation of the mean and median would be meaningless (because it is a rarely occurring variate). In this case the mode is apposite.

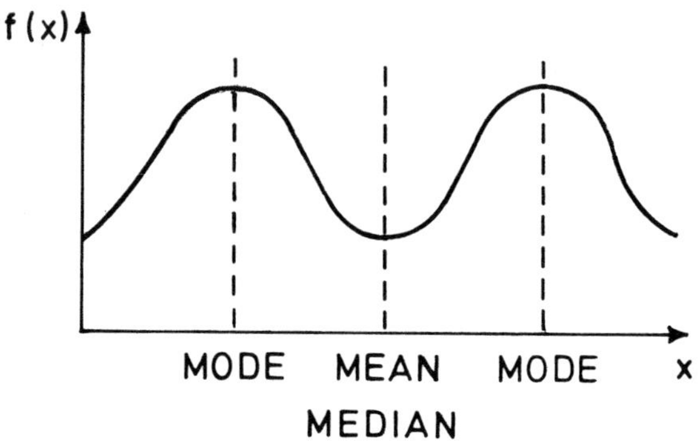


Figure I.10 A BIMODAL FREQUENCY DISTRIBUTION

To date, a number of simple, albeit widely used, measures of central tendency have been introduced; in terms of geography per se, they are very useful for describing many data sets, but it must be appreciated that they are aspatial. However, the concepts of spatial mean, spatial median and spatial mode do exist and can be used to describe spatial distributed data (these statistics are introduced in the next appendix).

### I.4.3 *Measures of Dispersion*

It has been illustrated already that frequency distributions may be different even though they have the same value of central tendency. To give additional description of a data set, some indication of this variation or dispersion around the measure of central tendency is required. The following three samples,

| | |
|---|---|
| Sample A | 7; 7; 7; 7; 7; |
| Sample B | 1; 1; 7; 13; 13; |
| Sample C | 5; 6; 7; 8; 9; |

all have identical means ($\bar{x}$ = 7) but, clearly, the dispersion of the variates around this mean is dissimilar.

As with the concept of central tendency, no single or unambiguous definition of a measure of dispersion exists. Accordingly, a number of the most frequently employed statistics are described. Again, the data's measurement level is used to differentiate between the alternative statistics.

### I.4.3.1 *Standard Deviation*

For any data set measured on either an interval or ratio scale, the measure of dispersion around the sample mean ($\bar{x}$) for any variate $x_i$; is defined as

$$x_i - \bar{x}$$

and, clearly, this deviation can be positive or negative. One intuitive measure of dispersion around the sample mean would be the mean signed deviation (M.S.D.), that is, the 'average' of the deviations. Formally, it is as given by the following equation,

$$\text{M.S.D.} = \frac{\sum_{i=1}^{n} (x_i - \bar{x})}{n}$$

However from the definition of the mean, the numerator is always zero, and therefore, this statistic is useless (because it always equals zero). In fact, the sum of the positive and negative deviations are balancing out each other.

By a simple modification, it is possible to consider the mean absolute deviation (M.A.D.), which as the name suggests, disregards the signs of the deviations and is the mean of the absolute values of deviations around the sample mean. Formally this is,

$$\text{M.A.D.} = \frac{\sum_{i=1}^{n} |x_i - \bar{x}|}{n}$$

where the usual notation for an absolute value, two straight lines, has been used. This statistic involves a simple calculation, but it is rarely applied, because the so-called standard deviation is preferred as it possesses very useful properties for many inferential statistics. An alternative name for the standard deviation, which indicates its actual derivation, is the root mean squared deviation.

That is, a deviation is,

$$x_i - \bar{x}$$

and a squared deviation is,

$$(x_i - \bar{x})^2$$

Following the earlier definition of a mean, the mean squared deviation is

$$\frac{\sum_{i=1}^{n} (x_i - \bar{x})^2}{n}$$

and the root mean squared deviation or standard deviation (s) is,

$$s = \sqrt{\frac{\sum_{i=1}^{n} (x_i - \bar{x})^2}{n}}$$

It is noted that the square of the standard deviation is called the variance, but it is only the standard deviation that is measured in the original units, and this is a major reason for selecting the standard deviation in preference to the variance. The standard deviatiion for a population is represented by $\rho$ and for a sample it is represented by s. In addition, when calculating a sample standard deviation, usually for small samples, the denominator is often given the value (n-1) rather than n. That is,

$$s = \sqrt{\frac{\sum_{i=1}^{n} (x_i - \bar{x})^2}{(n-1)}}$$

Whilst the results are practically the same for large sample sizes, it does introduce an added conservatism for small samples, which could be important when using inferential statistics.

Whilst the above derivation of the equation shows explicitly why the phrase 'root mean squared deviation' is frequently used, for computational efficiency, an alternative formulation of the standard deviation is employed when data sets are large. It gives an identical answer and is much quicker, because it involves less arithmetic operations. Formally, it is given by,

$$s = \sqrt{\frac{\sum_{i=1}^{n} x_i^2}{n} - \left[\frac{\sum_{i=1}^{n} x_i}{n}\right]^2}$$

As a simple example, the standard deviations of the three samples presented at the beginning of this sub-section are calculated. As they are all small samples (n = 5), the equation with a modified denominator is used. For the first sample, it is clear that there is no deviation around the mean, and therefore, the standard deviation is zero. For the second case,

$$n = 5$$

$$\sum_{i=1}^{5} x_i = 35$$

$$\sum_{i=1}^{5} x_i^2 = 389$$

$$\left[\sum_{i=1}^{5} x_i\right]^2 = 1225$$

and, thus

$$s = \sqrt{\frac{1}{4}\left(389 - \frac{1225}{5}\right)} = 6.0$$

For the third case,

$$n = 5$$

$$\sum_{i=1}^{5} x_i = 35$$

$$\sum_{i=1}^{5} x_i^2 = 225$$

$$\left[\sum_{i=1}^{5} x_i\right]^2 = 1225$$

and, thus

$$s = \sqrt{\frac{1}{4}\left(225 - \frac{1225}{5}\right)} = 1.6$$

As to be expected, the standard deviation for the second sample is bigger than that for the third sample.

Standard deviations can be calculated for classified data by using either of the following formulae, where $x_k$ is the mid-value of the $k^{th}$ class.

$$s = \sqrt{\frac{\sum_{k=1}^{c} (x_k - \bar{x})^2 f_k}{n}}$$

Or

$$s = \sqrt{\frac{1}{n}\left[\sum_{k=1}^{c} x_k^2 f_k - \frac{\left[\sum_{k=1}^{c} x_k f_k\right]^2}{n}\right]}$$

where,

$$n = \sum_{k=1}^{c} f_k$$

for c classes.

One property of the standard deviation gives especially useful information about any data set. Independent of the data's frequency distribution,

(i) at least 56 per cent of the variates are within 1.5 standard deviations either side of the mean (that is, within the range of values, $\bar{x}$ - 1.5s to $\bar{x}$ + 1.5s).

(ii) at least 75 per cent of the variates are within 2.0 standard deviations either side of the mean.

(iii) at least 89 per cent of the variates are within 3.0 standard deviations either side of the mean,

and (iv) at least 94 per cent of the variates are within 4.0 standard deviations either side of the mean.

It is noted that whilst these properties characterise any frequency distribution, it is possible to be more specific when a frequency distribution is known.

### I.4.3.2 *Dispersion around the median*

The standard deviation is a measure of dispersion around the mean. Measures of dispersion around the median, such as the range, the inter-quartile range and quartile deviation, also exist, and they are described briefly in this subsection.

The range is a very straightforward statistic; as the name suggests, it is the difference between the highest and lowest variates. For an ordinal data set, such as,

7; 10; 10; 13; 16; 18; 24

the range, R, is

$$R = 24-7 = 17$$

Due to its complete dependence on extreme variates, it does not offer a particularly useful description, although it may have meaning for 'standards' or bands of control. For classified data, the range is defined as the difference between the upper boundary of the highest class and the lower boundary of the lowest class.

Remembering that for a set of n (n is an odd number) ordered variates, $x_i$, the median is the $(\frac{n+1}{2})^{th}$ variate, the interquartile range (I.Q.R.), is defined as the numerical difference between the upper quartile, $x_{(\frac{3n}{4})}$, and the lower quartile $x_{(\frac{n}{4})}$; that is

$$I.Q.R. = x_{(\frac{3n}{4})} - x_{(\frac{n}{4})}$$

and if the subscripts, $(\frac{3n}{4})$ and $(\frac{n}{4})$, are not integers, it is usual to approximate them by the nearest whole number. The quartile deviation, Q.D, is

$$Q.D. = \frac{I.Q.R.}{2}$$

and, by comparing the values of these three statistics, it is possible to gain insights into the deviations around the median.

### I.4.3.3 *Dispersion around the mode*

The variation ratio (V.R.) is a statistic that measures the degree to which the mode of a given frequency distribution is representative of all the variates. It is a crude, rarely used, statistic, but it is the only measure of dispersion for nominal data. Formally, it is defined as,

$$V.R. = 1 - \frac{f_{modal}}{n}$$

where $f_{modal}$ is the frequency of the mode or modal class, and n is the sample size. Clearly, by definition, the more representative the mode is of a frequency distribution, the lower the value of the variation ratio. The variation ratio can take a range of values,

$$0 \leqslant V.R. < 1$$

## I.5 Measures of Association

### I.5.1 *Introduction*

To date, concern has been with single variable data sets. Geographers often want to analyse the association between two variables, such as rainfall and runoff, and, for this reason, correlation techniques have been applied widely. In examining the relationships between two variables (such as the number of children (x) and the number of schools (y), over a set of individuals (such as towns in England with a population of greater than 10,000), particular attention is given to how one variable alters as the other one changes. Thus, for each town, there is a particular number of children and a number of schools. In most statistical text-books, correlation and regression are considered together; they should be treated separately because the former is concerned with the strength and direction of associations between variables, whereas the latter is more specific, involving causal, functional relationships between variables. For example, knowledge, of the correlation between rainfall and runoff would be treated as an empirically observed association which would be used for prediction. On the other hand, the causal relationship, rainfall influences runoff, can be analysed with regression techniques. Whilst it is obviously important to attempt to understand and explain associations, empirically observable associations can act as a basis for prediction by themselves. In regression analysis, the specification of the independent (explanatory) and dependent (response) variables is of fundamental importance, but in simple linear correlation, this is not essential.

A number of statistics are available to measure the strength and direction of an association between two variables. It should be appreciated, at the outset, that attention focuses on *linear* associations. Continuing the explicit recognition of the fact that statistical techniques are specific to particular measurement levels, the various correlation coefficients are distinguished on this basis and are described systematically. The most frequently applied correlation coefficient is Pearson's so-called product moment correlation coefficient which is only appropriate for data available at the interval and ratio scales. For ordinal level data, a choice of statistics is available: Spearman's rank correlation coefficient and Kendall's tau. Finally, for nominal level data, the contingency coefficient or the phi-coefficient can be used.

The association between variables, x and y, is visually portrayed in the form of scatter diagrams.

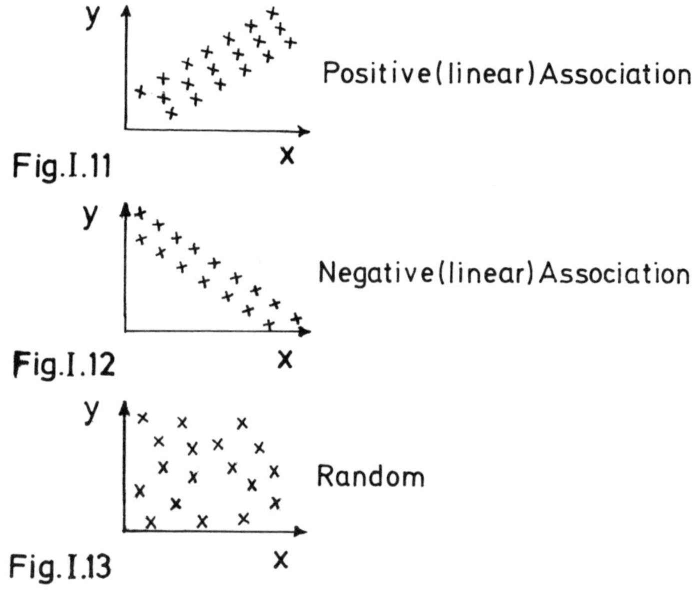


Correlation coefficients are single statistics which present a numerical value of the strength and direction of an association. The most commonly employed coefficients are discussed below.

I.5.2 *Pearson's Product-Moment Correlation Coefficient*
The most widely applied correlation coefficient, Pearson's product-moment correlation coefficient (r), is only applicable when interval or ratio (high-level) data are available. Like the majority of correlation coefficients, r ranges between

$$-1 \leqslant r \leqslant 1$$

The two extremes are a perfectly positive linear association (r = +1) and a perfectly negative linear association (r = -1); when r equals zero there is no linear association between the two variables. r is the sample product-moment correlation coefficient, and $\rho$ (rho) is the bivariate population's product-moment correlation coefficient. (An underlying assumption of the statistic is that both variables are from normally distributed populations).

The term 'product-moment' describes the derivation of this statistic. From the discussions of standard deviation, the concept of moment or deviation around the mean is familiar to the reader. Indeed, for example, the sample variance, $s^2$, is the mean squared deviation (or second moment).

$$s^2 = \frac{1}{n} \sum_{i=1}^{n} (x_i - \bar{x})^2 = \frac{1}{n} \sum_{i=1}^{n} (x_i - \bar{x})(x_i - \bar{x})$$

For two variables, $x_i$ and $y_i$, rather than one variable $x_i$, the mean of the product of their deviations around their means is known as the covariance, cov (x, y).

$$\text{Cov}(x, y) = \frac{1}{n} \sum_{i=1}^{n} (x_i - \bar{x})(y_i - \bar{y})$$

The fact that the covariance is a measure of the linear association between the two variables x and y is graphically demonstrated below. A scatter diagram is divided into four quadrats on the basis of the sample means of both variables, $\bar{x}$ and $\bar{y}$, around which the deviations of the variates are measured. Given that an association relates to how x changes when y changes (or vice-versa), if either x or y has a constant value, no linear association exists.

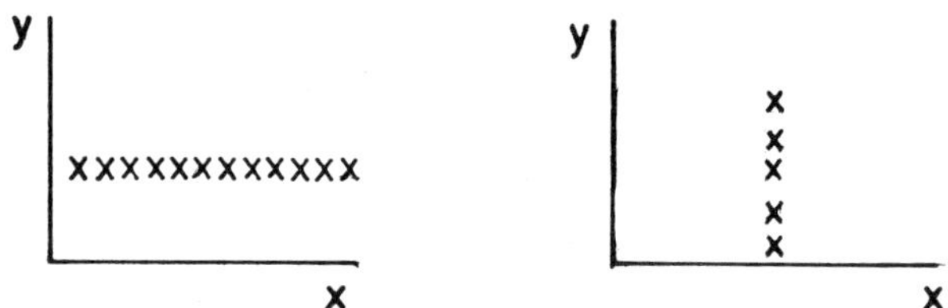


Figure I.14 NO LINEAR ASSOCIATION

Removing these two situations leaves two other examples of linear associations. The first when the points are primarily located in quadrants I and III, and the second when the points are primarily located in quadrants II and IV. In the former, the product

$$\sum_{i=1}^{n} (x_i - \bar{x}) \ (y_i - \bar{y})$$

is positive, and, in the latter, this product is negative. Thus, the sign of the covarience indicates the direction of the association, and the strength is dependent on the absolute value. As the units of measurements are those of the x and y variables, the value of the covariance cannot be easily used for comparisons. Standardisation of the covariance into a dimensionless quantity to facilitate direct comparison is achieved by dividing it by the sample standard deviations of x and y, $s_x$ and $s_y$, respectively. This gives Pearson's product moment correlation coefficient, r,

$$r = \frac{\text{Cov}\ (x,\ y)}{s_x \ s_y}$$

which is clearly the proportion of the products of the two standard deviations represented by the covariance. To assist calculations, a computational formula is available which reduces the number of arithmetic operations required.

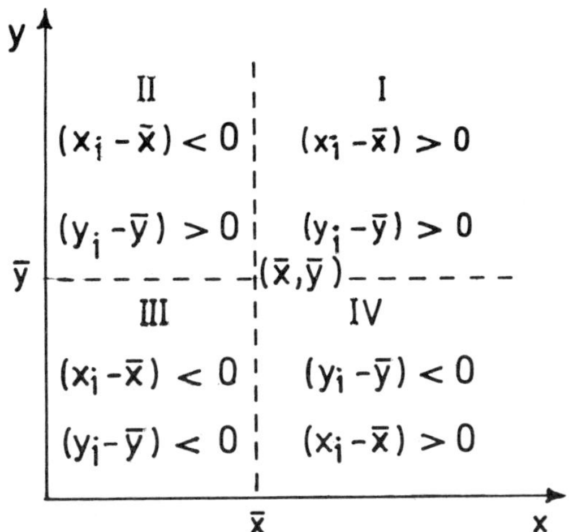


Figure I.15 A DEMONSTRATION THAT THE COVARIANCE IS A MEASURE OF THE LINEAR ASSOCIATION BETWEEN x AND y

$$r = \frac{\sum_{i=1}^{n} x_i y_i - \frac{\sum_{i=1}^{n} x_i \sum_{i=1}^{n} y_i}{n}}{\sqrt{\left[\sum_{i=1}^{n} x_i^2 - \frac{\left[\sum_{i=1}^{n} x_i\right]^2}{n}\right]\left[\sum_{i=1}^{n} y_i^2 - \frac{\left[\sum_{i=1}^{n} y_i\right]^2}{n}\right]}}$$

The square of this correlation coefficient, $r^2$, is defined as the coefficient of determination. Simply stated, it is the proportion of the total variations (or variance) on one variable, say y, which can be explained by the linear association between x and y (whether it be positive or negative), the variation in one variable accounts for the total variability in the other.

To illustrate this correlation coefficient, the association between rainfall and sunshine levels is examined (see Table I.8). The correlation is

$$r = \frac{\sum_{i=1}^{36} x_i y_i - \frac{\sum_{i=1}^{36} x_i \sum_{i=1}^{36} y_i}{n}}{\sqrt{\left[\sum_{i=1}^{36} x_i^2 - \frac{\left[\sum_{i=1}^{36} x_i\right]^2}{n}\right]\left[\sum_{i=1}^{36} y_i^2 - \frac{\left[\sum_{i=1}^{36} y_i\right]^2}{n}\right]}}$$

$$r = \frac{8847 - \frac{(86.58)\ (4199.7)}{36}}{\sqrt{\left(272.12 - \frac{7496.1}{36}\right)\left(652031.99 - \frac{17637480.09}{36}\right)}}$$

$$r = \frac{8847.4 - 10100.3}{\sqrt{(63.9)\ (162101.9)}}$$

$$r = -0.39.$$

The correlation coefficient shows a negative association between rainfall and sunshine, which is, obviously, expected.

Thus a correlation coefficient indicates the magnitude (by its absolute value) and the direction (by its sign) of an association between two variables.

It is of paramount importance to remember that attention is restricted to linear association; it is impossible to say whether a non-linear association exists or not.

| $x_i$ | $y_i$ | $x_i^2$ | $y_i^2$ | $\Sigma x_i y_i$ |
|---|---|---|---|---|
| 3.50 | 53.4 | 12.25 | 2851.56 | 186.9 |
| 2.67 | 45.9 | 7.13 | 2106.81 | 122.5 |
| 1.66 | 107.4 | 2.76 | 11434.76 | 178.3 |
| 0.38 | 161.8 | 0.14 | 26179.24 | 61.5 |
| 1.82 | 184.9 | 3.31 | 34188.01 | 336.5 |
| 2.33 | 175.3 | 5.43 | 20730.09 | 408.5 |
| 3.50 | 161.3 | 12.25 | 26017.69 | 563.5 |
| 3.10 | 177.8 | 9.61 | 31612.84 | 551.2 |
| 3.90 | 82.9 | 15.21 | 6872.41 | 323.3 |
| 4.14 | 50.9 | 17.14 | 2590.81 | 210.7 |
| 1.97 | 53.9 | 3.88 | 2905.21 | 106.2 |
| 3.49 | 33.5 | 12.18 | 1122.25 | 116.9 |
| 1.07 | 52.1 | 1.14 | 2714.41 | 55.7 |
| 1.93 | 86.0 | 3.72 | 7396.0 | 165.9 |
| 3.29 | 111.4 | 10.82 | 12409.96 | 366.5 |
| 1.42 | 208.0 | 2.02 | 43264.0 | 295.4 |
| 0.61 | 247.8 | 0.37 | 61404.84 | 151.2 |
| 4.31 | 190.6 | 18.57 | 37328.36 | 821.5 |
| 2.22 | 221.9 | 4.93 | 49239.61 | 492.6 |
| 2.09 | 147.6 | 4.37 | 21785.76 | 308.5 |
| 1.15 | 96.1 | 1.32 | 9235.21 | 110.5 |
| 2.13 | 79.4 | 4.54 | 6304.36 | 169.1 |
| 2.73 | 29.9 | 7.45 | 894.01 | 81.6 |
| 2.03 | 39.6 | 4.12 | 1568.16 | 80.4 |
| 2.02 | 51.2 | 4.08 | 2621.44 | 103.4 |
| 2.18 | 88.4 | 4.75 | 7814.56 | 192.7 |
| 0.93 | 116.9 | 0.86 | 13665.61 | 108.7 |
| 3.39 | 133.5 | 11.49 | 17822.25 | 452.5 |
| 0.73 | 219.1 | 0.53 | 48004.81 | 159.9 |
| 0.49 | 230.0 | 0.24 | 52900.0 | 112.7 |
| 0.20 | 226.7 | 0.04 | 51392.89 | 45.3 |
| 4.87 | 73.9 | 23.72 | 5461.21 | 359.9 |
| 5.23 | 50.5 | 27.35 | 2550.25 | 264.4 |
| 1.78 | 60.4 | 3.17 | 3648.16 | 107.5 |
| 2.17 | 31.9 | 4.71 | 1017.61 | 69.2 |
| 5.15 | 117.8 | 26.72 | 13876.84 | 606.6 |
| 86.58 | 4199.7 | 272.12 | 652031.99 | 8847.1 |

TABLE I.8: MONTHLY RAINFALL ($x_i$) and SUNSHINE ($y_i$) FIGURES FOR KEELE 1974-1976[1]

### I.5.3 *Correlation Coefficients for Ordinal Data*

A number of different correlation coefficients are appropriate statistics for ordinal level data. The most commonly applied statistic is Spearman's rank correlation coefficient, (and it is the only one 'A' level geographers need to understand), although Kendall's tau is also used. The latter is based on changes in rank of the two variables for the individuals, whereas the former emphasises the differences between the ranks by squaring these differences. Whilst their particular values may be different when applied to the same data set, it is unlikely that the results will be significantly different. Spearman's rank correlation coefficient is illustrated for a simple example which is concerned with the association between rainfall ($x_i$) and sunshine ($y_i$) levels at Keele in 1954. This monthly data is ranked in terms of increasing rainfall and sunshine.

| 1954 | $x_i$ | $y_i$ | $(x_i - y_i)$ or $d_i$ | $d_i^2$ | |
|---|---|---|---|---|---|
| Jan | 2 | 1 | -1 | 1 | |
| Feb | 5 | 4 | 1 | 1 | |
| Mar | 3 | 6 | -3 | 9 | |
| April | 1 | 12 | -11 | 121 | |
| May | 4 | 9 | -5 | 25 | $\sum_{i=1}^{12} d_i^2 = 328$ |
| June | 9 | 10 | -1 | 1 | |
| July | 6 | 7 | -1 | 1 | |
| Aug | 10 | 8 | 2 | 4 | |
| Sept | 7 | 11 | -4 | 16 | |
| Oct | 12 | 5 | 7 | 49 | |
| Nov | 11 | 3 | 8 | 64 | |
| Dec | 8 | 2 | 6 | 36 | |

TABLE I.9 RAINFALL AND SUNSHINE LEVELS AT KEELE: 1954

The n (n=12) observations for both variables have been ranked separately. The difference between their ranks, $d_i$, is squared, and summed over the n individuals. It is then placed into the formula for Spearman's rank correlation coefficient, $r_s$,

$$r_s = 1 - \frac{6 \sum_{i=1}^{n} d_i^2}{n^3 - n}$$

for example,

$$r = 1 - \frac{6 \times 328}{1728 - 12} = -0.15$$

It is noted that this formula is only correct when there are no tied ranks (that is, all the differences must be non-zero). However, if there are only a small number of tied ranks, the result of the 'corrected' equation is usually not very different.

### I.5.4 *Correlation Coefficients for nominal level data*

At the lowest level of measurement, the nominal, it is possible to determine the association between two variables. Two statistics, the contingency coefficient, c, and the phi coefficient, $\phi$, can be used. Both incorporate the chi-square ($\chi^2$) statistic which is described separately below. A problem with the contingency coefficient is that, whilst it has a value of zero to indicate no linear association, its upper limit varies and is only asymptotic to one. Indeed, for the common 2 by 2 contingency table case, its maximum value is 0.707, and therefore interpretation of this statistic is not straight forward.

Formally, the contingency coefficient is defined as,

$$c = \sqrt{\frac{\chi^2}{N + \chi^2}}$$

where N is the total frequency. To illustrate this statistic the following problem of shopping preferences among social social classes is considered.

Formally, the null and alternative hypotheses are

$H_o$ : no association between social class and shopping habits

$H_1$ : an association between social class and shopping habits

To calculate the $\chi^2$ statistic we make use of the fact that the expected frequencies under such an hypothesis are:

$$E_i = \frac{\text{Row total x column total}}{\text{Grand total}}$$

| Choice of Shopping area | SOCIAL CLASS | | | | TOTAL |
|---|---|---|---|---|---|
| | I-II | III | IV | V | |
| City centre | 23 | 40 | 16 | 2 | 811 |
| District area | 11 | 75 | 107 | 14 | 207 |
| Neighbourhood | 1 | 31 | 60 | 10 | 102 |
| | 35 | 146 | 183 | 26 | 390 |

TABLE I.10: SHOPPING PREFERENCES AMONG SOCIAL CLASSES

| Observed Frequency ($O_i$) | **Expected** Frequency ($E_i$) | $O_i - E_i$ | $(O_i - E_i)^2$ | $\frac{(O_i - E_i)^2}{E_i}$ |
|---|---|---|---|---|
| 23 | 7.3 | 15.7 | 246.49 | 33.77 |
| 11 | 18.6 | -7.6 | 57.76 | 3.11 |
| 1 | 9.1 | -8.1 | 65.61 | 7.21 |
| 40 | 30.3 | 9.7 | 94.09 | 3.11 |
| ' | ' | ' | ' | ' |
| ' | ' | ' | ' | ' |
| ' | ' | ' | ' | ' |
| ' | ' | ' | ' | ' |
| 14 | 13.8 | 0.2 | 0.04 | 0.003 |
| 10 | 6.8 | 3.2 | 10.24 | 1.51 |
| | | | | $\chi^2 = 69.2$ |

TABLE I.11: SHOPPING PREFERENCES: $\chi^2$ calculations

Substituting the cbi-square statistic into the formula we get,

$$c = \sqrt{\frac{69.2}{390 + 69.2}} = 0.39$$

It should be noted that for a 3 by 3 table, the maximum value of c is 0.816 not 1.0. The difference between $\chi^2$ - statistic and the contingency coefficient is that the former tests whether there is a difference and the latter tests the degree of difference.

In many respects, for nominal data, the phi coefficient is more useful. Not only does it range between +1 and -1 with zero indicating no association, but also, for the analysis of two binary variables, a very straight forward computational equation exists.

For the particular 2 by 2 contingency table situation, where the first variable has two possibilities I and II, and the second variable has two possibilities, x and y, the frequencies of the different occurences can be tabulated in the following form.

FIRST VARIABLE

| SECOND VARIABLE | I | II | TOTAL |
|---|---|---|---|
| Y | A | B | A + B |
| Y | C | D | C + D |
| TOTAL | A + C | B + D | |

TABLE I.12: TABULATION OF FREQUENCIES OF DIFFERENT OCCURENCES

The phi-coefficient, $\phi$ is defined as

$$\phi = \frac{AD - BC}{\sqrt{(A + B)\ (C + D)\ (A + C)\ (B + D)}}$$

If, for instance, we were interested in the association between the sex of students and those selecting to read geography, a random sample of individuals may provide the following data set.

| SUBJECT | SEX: Male | Female | TOTAL |
|---|---|---|---|
| Geography | 67 | 46 | 113 |
| Not Geography | 99 | 72 | 171 |
| TOTAL | 166 | 118 | |

Thus,

$$\phi = \frac{(67 \times 72) - (99 \times 46)}{\sqrt{(67+46)\ (99+72)\ (67+99)\ (46+72)}}$$

$$\phi = \frac{4824 - 4554}{\sqrt{(113)\ (171)\ (166)\ (118)}}$$

$$\phi = \frac{270}{378498924}$$

$$= 0.0000007$$

This indicates no association.

The more general forumula for calculating the phi-coefficient is

$$\phi = \frac{\chi^2}{N\ (m - 1)}$$

Where N is total number of individuals and m is the number of rows or columns (whichever is smaller).

### I.5.5 *Some Difficulties in Interpreting Correlation Coefficients*

It should be emphasised that the correlation coefficients only measure the strength of linear associations and therefore, whilst a scatter diagram such as the figure below indicates a strong non-linear association between the variables x and y, the correlation coefficient is approximately zero. Whilst it may be possible to transform the variables to get a linear association, difficulties of interpretation often occur.

In addition, if the correlation coefficient is approximately zero, it does not mean that a non-linear relationship exists. For example, the following figures shows two scatter diagrams in which there are no association between the two variables.

Particular care must be given to the interpretation of results. The value of a correlation coefficient gives the strength and direction of a linear association, and unlike regression analysis, no direct indication of causality is given although it may be possible to infer which processes are operating as in the rainfall-run off example.

Moreover, two variables, may be associated because both are adjusting to a third variable. It is necessary to think about the specific association one is considering; it is no good selecting any two variables!

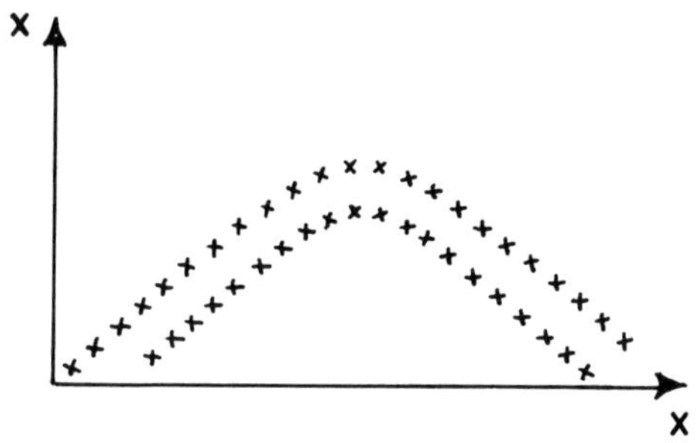


Figure I.16 NON-LINEAR ASSOCIATION

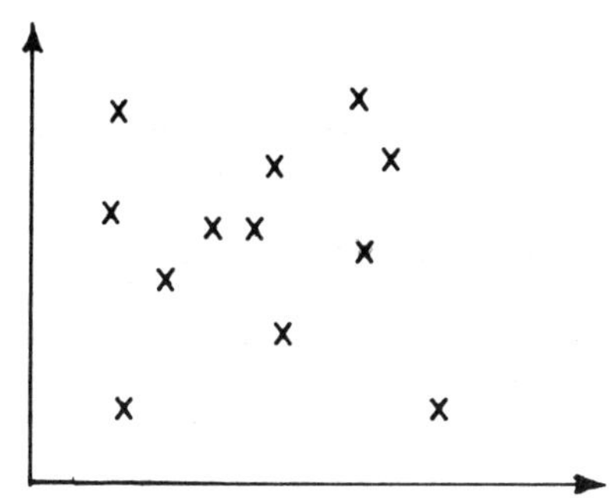

Figure I.17 A RANDOM PATTERN

Other spurious correlations may arise from the definition of variables. For example, this is especially true when ratio variables with a common denominator are analysed from the census.

Finally, and of particular interest to geographers, there is the difficulty of so-called ecological correlation. This problem arises from modifiable areal units; correlation coefficients are (areal) scale-specific and any inferences made at a particular level of resolution should not be transferred directly to another one. A much cited example of this problem is the study of the linear association between wheat and potato yields in the English counties in 1936. As Table I.13 indicates, different values of correlation coefficients exist for different spatial scales of analysis.

| Number of Areas | Correlation Coeff. |
|---|---|
| 48 | + 0.2189 |
| 24 | + 0.2963 |
| 12 | + 0.5757 |
| 6 | + 0.7649 |
| 3 | + 0.9902 |

TABLE I.13: CORRELATIONS BETWEEN WHEAT AND POTATO YIELDS

Following Silk's (1979) illustration, enormous care must be given to the interpretation of data which is in the form of a closed number system, such as percentages. Table I.14 (a) gives the area of land under three types of land use for four regions. Table I.14 (b) gives the corresponding percentage of each type of land use for each region. The correlation coefficient between the land-use types are given for both examples. Clearly, there is a change from a strong positive association to a strong negative association for two of the correlation coefficients.

### I.5.6 *Chi-Square Statistic*

The chi-square statistic is the basis of correlation coefficients for nominal data, and, consequently, students should be familiar with it. In addition, in its own right, it is a very useful statistic for hypothesis testing when data are in the form of a contingency table. Objects are classified according to two criteria, one criterion being entered in the

TABLE I.14 (a) Open Number System

| | Land Type | | |
|---|---|---|---|
| Region | Arable (1) | Pasture (2) | Scrub (3) |
| A | 5 | 2 | 1 |
| B | 7 | 4 | 2 |
| C | 8 | 6 | 4 |
| D | 10 | 8 | 5 |

$r_{12}$ = C.99 $r_{13} = 0.96$ $r_{23} = 0.89$

(b) Closed Number System

| | Land Type | | | |
|---|---|---|---|---|
| Region | Arable (1) | Pasture (2) | Scrub (3) | TOTAL |
| A | 63 | 25 | 12 | 100 |
| B | 54 | 31 | 15 | 100 |
| C | 44 | 33 | 23 | 100 |
| D | 43 | 35 | 22 | 100 |

$r_{12} = {}^{-}0.97$ $r_{1} = {}^{-}0.98$ $r_{23} = 0.89$

| | Sex | | |
|---|---|---|---|
| | Male | Female | TOTALS |
| Working age | 1605,000 | 2370,000 | 3975,000 |
| Not of working age | 1045,000 | 1630,000 | 2675,000 |
| TOTALS | 2650,000 | 4000,000 | 6650,000 |

SOURCE: Townsend (1979)

TABLE I.15: INCIDENCE OF DISABLEMENT: U.K.

rows and the other in the columns. (The number of rows of a contingency table is denoted by i and the number of columns is denoted by j; the data is presented as i x j contingency table). The basic hypothesis that is tested is whether the two criteria are independent of each other. The expected frequencies under this hypothesis are compared with the observed frequencies from collected data.

As an illustration of this form of hypothesis testing and as a means of introducing the mechanics of calculating the statistic, information on the incidence of disablement is provided in a 2 by 2 contingency table involving sex and working age. (See Table I.15)

Simply stated, is there any indication that women are more susceiptible to disablement than men (at the 95% probability level - that is, there is only 5% probability that the outcome is the result of chance)? Cursory inspection of the table suggests that women are more suceptible but it is sensible to undertake a formal statistical test. Given this independence hypothesis, the proportion of disabled men from all men is expected to be equal approximately to the proportion of disabled women from all women, and also the same as the proportion of all disabled people from the total number of people.

The test involves a comparison of expected and observed frequencies of each category, and expected frequencies in a $\chi^2$ test are determined by taking the sum of the column of interest, multiplying it by the sum of the row of interest and dividing this product by the total number of observations. For example, the expected frequency of disabled males of working age (which would be compared with the observed frequency of 1605,000) is,

$$\frac{2650{,}000 \times 3975{,}000}{6650{,}000} = 1584022.6$$

Formally, the $\chi^2$ statistic is given as

$$\chi^2_{\text{d.f.}} = \sum_{i=1}^{k} \frac{(O_i - E_i)^2}{E_i}$$

where d.f. are the degrees of freedom (see below), $O_i$ and $E_i$ are the observed and expected frequency for category i and there are k categories; for an i x j contingency table, the number of degrees of freedom is,

$$\text{d.f.} = (i - 1) \times (j - 1)$$

Thus, for this specific example there is one degree of freedom. To actually calculate the statistic, it is useful to construct the following table.

| $O_i$ | $E_i$ | $O_i-E_i$ | $(O_i-E_i)^2$ | $(O_i-E_i)^2/E_i$ |
|---|---|---|---|---|
| 1605,000 | 1584022.6 | 20977.4 | 440051310.8 | 277.8 |
| 2377,000 | 2390977.4 | -13977.4 | 195367710.8 | 81.7 |
| 1045,000 | 1065977.4 | -20977.4 | 440051310.8 | 412.8 |
| 1630,000 | 1609022.6 | 20977.4 | 440051310.8 | 273.5 |
| 6650,000 | 6650,000 | | | 1046.1 |

TABLE I.16: CALCULATION OF CHI-SQUARED STATISTIC

It is noted that the total expected frequency equals the total observed frequency and this can be used as a simple check to ensure correct calculations.

The calculated chi-squared statistic

$$\chi^2_1 = 1046.1$$

is then compared with the critical value obtained from widely available statistical tables. Using a non-directional test at the 95% significance level (0.05), the critical value is 3.84. As the calculated value is greater than the critical value, the hypothesis that sex and age are independent is rejected.

For completeness, it is noted that, in the calculation of the $\chi^2$ statistic, the expected frequency for each category must be greater than five. If this is not the case, categories have to be joined together.

## I.6 Simple Linear Regression

### I.6.1 *Introduction*

In the last section, particular attention was given to the strength and direction of the association (or correlation) between two variables. If a significant correlation exists, it would be useful to know more about the functional relationship between the two variables. If prior theoretical knowledge indicates the direction of causality between two variables (say rainfall affects run-off, but not vice-versa), it is possible to specify which is the independent or explanatory variable (x) and which is the dependent or response variable (y). In the above example, rainfall is the independent variable and run-off is the dependent variable. A functional relationship is written as

$$y = f(x)$$

Regression analysis has been developed to determine such functional relationships. In this section, simple (bivariate) linear regression is described in detail, and some extensions are briefly mentioned.

Simply stated, regression analysis indicates the rate of change of a dependent variable for unit changes of the independent variable. Such mathematical statements are not used wholly for description; predicting and forecasting are also important and inferential tests can be undertaken to see how closely the regression model fits the data. In general, regression analysis requires data measured at either the interval or ratio levels (although Hammond and McCullagh (1978) present simpler methods to be applied at lower levels of measurement).

Restricting ourselves to the bivariate situation, the functional relationship may often result in the neglect of other important influences, say $x_1$, $x_2$ and $x_n$. These factors and measurement error are assumed away under an error term, $\varepsilon$. Accordingly, total specification of the functional relationship is

$$y = f(x,\varepsilon)$$

and, as it is assumed that the relationship is linear, the population's regression line is

$$y_i = \alpha + \beta x_i + \varepsilon_i$$

$\alpha$ and $\beta$ are regression parameters representing the intercept and slope of the line, respectively. Figure I.18, for example, shows two types of linear relationship. In both cases, the intercept, $\alpha$ is the value of y when x equals zero. $\beta$, the gradient of the line, measures the rate of change of y for a unit change in x. The downward-sloping line (case (i)) has a negative $\beta$ value and the upward sloping line (case (ii)) has a positive $\beta$ value; the sign of the $\beta$ coefficient is the same as that of the correlation coefficient. The regression line is, therefore, completely specified by

$$y_i = \alpha + \beta x_i$$

and, accordingly, regression analysis involves the determination of the values of $\alpha$ and $\beta$.

To date, the regression model has been describing the population. Invariably, however, analysis is concerned with a sample. In this situation, the regression coefficients are estimates, and this fact is represented by including a circumflex, $\hat{\alpha}$ and $\hat{\beta}$. Clearly, from the regression model, the dependent (or response) variable is also an estimate ($\hat{y}_i$) of the actual value ($y_i$) for a particular value of the independent (or explanatory) variable ($x_i$). The regression model for a sample is

$$\hat{y}_i = \hat{\alpha} + \hat{\beta}x_i$$

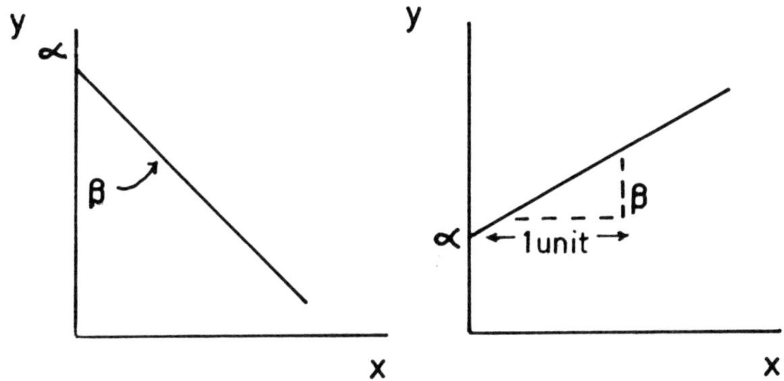


Figure I.18 REGRESSION LINES

I.6.2 *Determination of the regression coefficients*

The determination of the regression coefficients is obviously dependent on the criterion employed to fit a straight line to a scattergram. There is only a unique straight line when there is a perfect relationship between the two variables (that is, the correlation coefficient has an absolute value of one).

One possible criterion is to place the straight line so that there is as many points above the line as there is below it. Such a procedure does not define one unique line for a particular data set; a number of different lines can be drawn which satisfy this criterion.

Attention focuses on the deviations between the actual and the estimated (or predicted) values of the dependent variable, $y_i$ and $\hat{y}_i$, respectively. Figure I.19 shows these vertical deviations around the regression line (and

conceptually, the idea is analogous

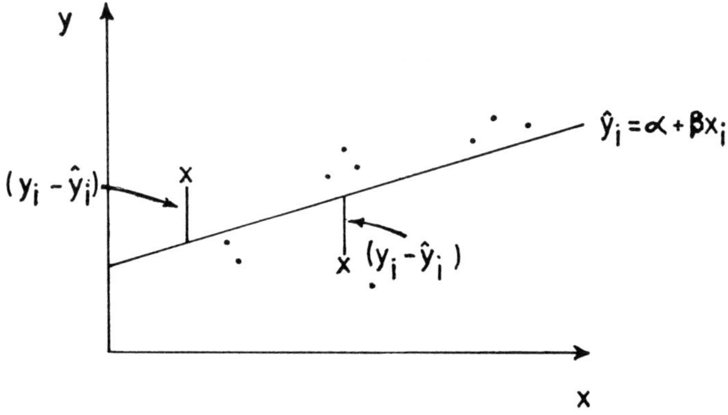


Figure I.19 LEAST SQUARES METHOD

to considering the various deviations around the mean such as the standard deviation, skewness, and kurtosis). Clearly, the "best" line is one in which such deviations between the estimated and actual value of the dependent variables are minimised. The minimisation of the sum of these deviations for n observations

$$\min \sum_{i=1}^{n} (y_i - \hat{y}_i)$$

does not produce a unique line, and the minimisation of the sum of the absolute deviations

$$\min \sum_{i=1}^{n} |y_i - \hat{y}_i|$$

proves computationally difficult. Instead, a straight line is fitted on the basis of minimising the sum of the squared deviations - the so called least sqaures method

$$\min \sum_{i=1}^{n} (y_i - \hat{y}_i)^2$$

The regression coefficients for a unique line are readily determined using simple differential calculus (although it is sufficient to know the equations rather than their derivation). It is noted that, by minimising the sum of square deviations, particular weight is given to the removal of large deviations.

$$\hat{\beta} = \frac{\sum^{n} x_i y_i - n \bar{x} \bar{y}}{\sum_{i=1}^{n} x_i^2 - n \bar{x}^2}$$

Given $\hat{\beta}$, $\hat{\alpha}$ is easily determined,

$$\hat{\alpha} = \bar{y} - \hat{\beta} \bar{x}$$

I.6.3 *A Final Comment*

The use of this simple regression analysis is based on the fundamental assumption that a linear functional relationship exists between two variables, an independent variable ($x_i$) and a dependent variable ($y_i$) which are measured on either the interval or ratio scales.

It should, however, be noted that the variables of a non-linear (or curvilinear) relationship can often be transformed so that the newly defined variables are linearly related (see Chapter Seven). If a curvilinear relationship is described on a scattergram, and the (linear) correlation coefficient is calculated, this statistic will under estimate the association between two variables.

Appendix II

# AN INTRODUCTION TO SPATIAL STATISTICS FOR ADVANCED LEVEL GEOGRAPHY

## II.1 Introduction

Whilst the statistics discussed in the last appendix are of enormous use in Geography, they are not spatial statistics *per se*. During the last two decades, however, a range of important methods for spatial analysis have been developed. Although the majority of the methods are outside the scope of this discussion, it is appropriate to introduce a number of the elementary, albeit potentially useful, methods. Following the structure adopted in the preceding appendix, specific attention focuses on measures of spatial central tendency, spatial dispersion and associations over space. In addition, given the obvious importance of flow data, a method of summarising its structure is presented.

## II.2 Centrographic measures

The concepts of spatial mean, spatial median, and spatial mode do have meaning, and can be applied to describe the central tendency of a given spatial distribution. The derivation of these spatial analogues or so-called centrographic measures follow their respective aspatial statistic. It is noted that the analyses can be undertaken for data based on both points or areas. Areal data, for example, can be summarised as point data using the centroids of the areas.

Simply stated, at the outset, a spatial referencing system must be defined. Whilst the National Grid provides a suitable referencing system, it is often more convenient to define an arbitrary Euclidean coordinate system because laborious calculations using six- or eight-figure references are avoided. As Figure II.1 exemplifies, it is a straightforward task to construct an (x, y) coordinate system from which the location coordinates of each settlement can be derived. From this information, the spatial mean and the spatial median can be determined. Given the location of n points, the spatial mean is $(\bar{x}, \bar{y})$, where

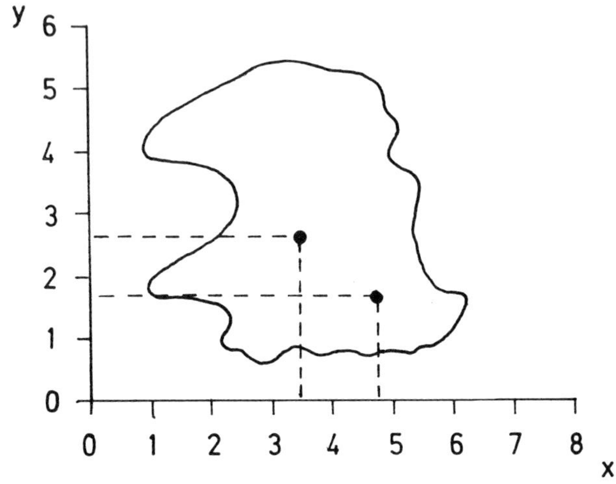


Figure II.1 A COORDINATE SYSTEM

$$\bar{x} = \frac{\sum_{i=1}^{n} x_i}{n} \quad \text{and} \quad \bar{y} = \frac{\sum_{i=1}^{n} y_i}{n}$$

Often, however, it is unrealistic to consider each point as identical. Central place theory, for instance, suggests that settlements are of different sizes and provide different ranges of goods and services. It is possible to calculate a so-called weighted spatial mean, which explicitly takes account of the differential characteristics of particular locations. By defining a weight for each location i as $w_i$, the weighted spatial mean $(\bar{x}, \bar{y})$ is calculated using the following equations,

$$\bar{x} = \frac{\sum_{i=1}^{n} w_i x_i}{\sum_{i=1}^{n} w_i} \quad \text{and} \quad \bar{y} = \frac{\sum_{i=1}^{n} w_i y_i}{\sum_{i=1}^{n} w_i}$$

The weights would be defined in relation to the specific problem of interest, and, for example, with regard to settlement patterns, population, number of services and so on could be used. To facilitate interpretation, it is often appropriate to compare the relative locations of the unweighted and the weighted spatial means. Moreover, a description of their changing locations over time (using say census data which is compiled every ten years) can be also informative.

Correspondingly, a so-called spatial median can be defined. Again, the x and y coordinates are considered separately. Both sets of coordinates are ranked (say in an east-west and north-south direction, respectively), and the median is calculated in the same way as described in the preceding appendix. However, it should be noted that the definition of the orthagonal axes of the coordinate system, specifically its orientation (for example, east-west and north-south), directly affects the location of the spatial median. In most cases, no single, unique spatial median exists. (Ebdon (1977) p.111), for example, presents a simple illustration of this feature).

## II.3 Point pattern analysis

Geographers' interest in the description of the spatial distribution of points can be thought to be the spatial analogue of the measures of dispersion discussed in Appendix I. Whilst the rigour of the description is useful, it can also be argued that point patterns are a geometrical expression of location theory (Rogers, 1974), (and it may be possible to infer the type of process or processes that generated such a pattern). Two methods of analysis are applied frequently: nearest neighbour analysis and quadrat analysis. The former is a distance-based technique, and the latter is an area-based technique. Whilst quadrat analysis is really outside the scope of Advanced Level students, it is noted that a useful pedagogic introduction is provided by Thomas (1977). In contrast, nearest neighbour analysis is the one method that has been introduced widely in schools, and, therefore, only a brief discussion is presented here (problems relating to boundary definitions and so on are not considered; CATMOG reference should be consulted). First, however, it is emphasised that both techniques describe a spatial distribution of points in terms of the degree of clustering, randomness or regularity (see Figure II.2); in fact, it can be suggested that a random pattern is of no intrinsic interest, but it provides an operational benchmark to distinguish between clustering and regularity.

Simply stated, nearest neighbour analysis involves calculating the mean of the distance between each point in the study area and its nearest neighbour. That is, in practice, each point is considered in turn, its nearest neighbour is determined, and the distance between them is measured. Formally, the observed mean nearest neighbour distance, $\bar{d}_{obs}$, is represented by

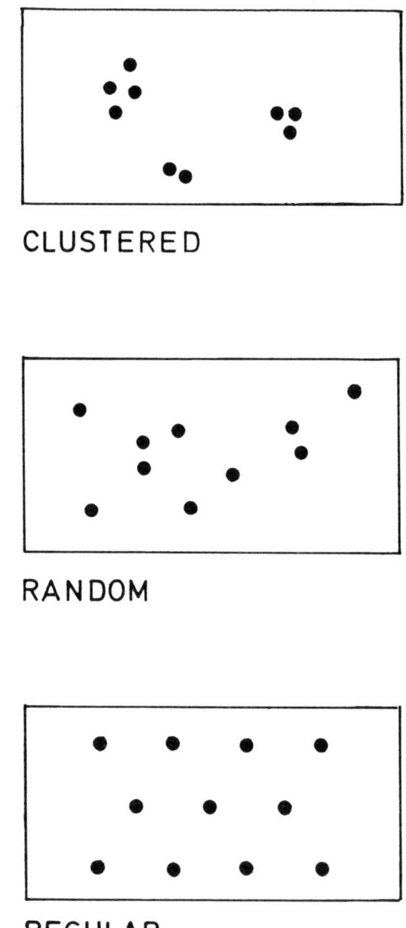


Fig. II.2 Spatial patterns

$$\bar{d}_{obs} = \frac{\sum_{i=1}^{n} \min d_{ij}}{n}$$

where $\min d_{ij}$ is the distance between a point located at i and its nearest neighbour located at j, and there are n points in the study area. (It is noted that a specific point need not be the nearest neighbour of any point, although it can be the nearest neighbour of more than one point).

To permit interpretation, this observed mean nearest neighbour distance is compared with the expected mean nearest neighbour distance if the same number of points were randomly distributed over an area of the same size. It can be shown that this expected mean distance, $\bar{d}_{exp}$, is

$$d_{exp} = \frac{1}{2\sqrt{\frac{n}{a}}}$$

where a is the area of study and, therefore, the term, n/a, is density. (The distances and area should be measured in the same basic unit, say kilometres and square kilometres or centimetres and square centimetres). It is noted that no unique spatial pattern is described by a specific value of $\bar{d}_{obs}$ or $\bar{d}_{exp}$.

Given both the expected and the observed mean nearest neighbour distances, $\bar{d}_{exp}$ and $\bar{d}_{obs}$, respectively, the nearest neighbour index, R, is defined as the ratio of these two values. Specifically,

$$R = \frac{\bar{d}_{obs}}{\bar{d}_{exp}}$$

and, clearly, by definition, a perfectly random point pattern has an R value of one (because $\bar{d}_{obs}$ equals $\bar{d}_{exp}$). Intuitively, if a particular point pattern possesses a degree of clustering, the observed mean nearest neighbour distance would be less than the expected mean distance for a random pattern, and, if a particular point pattern possesses a degree of regularity, the observed mean nearest neighbour distance would be greater than the expected mean distance for a random pattern. In terms of the nearest neighbour index, R, three different cases are distinguished:

(i) R less than one - clustering
(ii) R equal to one - perfectly random
(iii) R greater than one - regularity

These conditions are qualified by the fact that the value of R can range from 0 (perfect clustering) to 2.1491 (perfect regularity). Thus, the nearest neighbour index is a descriptive statistic that offers some precision to the terms 'clustered' and 'regular'. However, as stated earlier, it does not describe a unique point pattern. Indeed, an R value of zero does not necessarily describe a point pattern of one perfect cluster; multiple, perfect clusters, each with at least two points, would have the same value. A natural extension would be to consider distances from points to their second nearest neighbours, to their third nearest neighbours, and so on. Obviously,

$$\bar{d}_{obs}(1) \leqslant \bar{d}_{obs}(2) \leqslant \bar{d}_{obs}(3) \leqslant \ldots . \leqslant \bar{d}_{obs}(m)$$

where $\bar{d}_{obs}(m)$ is the observed mean distance between points and their $m^{th}$ nearest neighbour. Such an extension would help in the description of single and multiple clusters, and, for a perfectly regular point pattern (the lattice of equilateral triangles behind central place theory), the following situation would be present

$$\bar{d}_{obs}(1) = \bar{d}_{obs}(2) = \bar{d}_{obs}(3) = \bar{d}_{obs}(4) = \bar{d}_{obs}(5) = \bar{d}_{obs}(6)$$

## II.4 Spatial autocorrelation

Recently, new developments in spatial statistics have been concerned with spatial autocorrelation (see, for example, Cliff and Ord (1981)). That is, rather than being interested in whether the values of two variables are associated (as in conventional correlation analysis described in Appendix I), attention focuses on whether the spatial distribution of a variable's values are associated with their neighbouring values; hence the term, spatial autocorrelation.

Different statistical tests are available to test whether spatial autocorrelation exists or whether the pattern is random. As usual, these tests can be distinguished on the basis of the level of measurement of the data. In this subsection, a simple example from electoral geography is presented to describe the spatial autocorrelation test for a nominal data set. More specifically, it is a binary (two categories) data set, and the study involves an examination of the pattern of Conservative and Labour representation in Staffordshire constituencies in the 1979 General Election. Although the method can be extended to more than two categories, it is especially useful because many areal studies are based on two categories - presence and absence of some phenomenon, above and below average values, positive and negative residuals, and so on.

The statistic used is called a black/white joint count statistic, because, for convenience and generality, the two categories are black and white and spatial autocorrelation is interpreted in terms of the similarity of contiguous areas. In practice, it involves counting the number of black and white joins. The basic test involves determining whether the spatial distribution of Conservative (white) and Labour (black) constituencies is random or not random. The observed number of black/white joins is compared with the expected number if the pattern is random. If these values are blatantly different, the null hypothesis of a random pattern can be rejected; a significant level of spatial autocorrelation exists. If the values are only slightly different, interpretation is difficult and a simple significance test should be undertaken.

In the example from Staffordshire (see Figure II.3), the observed number of black/white joins, $O_{BW}$, is nine ($O_{BW} = 9$). Using a so-called non-free sampling hypothesis, because it would be inappropriate to calculate the probability of a Conservative or Labour constituency in Staffordshire using known national figures, the expected number of black/white joins for a random pattern $E_{BW}$, is calculated using the following equation,

$$E_{BW} = \frac{2JBW}{(n-1)n}$$

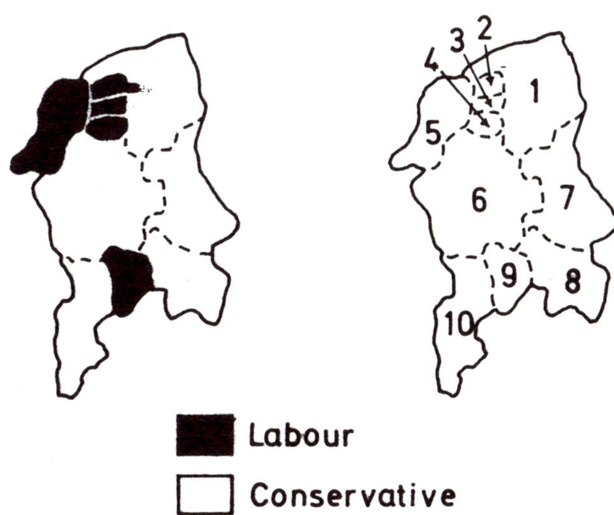


Figure II.3 STAFFORDSHIRE: 1979 ELECTION RESULTS

where J is the total number of joins in the study area, B and W are the total number of black and white areas, respectively, and n is the total number of areas; obviously,

$$n = B + W$$

The fact that many of the constituencies have adjoining constituencies with the same political representation suggests that spatial autocorrelation may exist. In this example,

$$\begin{aligned} J &= 20 \\ B &= 5 \\ W &= 5 \\ n &= 10 \end{aligned}$$

and, therefore,

$$E_{BW} = \frac{2 \times 20 \times 5 \times 5}{(10-1)\ 10} = \frac{1000}{90} = 11.1$$

Is this value really very different from the observed value of nine? To test the significance of this difference, it is necessary to calculate the expected standard deviation of the number of black/white joins for a random pattern. Whilst the equation is very daunting, in practice, the calculation only involves putting the appropriate values into the equation and undertaking some basic arithmetic operations. This standard deviation, $\sigma_{BW}$, is

$$\sigma_{BW} = \sqrt{E_{BW} + \frac{\sum_{i=1}^{n} L_i(L_i - 1)\ BW}{n(n-1)} + \frac{4\{J(J-1) - \sum_{i=1}^{n} L_i(L_i - 1)\}\ B(B-1)W(W-1)}{n(n-1)\ (n-2)\ (n-3)} - E_{BW}^2}$$

$L_i$ is the number of joins between area i and all its contiguous or adjoining areas in the study region. For instance, in this example, $L_{Leek}$ equals six and $L_{Buxton}$ equals three; it is noted that joins with areas not in the study region are disregarded. This information is summarised in Table II.1.

In the calculation of the standard deviation, the value of the term, $\sum_{i=1}^{n} L_i(L_i - 1)$, must be determined. If

| | Area i | $L_i$ | $(L_i - 1)$ | $L_i(L_i - 1)$ |
|---|---|---|---|---|
| 1. | Leek | 6 | 5 | 30 |
| 2. | Stoke-on-Trent (North) | 3 | 2 | 6 |
| 3. | Stoke-on-Trent (Central) | 4 | 3 | 12 |
| 4. | Stoke-on-Trent (South) | 4 | 3 | 12 |
| 5. | Newcastle-under-Lyme | 5 | 4 | 20 |
| 6. | Stafford and Stone | 7 | 6 | 42 |
| 7. | Burton | 3 | 2 | 6 |
| 8. | Lichfield and Tamworth | 3 | 2 | 6 |
| 9. | Cannock | 3 | 2 | 6 |
| 10. | South West Staffordshire | 2 | 1 | 2 |
| TOTAL | | | | 142 |

TABLE II.1: THE NUMBER OF ADJOINING AREAS

calculations are tabulated following Table II.1, this value is determined readily. Consequently,

$$\sigma_{BW} \sqrt{11.1 + \frac{142 \times 5 \times 5}{10 \times 9} + \frac{4\ \{(20 \times 19) - 142\} \times 5 \times 4 \times 5 \times 4}{10 \times 9 \times 8 \times 7} - (11.1)^2}$$

Therefore,

$$\sigma_{BW} = 1.69$$

As part of the significance test, it is necessary to calculate the standard normal deviate, $Z$, which is defined as,

$$Z = \frac{O_{BW} - E_{BW}}{\sigma_{BW}} = \frac{9 - 11.1}{1.69} = -1.24$$

From statistical tables, at the 95 per cent significance level for a two-way test, the critical value of $Z$ is 1.96. Hence, if spatial autocorrelation exists significantly, either of the following conditions must be satisfied

$$Z \leq -1.96 \quad \text{or} \quad Z \geq 1.96$$

Thus, in this example, significant spatial autocorrelation does not exist.

## II.5 Primary linkage analysis

Often flow data are so complicated that simplification is necessary to obtain a comprehension of the general underlying pattern. In this subsection, Nystuen and Dacey's (1961) primary linkage analysis is introduced using their original example. As a further illustration, the result of the simplification of a large transportation trip matrix for Devon is presented diagrammatically. It is noted that this simple approach can be extended to take account of multiple linkages.

In a study of inter-city telephone calls, Nystuen and Dacey (1961, p. 7) argued that '...within the myriad relations existing between cities, the network of largest flows will be the ones outlining the skeleton of the urban organisation within the entire region'. Their method simplifies node-specific, flow data to indicate the underlying hierarchical structure of flow movements. (In terms of applications, obvious links to central place theory could be examined). As the name, primary linkage analysis, suggests, the basis of the method is the dominant outflow from each node.

Table II.2 presents the hypothetical data applied by Nystuen and Dacey. It is a matrix of telephone calls from one city to another. The relative magnitude or hierarchical order/rank of a city is measured by its total *incoming* flow (which is determined by summing the elements in columns and ordering the column totals). As stated earlier, the hierarchical structure of the nodes is based on the largest *outgoing* flow; in primary linkage analysis, only one flow from each node is included, the nodal flow. A terminal point of the structure is defined when the nodal flow of a city is to a lower-order city. In this example, cities two, five, seven and ten are terminal (or dominant) points. To construct the hierarchical structure of the flow, it is necessary to begin with the terminal points, because the remaining nodes are linked, directly or indirectly, to one of them according to their nodal flow. As Figure II.4 illustrates, a summary of this data set highlights four district clusters, which can be interpreted in terms of the area over which a particular city exerts an influence. Figure II.5 presents a similar kind of summary for heavy goods transport data in Devon. Whilst a number of indirect linkages are present, the nodal structure clearly highlights the local significance of the larger settlements, such as Exeter, Plymouth, Barnstaple, Torbay and Newton Abbott.

DESTINATION

| ORIGIN | 1 | 2 | 3 | 4 | 5 | 6 | 7 | 8 | 9 | 10 | 11 | 12 | |
|---|---|---|---|---|---|---|---|---|---|---|---|---|---|
| 1 | | 75 | 15 | 20 | 28 | 2 | 3 | 2 | 1 | 20 | 1 | 0 | Satellite |
| 2 | 69 | | 45 | 50 | 58 | 12 | 20 | 3 | 6 | 35 | 4 | 2 | Dominant |
| 3 | 5 | 51 | | 12 | 40 | 0 | 6 | 1 | 3 | 15 | 0 | 1 | Satellite |
| 4 | 19 | 57 | 14 | | 30 | 7 | 6 | 2 | 11 | 18 | 5 | 1 | Satellite |
| 5 | 7 | 40 | 48 | 26 | | 7 | 10 | 2 | 37 | 39 | 12 | 6 | Dominant |
| 6 | 1 | 6 | 1 | 1 | 10 | | 27 | 1 | 3 | 4 | 2 | 0 | Satellite |
| 7 | 2 | 16 | 3 | 3 | 13 | 31 | | 3 | 18 | 8 | 3 | 1 | Dominant |
| 8 | 0 | 4 | 0 | 1 | 3 | 3 | 6 | | 12 | 38 | 4 | 0 | Satellite |
| 9 | 2 | 28 | 3 | 6 | 43 | 4 | 16 | 12 | | 98 | 13 | 1 | Satellite |
| 10 | 7 | 40 | 10 | 8 | 40 | 5 | 17 | 34 | 98 | | 35 | 12 | Dominant |
| 11 | 1 | 8 | 2 | 1 | 18 | 0 | 6 | 5 | 12 | 30 | | 15 | Satellite |
| 12 | 0 | 2 | 0 | 0 | 7 | 0 | 1 | 0 | 1 | 6 | 12 | | Satellite |
| TOTAL | 113 | 337 | 141 | 128 | 290 | 71 | 118 | 65 | 202 | 311 | 91 | 39 | |
| RANK | 8 | 1 | 5 | 6 | 3 | 10 | 7 | 11 | 4 | 2 | 9 | 12 | |

TABLE II.2: NYSTUEN AND DACEY'S (1961) DATA MATRIX OF TELEPHONE CALLS

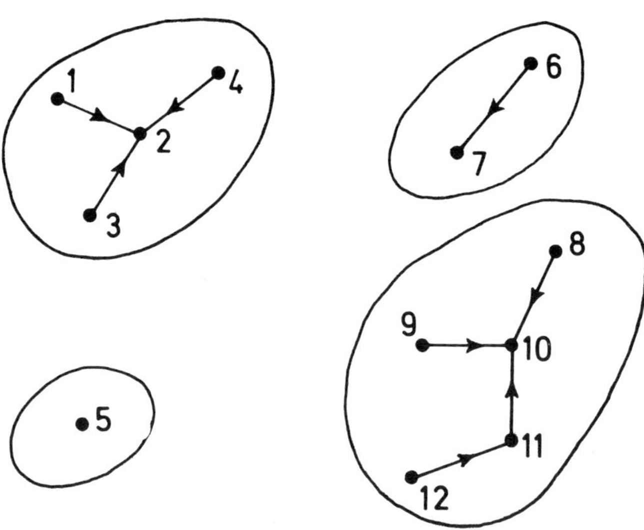


Figure II.4 NODAL STRUCTURE OF NYSTUEN AND DACEY'S (1961) DATA SET

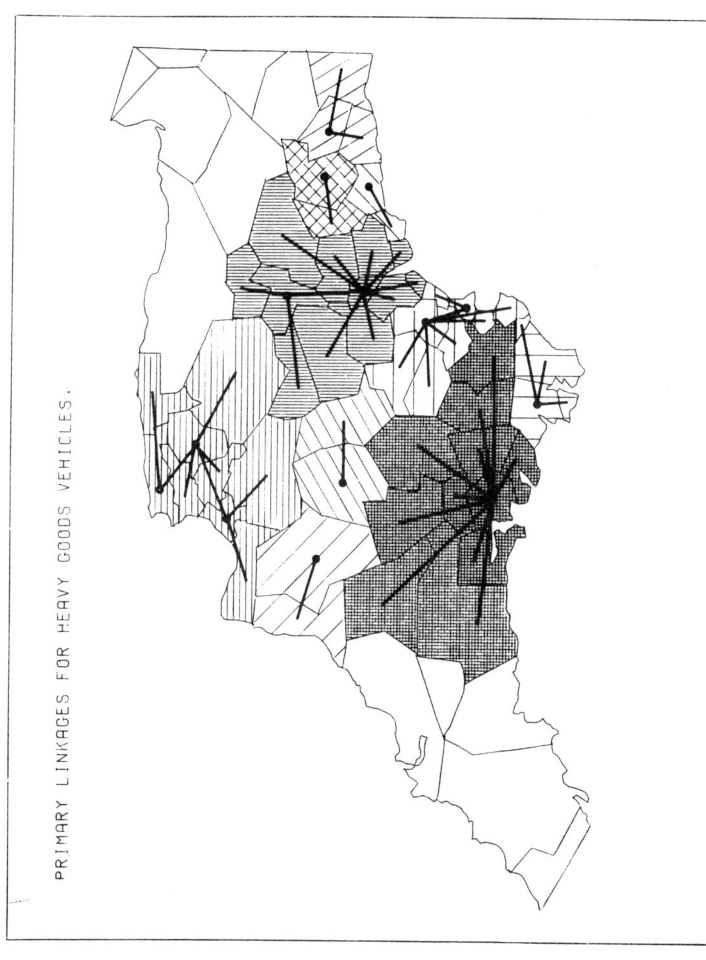


Figure II.5 PRIMARY LINKAGE ANALYSIS: HEAVY GOODS TRANSPORT IN DEVON

Appendix III

# THE DESIGN OF QUESTIONNAIRES

## III.1 Introduction

For many students, the attraction of a questionnaire survey is immediate; rapidity of design and execution, together with the consequent availability of readily accessible data for analysis. However, the apparent advantages of questionnaire surveys obscures the fact that, very often, questionnaires are used as a last resort. The purpose of this appendix is to emphasise that a questionnaire is an instrument for the measurement and collection of particular kinds of data. As such the design and implementation of a questionnaire requires careful attention if the final project is to be successful. Moreover, from a practical viewpoint, it requires access to duplicating facilities, which could be costly if a large number of students are undertaking questionnaire surveys.

In the following sections some of the more important problems associated with questionnaire design are considered. These include, the choice of survey, construction and wording of questions, definition of background variables and a discussion of attitude measurement and perception approaches.

## III.2 Choosing the type of survey.

There are essentially two types of survey research; the personal interview and the impersonal survey. The personal interview involves an interview schedule which is filled in by the interviewer in a face-to-face situation with the respondent. In contrast, the impersonal survey requires the respondent to answer the questions without the interviewer being present. This type of survey frequently involves the use of postal questionnaires or the 'drop and collect' method. Strictly speaking, the term 'questionnaire' only applies to impersonal surveys rather than personal interviews. However, in common usage this term is applied to both types of survey, and, to avoid confusion, this convention is used throughout the following discussion.

The initial decision, therefore, involves a choice between either the personal interview or the impersonal survey. In general, the use of postal questionnaires is precluded because of the excessive cost and relatively poor response rate, and in most instances the personal survey is the obvious choice. However, the 'drop and collect' survey method may be used if the study area is reasonably compact.

### III.2.1 *The Personal Interview*

This type of survey takes place on a face-to-face basis, and can range from a structured interview to a totally unstructured interview. The more structured interview involves the use of an interview schedule and is most commonly used by geographers. The questions, their wording and sequence are identical for every respondent. This strict format permits comparisons to be made directly, because it ensures that when variations appear between responses they can be attributed to actual differences between respondents, rather than to the way in which the interview was conducted. The personal interview using an interview schedule is based on three important assumptions. First, that the respondents have a sufficiently common vocabulary, insofar as it is possible to formulate questions which have the same meaning for each of them. Second, that its possible to phrase all questions in a form that is equally meaningful and straightforward to each respondent. Third, the sequence of questions must be identical.

In geography, unstructured interviews are usually used in conjunction with structured interviews rather than in isolation. For example, prior to a formal survey, a number of unstructured, pilot interviews could be undertaken to make sure that the correct issues are being focused on.

## III.3 The construction of questions.

The basis of all interviews is the question, and the interview itself serves two essential purposes. First, it must translate the objectives of the research project into specific questions whose answers will provide the necessary data for hypothesis testing. Second, the interview must motivate the respondent so that the necessary information is obtained. The construction of specific questions involves four principal considerations:

1. *Wording of the question*

   Short simple questions are preferable, which also avoid vague or technical terms. For example, two short questions are more easily understood than one long question.

   Double negatives should be avoided. The following

attitude question, for example, could give rise to ambiguity: 'Do you agree or disagree with the view that the town centre should not be open to traffic on a Saturday?'

Avoid apologetic wording. For example, the question 'would you mind telling me your weekly income?' may invite refusal.

Embarassing or personal questions should be left to the end of the interview.

If a question is potentially embarassing then a 'prompt card' could be used. These cards list possible responses to particular questions. Age, for example, may be divided into five classes (A-E) and the respondent simply indicates the relevant category. Prompt cards can also be used when a number of questions have the same range of responses (for example, attitude statements).

2. *Open and closed questions*
Questions can be either open-ended or fixed-alternative and both types are found usually on a single questionnaire.

In the case of fixed-alternative questions, respondents are presented with a set of answers and asked to select the one that best represents their views or fits their case. For example, in a study which focuses on the reasons for migration the following categories might be relevant: jobs, health, retirement, personal. Fixed-alternative questions are also used frequently in attitude surveys, and the following example illustrates. Over the last year, has traffic noise along this road: Decreased a lot; Decreased a little; Stayed about the same; Increased a little; Increased a lot.

Fixed-alternative questions are easy to ask, quick to answer and straightforward to analyse. However, they may introduce bias by forcing a respondent to make a choice.

In contrast, in open-ended questions the respondent's answers are recorded in full. The advantage of this approach is that the respondent is given freedom to express his or her views without being forced into pre-conceived categories. A major drawback, however, with open-ended questions is that they can be difficult to answer and even more difficult to analyse.

3. *Leading questions*
A leading question is phrased in such a way that the respon-

dent is led to believe that the interviewer expects a certain answer. For example, the question 'do you like shopping in the new pedestrian precinct?', could be rephrased in a more neutral manner to read 'do you like or dislike shopping in the new pedestrian precinct?'.

4. *Sequence of questions*
In general, questions which are used to classify respondents (for example, age, occupation) should be asked last. Hopefully, by placing the more interesting questions at the beginning the interest of the respondent should also be captured. In addition, the earlier questions should be straightforward and relaxing, while the more complex questions should be left until later. In certain instances, however, it will be necessary to ask some personal questions at the beginning of the questionnaire. This situation might arise when particular types of respondents are being sought (for example, pensioners, the unemployed), or in quota samples where a certain number of respondents in particular age/occupation categories are required.

Questions should be in a logical sequence and an interview should flow naturally like a conversation. If the questionnaire is divided into sections which cover distinct topics, then some explanation should be provided. Many questions will only apply to some respondents. To avoid irritating the respondent it is often useful to use a filter question which allows the interviewer to skip irrelevant questions, for example:

Question A. Do you go out to do any of your own shopping for groceries?

Yes, sometimes
No, never

If NO skip next question (B)

Question B. Do you use the shop on the estate?

## III.4 Definition of background variables.

At some point, all questionnaire surveys will require the unambiguous definition of key or background variables. The following list contains the most commonly used background variables.

1. *Income.* This is a difficult question both because of its sensitivity and because of the problem of deciding what should count as income. In the context of 'A' level projects this question could be deleted.

2. *Occupation.* The Registrar General for the Population Census has produced a list of occupations called the *Classification of Occupations* (HMSO) which should be available in the local main Library. This allows respondents to be classified into five social classes or eighteen socio-economic groups (see Chapter ). The relevant information can be obtained from the following questions:

Are you working at present?

What is your present/was your last job?
This question requires a full job description, in some cases involving numbers of people employed, number supervised, whether the respondent is self-employed and in which industry does he or she work.

3. The *Age, Sex* and *Marital Status* of respondents are usually collected. In conjunction with the age question it is usually wise to use a prompt card listing age categories.

4. In some studies details relating to educational qualifications will be necessary. Information should be collected on age at which full-time education finished and qualifications obtained.

5. Type of accommodation. This information can easily be pre-coded (for example, flats, detached, semi-detached, terraced houses).

6. When data is collected on units other than households (for example, industrial or commercial concerns) the classification used should be comparable with government statistics. Industrial studies, for example, should use the Standard Industrial Classificationl while agricultural studies should use standard classifications such as land use maps and the Agricultural Census.

7. In the specific context of geographical studies, some difficulty might be encountered by respondents with the definition of spatial terms (for example, locality, neighbourhood, region and area). If this is expected some prior definition of terms may be required.

## III.5 Attitude measurement and perception

Geographers very often wish to obtain information on the respondent's perception of, and attitudes to, various environmental phenomena (see, for example, Chapters Two and Four). Attitudes are measured by attitude scales each consisting of five to two dozen attitude statements with which the respondent is asked to agree or disagree. It is important to stress that an attitude cannot be gauged by a single statement; this would approach the attitude from only one direction and, therefore, might give unreliable results.

Attitudes are usually measured by a rating scale. The simplest form of rating merely requires the respondent to state an opinion:

Over the last year has traffic noise along this road

... decreased a lot
... decreased a little
... stayed about the same
... increased a little
... increased a lot

However, as noted above it is preferable for an attitude to be measured by more than one statement. In addition, attitude statements should be phrased in a way such that they do not have a right or wrong answer. The statements contained in Table III.1 are taken from a research project which examined various aspects of agricultural cooperation amongst a group of farmers. The statements are designed to form a scale which measures a farmer's attitude towards certain aspects of life in a rural area. In particular, the statements examine the degree to which a farmer is either individualistic in outlook or 'community oriented'. A respondent's attitude to a particular idea can be quantified by assigning a score to each of the response categories (the figures in brackets in Table III.1). In the example, a high score represents a 'community oriented' response, while a low score represents a more individualistic outlook (whether a high score corresponds to 'agree' or 'disagree' will depend on the wording of the statement). A highly 'community oriented' farmer could, therefore, record a maximum aggregate score of 25 (5 x 5) while a score of 5 (5 x 1) would represent a totally individualistic farmer. These scores could eventually be used to discriminate between different groups in the sample, although each specified category may have slightly different interpretations for different people.

A more elaborate rating scale is the semantic differential (for greater detail see Chapter Three). The respondent is required to evaluate concepts with reference to a series of bipolar statements. One end of the scale consists of a positive response while the opposite end is negative. Again, these scales can be examined quantitatively by assigning appropriate scores to the response categories.

Perception studies in geography employ a variety of techniques. In Chapter Three two of the principal methods were discussed; the use of sketch maps and the verbal listing of features in the environment. A further method involves the respondent expressing a preference for a set of objects

Please assess the following statements about life in a rural area.

1. Country people should meet socially as often as possible

| Agree Strongly | Agree | Neutral or don't know | Disagree | Disagree Strongly |
|---|---|---|---|---|
| (5) | (4) | (3) | (2) | (1) |

2. Other people are usually more of a nuisance than anything else.

| Agree Strongly | Agree | Neutral or don't know | Disagree | Disagree Strongly |
|---|---|---|---|---|
| (1) | (2) | (3) | (4) | (5) |

3. People in rural areas should try to be independent of others.

| Agree Strongly | Agree | Neutral or don't know | Disagree | Disagree Strongly |
|---|---|---|---|---|
| (1) | (2) | (3) | (4) | (5) |

4. It is a good thing for farmers to share equipment

| Agree Strongly | Agree | Neutral or don't know | Disagree | Disagree Strongly |
|---|---|---|---|---|
| (5) | (4) | (3) | (2) | (1) |

5. Getting involved in local affairs is a waste of valuable time

| Agree Strongly | Agree | Neutral or don't know | Disagree | Disagree Strongly |
|---|---|---|---|---|
| (1) | (2) | (3) | (4) | (5) |

TABLE III.1 ATTITIDES TO RURAL LIFE

distributed in space. For example, in an attempt to examine the mental images of different areas in the British Isles, respondents could be asked to rank a list of towns or regions on the basis of their relative attractiveness as places in which to live and work (see Gould and White, 1974 for example).

## III.6 Some important practical considerations when interviewing

1. Each question must be asked clearly and uniformly to each respondent, no elaboration or rephrasing of questions should be allowed.

2. In some cases, when an answer is insufficient or could be expanded, it might be necessary to 'probe'. However, great care should be taken not to bias the respondent's answer.

3. Some respondents might try to falsify or mislead the interviewer. Most, however, try to help to please; sometimes to the extent that respondents try to impress or give what they think to be the 'right' answer.

4. Interview length will vary according to the location. It is suggested that in the street an interview should last for no more than ten minutes, while in the home an interview could range from twenty to forty-five minutes.

5. Usually interviews in the home can only be conducted from 5 - 9 p.m. or at the weekend.

6. When starting the interview some means of identification should be available, preferably on the headed paper of the school or college. In addition the aim of the survey and the length of time it will take should be firmly and clearly stated at the outset.

## III.7 The importance of a pilot survey

Before the main survey is implemented a pilot survey should always be conducted. The purpose of the pilot survey is to identify which questions are inappropriate, unclear, or cause difficulty for the respondent. A small sample consisting of fifteen to twenty questionnaires should be sufficient. On the basis of the results from the pilot survey the original questionnaire could then be amended. Whilst, at first sight, this might appear to be unnecessary, additional work, it must be stressed that it is a fundamental stage in a project based on questionnaires; it is essential to have a check on the quality of the data that is being collected.

Appendix IV

# A LAY GUIDE TO BASIC PROGRAMMING

## IV.1 Introduction

Given the increasing availability of micro computers in schools, a knowledge of computer programming can be useful for geography teachers (see also Chapter Ten). In this short introduction, the aim is to help familiarise the reader with the most simple elements of BASIC programming, and particular reference is made to the listing of the statistical package that is presented in Appendix V; this package was written to assist this pedagogic discussion, rather than an illustration of efficient programming. A computer can only undertake the tasks that it is told to perform. A computer program is defined as a set of instructions for carrying out one or more, well-defined tasks. To communicate with a computer, a number of precise languages are available, but, in this Appendix, the BASIC language is described because it is used most frequently on micro computers (and, whilst different variations exist, the subtleties are not considered in this introduction).

This introduction is divided into three sections. Sufficient, simple characteristics of the language are introduced to permit a program to be written. This provides the foundation for two very useful types of refinement: a conditional transfer statement and alternative looping; and functions and subroutines. Throughout, examples from the statistical package are considered.

First, however, it is important to note that in writing a program, it is sensible to adopt the habit of drawing a flow chart of the stages involved in deriving the solution of the problem of interest; this is termed the algorithm. Moreover, detailed documentation, to complement any remarks in the program, is important, particularly to ensure portability, wide use, and understanding at a later date.

A number is found at the beginning of each line of a program, and this provides the sequence in which the computer performs the instructions. It is noted that lines do not have to be typed in in order, because BASIC sorts the lines into ascending order. This property is especially important when adding to existing programs, although the situation could arise in which too many new lines have to be inserted between two existing lines. At this stage, it is sufficient to know that the RENUMBER command, which renumbers all the program's lines in increments of 10 commencing with 10. In fact, it is sensible to program in increments of 10's to facilitate future insertions. Finally, the command LIST lists the entire program, although, more generally, it is possible to list parts of a program or individual lines. For example,

LIST 20, 220

results in a display of all the lines between 20 and 220 inclusive.

## IV.2 Some Background Definitions

As with any other language, a symbol or combination of symbols can be given a precise meaning. Given that the terminal/key board is the major way of communicating with a computer, keyboard familiarity is beneficial. A common error exists, for example, because the number 0 is different from the letter O. The layout follows the QWERT format of standard typewriters, although additional special characters are found. Without doubt, the most fundamental key is the 'Return' key that causes the typed instruction to be entered into the computer. (Until this key is pressed, alterations to the line can be made, although, if a mistake is entered, it can be edited out later). (On the B.B.C. micro, both the 'Break' and 'Escape' keys allow new entries to be received. The important instruction to note is that with the former all the previous entries are lost and with the latter they are kept. Consequently, it is sensible to use the 'Escape' key when possible).

Whilst it would be wrong to think that a computer is only a glorified calculator, in essence, it performs operations on numbers. Numbers can be differentiated on the basis of two criteria: constants or variables; integers or real numbers. Numbers should be given names, and, in BASIC, a name must commence with a letter, and its permitted length is machine dependent. The names refer to locations in the computer memory; that is, reference to A really means the number stored in location A. It is sensible to use letters which form an acronym of the variable of interest, say SD for standard deviation.

*Constant*

A constant in BASIC is a number whose value does not alter throughout the running of a program. From an efficiency perspective, it is appropriate to differentiate between integer (or whole) constants (such as, 7, 10 and so on) and real constants with decimal points (such as 3.14 and 2.7). If a constant or variable is an integer, its name ends in a % sign. It is noted that 0.5 is a constant, but ½ is an expression because it is a combination of two constants and one operation.

*Variables*
A variable, therefore, is any quantity whose value alters within a program. Moreover, unless indicated otherwise, all variables are interpreted as numerical values. To hold test, the symbols for a variable must finish with a (dollor) $ sign. For example, in the bivariate statistics package, the names of the two variables of interest, are entered as XNAME$ and YNAME$.

*Expressions*
On the keyboard, the special symbols for arithmetic operations are obviously of paramount importance. As Table IV.1 indicates, there are six fundamental arithmetic operations, and it is noted that the symbol for the operator '...raise to the power of ...' is machine dependent.

| OPERATION | SYMBOL |
|---|---|
| Addition | + |
| Subtraction | - |
| Multiplication | * |
| Division | / |
| ... is replaced by the value of ... | = |
| Exponentiation | ↑ or ** or Λ |

TABLE IV.1 ARITHMETIC OPERATIONS

An expression is any combination of constants, variables, arithmetic operations and parantheses, such as

```
X  + Y
S1 / D
T  * 3
```

Care must be taken with complex expressions because the hierarchical order in which arithmetic operations are performed is:

exponentiation
multiplication or division (from left to right)
addition or subtraction (from left to right)

That is, an expression is examined from left to right and the computer carries out all the multiplications and divisions. Then it starts from the left again and carries out all the additions and subtractions. Any exponentiations are undertaken first. For example, the expression,

A + B Λ C * D - E

is interpreted as B to the power C, multiplied by D. A is added to this product, and then E is subtracted. One way to avoid potential confusion and errors is to use parantheses; arithmetic operations are performed beginning with the inner most pair of parantheses. For example, the expression,

A + (B * (C - D * E)) / F

is interpreted as the subtraction of the product D * E from C all multiplied by B. This value is divided by F, and A is added to the result.

*Assignments*
If an expression is related to a variable by an equals symbol, it is called an assignment. Clearly, for calculations, assignments are used frequently. In the statistical package, for instance,

R = (S3 - (S1 * S2) / N) / (D1 * D2)

Thus, to date, sufficient details have been given to indicate how individual calculations can be specified and how they can be linked together. In general, a program involves operations on data, and the presentation of results. Consequently, it is necessary to know how to input and output numerical information.

In terms of data input, three different commands are possible:

LET *statement*
The LET statement assigns a value to a variable. For example,

```
10 LET A  =  7
20 LET B  =  6 + A
```

Its general format is,

Linenumber LET variable name = expression.

A particularly useful, but perhaps counter-intuitive statement is

```
10 LET X%  =  X% + 1
```

which is inconsistent if the 'equal to' symbol is interpreted in the conventional way. This statement increases the value assigned to X% by 1, and it is especially useful for counting sample sizes and in loops when a set of manipulations are completed a specified number of times.

DATA *and* READ *statements*

A second statement for inputing data is the DATA statement, which is used in conjunction with the READ statement. These statements are particularly useful when large amounts of data are to be processed. Whilst DATA statements can be inserted anywhere in a program, a sensible convention to locate them at the end of a program has developed; it means that, if the number of DATA statements alters, the earlier line numbers are unaffected. Entries in a DATA statement must be constants separated by commas. Data are input from left to right on a line. If there is more than one DATA statement in a program, the one possessing the lowest line number is considered first, and the remaining DATA statements are processed in order of ascending line numbers. The computer creates a block of data from the order of the DATA statements, and it retrieves this data by a READ statement. Every occasion a READ statement occurs during a program's execution, the next number (or numbers) is supplied from the data block. Consequently, it is the order of the numbers that is of fundamental importance. For example, the order of the data list for

```
10  DATA  7, 8, 9
```

is 7, then 8, and then 9, and, for two DATA statements,

```
10  DATA  7, 8, 9
20  DATA  6, 5, 4
```

the order of the data list is 7, then 8, then 9, then 6, then 5, and then 4.

A READ statement takes numbers from a data list and places them into specified storage location. For example,

```
10  READ  X, Y, Z
20  DATA  12.1, 7.4, 8
```

assigns 12.1 to X, 7.4 to Y, and 8 to Z.

INPUT *statement*

The INPUT statement is a useful alternative to the DATA and READ statements, specifically when it is desirable to 'interact' with the computer during the processing of the program. With an INPUT statement, a user inserts a number into a program from the keyboard. For instance, information is given to the computer through INPUT statements for the statistical package, because different people can enter their data without modifying the program. When a program is being processed, a (question mark) ? symbol will appear on the screen to indicate it has come to an INPUT statement, such as

```
20  INPUT  X
```

The program stops running until a value for X is typed in (and entered by using the 'Return' key). For example, in the statistical package, the sample size is given by

```
100 INPUT N%
```

PRINT *statement*
Clearly it is important for the computer to be able to output solutions. The PRINT statement informs the computer to present the required information at the terminal. For example,

```
100  PRINT  A
```

would print out the current value of A. Moreover, the value of an expression can be given by putting the expression in a PRINT statement.

```
110  PRINT  A * B / C
```

By separating expressions by commas, it is possible to print out more than one number on a single line. For example, the line

```
120  PRINT  A, B, C, A * B / C
```

would cause a computer to print the value of A, the value of B, the value of C, and the value of the expression A * B / C on the same line. (It is noted that up to five variables can be printed on a line using commas). Finally, to facilitate interpretation, it is often useful to present a label and the variable. For text to be printed, it must be in inverted commas

```
750  PRINT  "THE INDEPENDENT VARIABLE (x)."
```

To obtain continuity in the printing of text and the associated numerical value, they are separated by a semi-colon. For example,

```
770  PRINT  "Mean  =  " ; S1/N
```

Finally, to improve the layout of the presentation of results, it is often useful to cause the computer to skip a line; this is achieved by a simple PRINT statement

```
30  PRINT
```

Paranthetically, it is noted that, within the constraints of the maximum line length, it is possible to have a single line containing more than one statement. In such multistatement lines, a colon indicates where one statement finishes and where the next one commences. A common example of this type of line is where a number of PRINT statements are on the same line so that more than one blank line is printed. This often occurs after the screen has been cleared using the CLS statement.

```
190   CLS : PRINT : PRINT : PRINT
```

Before considering a small program, two further statements - END and REM - should be defined briefly. END, as the name suggests, is the statement with the highest line number; all programs must have one (and only one) END statement (although its absence does not cause errors on most computers).

Even with the knowledge presented already, it would be possible to write very long programs, and, therefore, it is useful to be able to indicate briefly what different sections of a program do. The REM (remark) statement possesses the following format

```
10  REM  DESCRIPTION
```

and, whilst a computer stores the remarks, the line is ignored in a program's execution. This statement is often thought to be unnecessary by a new programmer, but it does not take long to forget the structure of a program. REM statements are, therefore, especially important in programs for widespread distribution; the only constraint on their use is that the REM statement requires memory space.

To illustrate how some of the above statements can be linked together, consider the following program.

```
10  REM CONVERT DEGREES FAHRENHEIT TO DEGREES CELSIUS
20  READ  X
30  Y  =  (5/9) * (X - 32)
40  PRINT X; "DEGREES FAHRENHEIT IS";Y;"DEGREES CELSIUS"
50  DATA  40
60  END
```

RUN *statement*
This command tells the computer to execute the program from the first line. This is obviously a fundamental instruction.

## IV.3 Some additional refinements

In this section, some additional statements are introduced to improve and extend the reader's ability to write computer programs. Specifically, two transfer statements, GO TO and IF ... THEN are introduced. Both statements can be used for the very common looping process, although it is more convenient to use a FOR ... NEXT statement. As both the IF ... THEN and the FOR ... NEXT statements permit a series of calculations to be repeated over and over again, it is necessary to give each variable a unique name; through a DIM (dimension) statement, it is possible to alter variable names within a program by using subscripts.

### GO TO *statement*

GO TO is an unconditional transfer statement, which is the simplest means of making a computer execute a program in an order other than ascending line numbers. It is possible to jump both forwards and backwards, although care must be taken to ensure that no endless loop exists which would go on for ever.

A frequent application of the GO TO statement is to make the computer repeat a series of statements without having a long repetitive program. For example, in the sample program, it might be necessary to calculate the degrees Celsius for six different values of degrees Fahrenheit.

```
10  REM CONVERT DEGREES FAHRENHEIT TO DEGREES CELSIUS
20  READ X
30  Y  = (5/9) * (x - 32)
40  PRINT X ;"DEGREES FAHRENHEIT IS";Y;"DEGREES CELSIUS"
50  GO TO 20
60  DATA 40, 45, 50, 55, 60, 65
70  END
```

Line 50 sends the computer back to line 20 to read another value of X. This iterative process continues until the data list has been exhausted, which is indicated by some 'out of data' message.

### IF ... THEN *statement*

In many situations, particularly when comparing two or more numbers, the GO TO transfer statement is cumbersome; consequently, the conditional transfer statement, IF ... THEN is used very frequently. When an IF ... THEN statement is reached the normal sequential processing by the order of the line numbers is only broken if a specified relationship is satisfied. The computer only executes what is indicated by the THEN statement if the condition given by the IF statement is satisfied; if it is not satisfied, the remainder of the line is ignored, and the computer progresses to the next line. In making the comparisons, TABLE IV.2 presents the six most important relations. Many examples of the uses of IF ... THEN statements are found in the statistical package. It is noted

| RELATION | SYMBOL |
|---|---|
| equals | = |
| less than | < |
| greater than | > |
| not equal to | <> |
| less than or equal to | < = |
| greater than or equal to | >= |

TABLE IV.2

that comparisons are not restricted to numerical values. For example, in the statistical package, in selecting the type of histogram to be drawn, the text string, CLASS$, which can be a A, B or C, is part of conditional transfer statements.

```
2920  IF  CLASS$ = "A" THEN GO TO 3260
2930  IF  CLASS$ = "C" GO TO 4310
2940  IF  CLASS$ <> "B" GO TO 2820
```

(It is noted that the THEN statement can be usually missed out). In line 2920, CLASS$ is tested to see whether it is A. If it is A, the computer goes to line 3260 to commence the process of drawing histograms with equal class intervals; if CLASS$ is not A, the computer proceeds immediately to the next line. Here, the computer tests whether CLASS$ is C, and, if this condition is satisfied the computer goes to line 4310 to commence the process of drawing histograms with user specified classes. If CLASS$ is not C, the computer goes to line 2940 which checks whether CLASS$ is not equal to B. Clearly, by logic, after the conditional statements in lines 2920 and 2930, CLASS$ must be B, because, to get to line 2940, it is not A or C (and the only other possibility is B). Line 2940, therefore, is a check to ensure the input letter is acceptable, and, if there is an error, the computer goes back to line 2820 to enable the user to make the correct selection. (Such general, consistency checks are particularly important when a range of users intend to run a specific program).

FOR ... NEXT *statement*
As indicated earlier, looping is a process that is found in many programs, and, to do this in BASIC, the FOR ... NEXT statement is used. The FOR ... NEXT statement is the easiest way of making a particular group of lines be repeated within a program. The general format of this statement is

```
linenumber FOR index variable = first value TO last value STEP
                                                        increment
           .
           .
           . lines to be repeated
           .
           .
linenumber NEXT index variable
```

For example, a program to print the first ten numbers squared is

```
10  REM PRINT SQUARED NUMBERS
20  FOR I% = 1 TO 10 STEP 1
30  X% = I% * I%
40  PRINT X%
50  NEXT I%
60  END
```

It is noted that the term, STEP 1, is redundant on most machines, because it increments automatically in this interval. If the index variables are all integer variables, it is appropriate to explicitly recognise this fact by using the % symbol. The STEP increment is neither restricted to integer values nor to positive values. A common application for the FOR ... NEXT statement is to add up numbers. For example,

```
10  REM PROGRAM TO SUM THE FIRST 50 NUMBERS
20  LET S% = 0
30  FOR I% = 1 TO 50
40  S% = S% + I%
50  NEXT I%
60  PRINT S%
70  END
```

In the statistical package, for instance, to calculate the mean and standard deviation of the independent (X) and dependent (Y) variables, it is necessary to calculate a number of summations of this type. It is noted that it is possible to leave a FOR ... NEXT loop before completion through either a GO TO statement or an IF statement; it is impossible to jump into the middle of a FOR ... NEXT loop. Moreover, it is possible to have more than one loop together, the so-called nesting of loops. An inner loop, however, must be enclosed completely within an outer loop.

```
10  REM ILLUSTRATION OF CORRECT NESTING
20  FOR I% = 1 TO 10
30  FOR K% = 1 TO 30
40  PRINT I% * K% / 2
50  NEXT K%
60  NEXT I%
70  END
```

In contrast,

```
10  REM ILLUSTRATION OF INCORRECT NESTING
20  FOR I% = 1 TO 10
30  FOR K% = 1 TO 30
40  PRINT I% * K% / 2
50  NEXT I%
60  NEXT K%
70  END
```

DIM *statement*
Both the IF ... THEN and FOR ... NEXT statements are used to complete the calculation a number of times. During this process, when variables are used which must be retained, each one must possess a unique name. Following the short-hand notation employed in statistics (see Appendix I), a list of similar variables can be summarised using a subscript,

$x_i$ where $(i = 1, 2, \ldots , n)$.

Thus, if it desired to add together ten numbers and also keep the information on the individual numbers, a suitable notation is $x_1, x_2, x_3, \ldots, x_{10}$ or $x_i$ (where $i = 1, 2, \ldots, 10$). BASIC language does not permit such lower case subscripts; instead, a subscripted variable (or array) is written as

X(1), X(2), X(3), ..., X(10) or X(I).

If subscripted variables are included, it is necessary to have a corresponding DIM (dimension) statement before the particular subscripted variable is used More specifically, if a subscripted variable ,X(I), has 20 elements, the corresponding DIM statement

```
10  DIM X(20)
```

is required to indicate the necessary number of distinct storage locations for the variable. Often, as in the statistical package, dimensions of arrays are dependent on the sample size, N% and, therefore, for generality, it is appropriate to have

```
105  DIM X(N%), Y(N%), XR(N%), YR(N%)
```

## IV.4 Functions and Subroutines

In the last section, attention was drawn to the fact that specified lines were repeated a number of times in many programs. In addition, many programs require the same sequence of lines to be repeated a number of times in different parts of a program. In such circumstances, it is obviously desirable to only include the sequence of lines once and to have the opportunity to go to the sequence when it is desired. In the BASIC language, two methods, functions and subroutines, are appropriate. Whilst the distinction between an intrinsic and a user-specified function is important and is described below, simply stated, functions are employed when the process to be repeated is simple, and subroutines are employed when the process is complex.

*Intrinsic functions*

Certain commonly used mathematical functions, such as square root, sine, logarithms, and so on, are found on many calculators. In BASIC, a range of intrinsic functions are available for use in programs. For example, if the following line is entered

```
PRINT  SQR (9)
```

the value 3, the square root of 9, would appear. Table IV.3 presents some of the intrinsic functions in BASIC.

| FUNCTION | DEFINITION |
|---|---|
| ABS(X) | The absolute value of X |
| COS(X) | Cosine of X (where X is in radians) |
| EXP(X) | Natural exponent of X (e = 2.7183) |
| INT(X) | Integer part of X |
| LOG(X) | Natural logarithm of X |
| RND(X) | Random number between 0 and 1 |
| SIN(X) | Sine of X (where X is in radians |
| SGN(X) | +1 if X > 0<br>0 if X = 0<br>-1 if X < 0 |
| SQR(X) | Square root of X (where X is non-negative) |
| TAB(X) | for formulating output (but its precise nature is often micro specific) |
| TAN(X) | Tangent of X (where X is in radians) |

TABLE IV.3   INTRINSIC FUNCTIONS

For example, in the statistics package, the line

```
1420 IF ABS (R) > (1.01 Λ 3)
```

is testing whether the absolute value of the correlation coefficient is greater than 1.0; clearly, if this condition is satisfied, there is an error!

*User-specified functions*
To extend the range of functions, users can specify their own functions within a program, and they operate in the same way as intrinsic functions. The general format of a user-specified function is

```
line number  DEF  FN (variables) = expression
```

where the variables in the function's argument (inside the brackets) are found in the expression. For example, in calculating moving-averages, it would be useful to have a function that would calculate the mean of three numbers, X, Y, and Z.

```
100  DEF  FN (X, Y, Z) = (X + Y + Z) / 3
```

If the function is complicated and extends over a single line, the last line of the function is indicated by it possessing an = (equals to) symbol.

```
100  DEF FN (X, Y, Z)
110  TOTAL = X + Y + Z
120  = TOTAL / 3
```

(Whilst there are no restrictions on multi-line functions with the B.B.C. microcomputer, other microcomputers are restricted to single line functions). In the statistics program, user-specified functions, FNXPLOT(I) and FNYPLOT(I) were employed in the construction of the scattergram in association with the bivariate regression analysis.

*Subroutines*
In general, the problem of restricting user-specified function to one line means that complicated program segments that are repeated frequently are put in a subroutine: once the subroutine is present, it can be used at any point in the program by using a GOSUB statement. For example,

```
line number  GOSUB  line number
```

or, more specifically,

```
1000  GOSUB  5240
```

means that the first line of the subroutine to be used is at line 5240. The computer transfers to this line, and executes each line of the subroutine in order until it reaches a RETURN statement, which indicates the end of the subroutine.

## IV.5 Concluding Comment

Sufficient statements and commands have been introduced to permit anybody to write a BASIC program. Moreover, it should be possible to comprehend the listing of the bivariate statistics package in Appendix V; any aspects not covered in this introduction are defined through the use of REM (remark) statements.

Appendix V

# BIVARIATE STATISTICS PACKAGE

## V.1 Background

In this appendix, a bivariate statistics package is listed. The range of options available was selected to reflect the types of analyses that are undertaken most frequently by students (and the different statistics are introduced formally in Appendix I). Given that the new B.B.C. microcomputer, which is approved under the Department of Industry scheme that allows schools to purchase a computer at half price, will become the most accessible microcomputer for school children, the package has been written particularly for this machine (specifically, the (32K) model B). The package could be modified easily for other microcomputers, and, for completeness, some BASIC keywords and instructions which are not defined either in Appendix IV or in the program itself are listed below.

To summarise, the structure of this interactive, bivariate statistics package, which, obviously, can be adapted to describe univariate data sets, is:

- data input
- calculation of descriptive statistics (such as, the mean, standard deviation and range)
- ranking of data (which is especially useful for calculating the Spearman's rank correlation coefficient) and the calculation of the median
- calculation of Pearson's product moment correlation coefficient
- calculate residuals
- determination of the linear regression equation (and of the associated standard errors)
- plotting of the scatter diagram and the regression line
- drawing histograms (with a choice of class definitions; it is noted that classes based on standard scores relate to one, two or three standard deviations above and below the mean value).

## V.2 Listing of the bivariate statistics package

```
10 MODE 4
20 VDU 23,224,&FF,&FF,&FF,&FF,&FF,&FF,&FF,&FF
30 REM*BIVARIATE STATISTICS PACKAGE
40 REM*DISCLAIMER*ALTHOUGH THIS PACKAGE HAS BEEN THOROUGHLY TESTED, NO WARRANTY,
50 REM* EXPRESSED OR IMPLIED, IS MADE BY THE AUTHOR (John R. Beaumont).
60 DIM XCL$(10),YCL$(10)
70PRINT
80PRINT"Please give the sample size."
90INPUT N
100 IF N > 50 GOTO 130
110 DIM X(50),Y(50),XR(50),YR(50)
120 GOTO 140
130 DIM X(N),Y(N),XR(N),YR(N)
140F$="F":R$="R":E$="E":Q$="Q":U$="U":N$="N":C$="C":Y$="Y":L$="L":AA$="A"
150T$="T":II$="I":V$="V"
160IF N>250.1THEN 190
170IF N>0 AND N<=250 THEN 220
180 PRINT:PRINT"Try again !":GOTO 70
190PRINT"The sample size is too large for this   program.";
200PRINT"The maximum sample size is 250."
210GOTO 70
220 CLS:PRINT:PRINT:PRINT
230 PRINT"Do you want a copy of the results sent  to the printer (Y or N)?"
240 INPUT SPR$:IFSPR$<>"Y"ANDSPR$<>"N" THEN GOTO 220
250 CLS:PRINT:PRINT:PRINT:PRINT
260PRINT"Name the x variable":INPUT XNAME$
270 CLS:PRINT:PRINT:PRINT:PRINT
280PRINT"Name the y variable":INPUT YNAME$
290 CLS:PRINT:PRINT:PRINT:PRINT
300PRINT"The data can be entered in two ways:"
310PRINT"(1)case by case (x,y)"
320PRINT"(2)variable by variable"
330PRINT
340PRINT"Which method (1 or 2)?"
350INPUT A1$ : A1$=LEFT$A1$,1)
360IF A1$="1" THEN M7=1: GOTO 390
370IF A1$<>"2" THEN PRINT"Incorrect value;try again"; : GOTO 350
380M7=0
390 CLS:PRINT:PRINT:PRINT
400PRINT:PRINT"Enter the data now!";
410 IF M7=0 THEN PRINT:PRINT "One value is entered per line.";
420PRINT " A total  of     ";2*N;
430PRINT"   values are needed."
440IF M7=1 THEN 510
450IF M7=1 THEN PRINT"Enter the data:" : GOTO 510
460PRINT"The independent variable (x):"
470FOR I=1 TO N
480INPUT X(I)
490NEXT I
500 IF M7=0 THEN 520
510 PRINT "The values can be entered case by case  (i.e. x,y on each line)."
520S1=0
530S2=0
540X2=0
550Y2=0
560S3=0
570IF M7=0 THEN PRINT:PRINT"The dependent variable (y):"
580FOR I=1 TO N
590IF M7=1 THEN 620
600INPUT Y(I)
610GOTO 640
620INPUT X(I),Y(I)
630GOTO 650
640X0=X(I)
650S1=S1+X(I)
660S2=S2+Y(I)
```

```
670X2=X2+X(I)*X(I)
680Y2=Y2+Y(I)*Y(I)
690S3=S3+X(I)*Y(I)
700NEXT I
710PRINT
720PRINT
730D1=SQR(X2-S1*S1/N)
740D2=SQR(Y2-S2*S2/N)
750D3=SQR(N)
760 CLS:IF SPR$="Y" THEN VDU 2
770 PRINT
780 PRINT "THE INDEPENDENT VARIABLE (x)  : ";XNAME$
790 PRINT
800 PRINT "Mean = ";S1/N
810 REM*MAXIMUM VALUE OF THE X VARIABLE
820 GOSUB 5250
830 PRINT
840 PRINT "The maximum value of the x variable"
850 PRINT"=       ";MX
860 GOSUB 5370
870 PRINT
880 PRINT "The minimum value of the x variable"
890 PRINT"=       ";MMX
900 PRINT
910 PRINT "The range = ",MX-MMX
920 PRINT
930PRINT "Standard deviation = ";D1/D3
940 PRINT
950 PRINT
960 PRINT "THE DEPENDENT VARIABLE (y)  : ";YNAME$
970 PRINT
980 PRINT "Mean = ";S2/N
990 GOSUB 5310
1000 PRINT
1010 PRINT "The maximum value of the y variable"
1020 PRINT "=       ";MY
1030 GOSUB 5430
1040 PRINT
1050 PRINT "The minimum value of the y variable"
1060 PRINT"=       ";MMY
1070 PRINT:PRINT"The range = ";MY-MMY
1080 PRINT
1090 PRINT "Standard deviation = ";D2/D3
1100 R=(S3-(S1*S2)/N)/(D1*D2)
1110 PRINT :IF SPR$="Y" THEN VDU 3
1120 GOSUB 5480
1130 CLS
1140 PRINT
1150 PRINT
1160 FOR I=1 TO N
1170 XR(I)=X(I)
1180 YR(I)=Y(I)
1190 NEXT I
1200 PRINT"Do you want the x and y variates ranked and their medians?(Y or N)"
1210 INPUT B$
1220 IF B$="N" THEN1580
1230 IF B$<>"Y" GOTO 1200
1240 CLS:IF SPR$="Y" THEN VDU 2
1250 PRINT
1260 PRINT
1270 GOSUB 5530
1280 PRINT "Ranked values of x"
1290 FOR I=1 TO N
1300  PRINT X(I),
1310  NEXT I:IF SPR$="Y" THEN VDU 3
1320 GOSUB 5480
```

```
1330 IF SPR$="Y" THEN VDU 2
1340 M7=N+1:M7=(M7 MOD 2):IFM7<>0THENGOTO1370
1350 CLS:PRINT:PRINT:PRINT:PRINT
1360 PRINT"The Median of ";XNAME$;" is ";X((N+1)/2):GOTO1390
1370 CLS:PRINT:PRINT:PRINT:PRINT
1380 PRINT"The Median of ";XNAME$;" is ";(X(N/2)+X((N/2)+1))/2
1390 IF SPR$="Y" THEN VDU 3
1400 PRINT:PRINT:PRINT:GOSUB 5480
1410 GOSUB 5670
1420 CLS:IF SPR$="Y" THEN VDU 2
1430 PRINT
1440 PRINT
1450 PRINT "Ranked values of y"
1460 FOR I=1 TO N
1470 PRINT Y(I),
1480 NEXT I:IF SPR$="Y" THEN VDU 3
1490 GOSUB 5480
1500 IF SPR$="Y" THEN VDU 2
1510 M7=N+1:M7=(M7 MOD 2):IFM7<>0THENGOTO1540
1520 CLS:PRINT:PRINT:PRINT:PRINT
1530 PRINT"The Median of ";YNAME$;" is ";Y((N+1)/2):GOTO1560
1540 CLS:PRINT:PRINT:PRINT:PRINT
1550 PRINT"The Median of ";YNAME$;" is ";(Y(N/2)+Y((N/2)+1))/2
1560 IF SPR$="Y" THEN VDU 3
1570 PRINT:PRINT:PRINT:GOSUB 5480
1580 CLS :IF SPR$="Y" THEN VDU 2
1590 PRINT
1600PRINT"PEARSON'S PRODUCT MOMENT CORRELATION    COEFFICIENT."
1610IF ABS(R)>(1.00001^3) THEN PRINT "Error in the correlation coefficient."
1620 PRINT
1630PRINT"      R =";R
1640REM*REGRESSION ANALYSIS*
1650M=(S3-S1*S2/N)/(X2-S1*S1/N)
1660C=(S2-M*S1)/N
1670PRINT
1680 PRINT
1690 PRINT :IF SPR$="Y" THEN VDU 3
1700 GOSUB 5480
1710 CLS
1720 PRINT
1730PRINT"Do you require a list of the residuals  (Y or N)?"
1740INPUT A1$ :PRINT
1750A1$=LEFT$A1$,1)
1760IF A1$="N" THEN 1950
1770 IF A1$<>"Y" GOTO 1730
1780REM*PRINT THE RESIDUALS
1790 CLS:IF SPR$="Y" THEN VDU 2
1800 PRINT
1810PRINT"INDEPENDENT VARIABLE    RESIDUAL":PRINT
1820FOR I=1 TO N
1830  E=XR(I)*M+C:R1=YR(I)-E
1840PRINT XR(I);TAB(24);R1
1850NEXT I:PRINT:IF SPR$="Y" THEN VDU3
1860 GOSUB 5480
1870 CLS:IF SPR$="Y" THEN VDU 2:PRINT
1880 PRINT"PREDICTED VALUE         RESIDUAL":PRINT
1890 FOR I=1 TO N
1900 E=XR(I)*M+C:R1=YR(I)-E
1910 PRINT  E;TAB(24);R1
1920 NEXT I:PRINT:IF SPR$="Y" THEN VDU3
1930REM*CALCULATE AND PRINT THE STATISTICS
1940 GOSUB 5480
1950 CLS:IF SPR$="Y" THEN VDU 2
1960 PRINT
1970PRINT    "BIVARIATE REGRESSION ANALYSIS: y=a+bx"
1980 PRINT
```

```
1990 PRINT
2000 PRINT "y = ";C;" + ";M;" x "
2010 PRINT
2020 IF (1-R*R)<0 THEN GOTO 2050
2030S6=SQR(1-R*R)*D2/D3
2040 GOTO 2070
2050 S6=0
2060 PRINT:PRINT:PRINT
2070PRINT:PRINT:PRINT"Standard error of the estimate  ":PRINT:PRINT TAB(15);S6
2080 IF S6=0 GOTO 2120
2090S7=S6/D1
2100 GOTO 2130
2110 PRINT
2120 S7=0
2130 PRINT
2140 PRINT
2150PRINT"Standard error of the slope (b)":PRINT:PRINT TAB(15);S7
2160S8=S7*SQR(X2/N)
2170 PRINT
2180 PRINT
2190 PRINT:PRINT:PRINT
2200PRINT"Standard error of the intercept (a)":PRINT:PRINT TAB(15);S8
2210 PRINT:PRINT:IF SPR$="Y" THEN VDU 3
2220 GOSUB 5490
2230 MODE 4
2240 PRINT
2250 PRINT
2260 REM*X-AXIS
2270 IF(MMY<0 AND MY>0) THEN GOTO 2350
2280 MOVE 234,80:DRAW 1038,80
2290 VDU 5:FOR I=0 TO 8 STEP 2
2300 MOVE 234+I*(800/8),80:DRAW 234+I*(800/8),92
2310 MOVE 234+I*(800/8),50:@%=131594:PRINT;MMX+I*((MX-MMX)/8) :@%=10
2320 NEXT I
2330 VDU 4
2340 GOTO 2440
2350 XB=740/(MY-MMY)
2360 SM=-MMY*XB
2370 FOR YY=80+SM TO 84+SM STEP 4
2380 MOVE 234,YY:DRAW 1038,YY
2390 NEXT YY
2400 VDU 5:FOR I=0 TO 10 STEP 10
2410 MOVE 234+I*(800/10),YY:DRAW 234+I*(800/10),YY+12
2420 MOVE 234+I*(800/10),YY-30:@%=131594:PRINT;MMX+I*((MX-MMX)/10):@%=10
2430 NEXT I:VDU 4
2440 REM*Y-AXIS
2450 IF(MMX<0 AND MX>0) THEN GOTO 2520
2460 MOVE 234,80:DRAW 234,820
2470 VDU 5:FOR I=0 TO 8 STEP 2
2480 MOVE 234,80+I*(740/8):DRAW 246,80+I*(740/8)
2490 MOVE 60,80+I*(740/8):@%=131594:PRINT;MMY+I*((MY-MMY)/8):@%=10
2500 NEXT I:VDU 4
2510 GOTO 2610
2520 XA=800/(MX-MMX)
2530 TM=-MMX*XA
2540 FOR XX=234+TM TO 238+TM STEP 4
2550 MOVE XX,80:DRAW XX,820
2560 NEXT XX
2570 VDU 5:FOR I=0 TO 10 STEP 10
2580 MOVE XX,80+I*(740/10):DRAW XX+12,80+I*(740/10)
2590 MOVE XX-170,80+I*(740/10):@%=131594:PRINT;MMY+I*((MY-MMY)/10) :@%=10
2600 NEXT I:VDU 4
2610 REM* LABEL AXES
2620 IF(MMY<0 AND MY>0) THEN PRINT TAB(32,((1024-(100+SM))/1024)*32);XNAME$
2630 IF(MMY<0 AND MY>0) THEN GOTO 2650
2640 PRINT TAB(32,29);XNAME$
```

```
2650 IF(MX>0 AND MMX<0) THEN PRINT TAB(((234+TM)/1280)*40,5);YNAME$
2660 IF(MX>0 AND MMX<0) THEN GOTO 2680
2670 PRINT TAB(7,5);YNAME$
2680 VDU 5
2690 XA=800/(MX-MMX)
2700 XB=740/(MY-MMY)
2710 MOVE 0,0
2720 FOR I=1 TO N
2730 XR(I)=234+(XR(I)-MMX)*XA
2740YR(I)=84+(YR(I)-MMY)*XB
2750 MOVE XR(I)-12,YR(I):DRAW XR(I)+12,YR(I)
2760 MOVE XR(I),YR(I)-12:DRAW XR(I),YR(I)+12
2770 NEXT I:VDU 4
2780 RANGEX=MX-MMX
2790 RANGEY=MY-MMY
2800 I=MMX
2810 REPEAT X=FNXPLOT(I):Y=FNYPLOT(I)
2820 I=I+RANGEX/100
2830 UNTIL Y>=80 AND Y <=820
2840 MOVE X,Y
2850 FOR I=I TO MX STEP RANGEX/100
2860 X=FNXPLOT(I): Y=FNYPLOT(I)
2870 IF Y<80 OR Y>820 THEN GOTO 2890
2880 DRAW X,Y
2890 NEXT I
2900 PRINT TAB(0,0)
2910 GOSUB 5480
2920 CLS
2930 PRINT:PRINT:PRINT:PRINT
2940 PRINT"Do you want to draw histograms? (Y or N)."
2950 INPUT HIST$
2960 IF HIST$ = "N" THEN GOTO 5120
2970 IF HIST$<>"Y" GOTO 2940
2980 CLS
2990 PRINT
3000 PRINT"          HISTOGRAMS."
3010PRINT: PRINT"Class intervals can be :"
3020 PRINT:PRINT"  A : equal intervals"
3030PRINT: PRINT"  B : standard scores"
3040 PRINT:PRINT"  C : user specified"
3050 PRINT:PRINT"Input A or B or C."
3060 INPUT CLASS$
3070 MEANX=S1/N
3080 MEANY=S2/N
3090 SDX=D1/D3
3100 SDY=D2/D3
3110 IF CLASS$="A" THEN GOTO 3450
3120IF CLASS$="C" GOTO4500
3130 IF CLASS$<>"B" GOTO 3010
3140 REM*STANDARD SCORES
3150 CLS
3160 PRINT:PRINT:PRINT
3170 PRINT"Histograms drawn using standard scores  have six classes."
3180 PRINT:PRINT:PRINT
3190 GOSUB 5480
3200 CLS
3210 PRINT
3220 FOR I=1 TO 6
3230 XCL%(I)=0
3240 YCL%(I)=0
3250 NEXT I
3260 FOR I=1 TO N
3270 XR(I)=(X(I)-MEANX)/SDX
3280 YR(I)=(Y(I)-MEANY)/SDY
3290 NEXT I
3300 FOR I=1 TO N
```

```
3310 IF(XR(I)<=-2.0) THEN XCL%(1)=XCL%(1)+1
3320 IF(YR(I)<=-2.0) THEN YCL%(1)=YCL%(1)+1
3330 IF(XR(I)<=-1.0 AND XR(I) >-2.0)THEN XCL%(2)=XCL%(2)+1
3340 IF(YR(I)<=-1.0 AND YR(I) >-2.0)THEN YCL%(2)=YCL%(2)+1
3350 IF(XR(I)<=0.0 AND XR(I)>-1.0)THEN XCL%(3)=XCL%(3)+1
3360 IF(YR(I)<=0.0 AND YR(I)>-1.0)THEN YCL%(3)=YCL%(3)+1
3370 IF(XR(I)<=1.0 AND XR(I)>0.0)THEN XCL%(4)=XCL%(4)+1
3380 IF(YR(I)<=1.0 AND YR(I)>0.0)THEN YCL%(4)=YCL%(4)+1
3390 IF(XR(I)<=2.0 AND XR(I)>1.0)THEN XCL%(5)=XCL%(5)+1
3400 IF(YR(I)<=2.0 AND YR(I)>1.0)THEN YCL%(5)=YCL%(5)+1
3410 IF(XR(I)>2.0)THEN XCL%(6)=XCL%(6)+1
3420 IF(YR(I)>2.0)THEN YCL%(6)=YCL%(6)+1
3430 NEXT I
3440 GOTO 3690
3450 REM*EQUAL INTERVALS
3460 RANGEX=MX-MMX
3470 RANGEY=MY-MMY
3480 CLS
3490 PRINT
3500 PRINT     "EQUAL INTERVAL SHADING"
3510 PRINT
3520 PRINT"How many classes?":PRINT" (minimum is 2,maximum is 10)."
3530 INPUT CN%
3540 IF CN%<2 OR CN%> 10 GOTO 3520
3550 XINT=RANGEX/CN%
3560 YINT=RANGEY/CN%
3570 FOR I=1 TO CN%
3580 XCL%(I)=0
3590 YCL%(I)=0
3600 NEXT I
3610 FOR K=0 TO CN%-1
3620 FOR I=1 TO N
3630IF(X(I)>=(MMX+(K*XINT))ANDX(I)<(MMX+((K+1)*XINT)))THENXCL%(K+1)=XCL%(K+1)+1
3640IF(Y(I)>=(MMY+(K*YINT))ANDY(I)<(MMY+((K+1)*YINT)))THENYCL%(K+1)=YCL%(K+1)+1
3650 NEXT I
3660 NEXT K
3670 YCL%(CN%)=YCL%(CN%)+1
3680 XCL%(CN%)=XCL%(CN%)+1
3690 CLS
3700 PRINT
3710 FOR PYY=80 TO 84 STEP 4
3720 MOVE 238,PYY : DRAW 1038,PYY
3730 NEXT PYY
3740 FOR PXX=234 TO 238 STEP 4
3750 MOVE PXX,80 : DRAW PXX,820
3760 NEXT PXX
3770 REM*DETERMINE THE CLASS WITH THE LARGEST FREQUENCY
3780 IF CLASS$="B" THEN CN%=6
3790 MAXX=XCL%(1)
3800 FOR I=1 TO CN%-1
3810 IF XCL%(I+1)>MAXX THEN MAXX=XCL%(I+1)
3820 NEXT I
3830 MAXY=YCL%(1)
3840 FOR I=1 TO CN%-1
3850 IF YCL%(I+1)>MAXY THEN MAXY=YCL%(I+1)
3860 NEXT I
3870 PRINT"The variable is "; XNAME$
3880 XXINT=800/CN%
3890 YYINT=700/MAXY
3900 XYINT=700/MAXX
3910 FOR K=0 TO CN%-1
3920 MOVE 234+XXINT*K,80+0
3930 DRAW 234+XXINT*(K+1),80+0
3940 DRAW 234+XXINT*(K+1),80+XYINT*XCL%(K+1)
3950 DRAW 234+XXINT*K,80+XYINT*XCL%(K+1)
3960 DRAW 234+XXINT*K,80+0
```

```
3970 NEXT K
3980 IF CLASS$="B" THEN GOTO4060
3990 VDU 5:FOR I=0 TO CN%
4000 MOVE 234+I*XXINT,50:@%=131594:PRINT;MMX+I*XINT :@%=10
4010 NEXT I
4020 FOR I=0 TO MAXX
4030 MOVE 150,80+XYINT*I:PRINT;I
4040 NEXT I:VDU 4
4050 GOTO 4120
4060 VDU 5:FOR I=0 TO CN%
4070 MOVE 234+I*XXINT,50:PRINT;-3+I
4080 NEXT I
4090 FOR I=0 TO MAXX
4100 MOVE 150,80+XYINT*I:PRINT;I
4110 NEXT I:VDU 4
4120PRINTTAB(1,8);F$:PRINTTAB(1,9);R$:PRINTTAB(1,10);E$:PRINTTAB(1,11);Q$
4130PRINTTAB(1,12);U$:PRINTTAB(1,13);E$:PRINTTAB(1,14);N$:PRINTTAB(1,15);C$
4140PRINTTAB(1,16);Y$:VDU 4:PRINTTAB(1,2)
4150 GOSUB 5480
4160 CLS
4170 PRINT
4180 PRINT"The variable is "; YNAME$
4190 FOR PYY=80 TO 84 STEP 4
4200 MOVE 238,PYY : DRAW 1038,PYY
4210 NEXT PYY
4220 FOR PXX=234 TO 238 STEP 4
4230 MOVE PXX,80 : DRAW PXX,820
4240 NEXT PXX
4250 FOR K=0 TO CN%-1
4260 MOVE 234+XXINT*K,80
4270 DRAW 234+XXINT*(K+1),80
4280 DRAW 234+XXINT*(K+1),80+YYINT*YCL%(K+1)
4290 DRAW 234+XXINT*K,80+YYINT*YCL%(K+1)
4300 DRAW 234+XXINT*K,80
4310 NEXT K
4320PRINTTAB(1,8);F$:PRINTTAB(1,9);R$:PRINTTAB(1,10);E$:PRINTTAB(1,11);Q$
4330PRINTTAB(1,12);U$:PRINTTAB(1,13);E$:PRINTTAB(1,14);N$:PRINTTAB(1,15);C$
4340PRINTTAB(1,16);Y$:VDU 4:PRINTTAB(1,2)
4350 IF CLASS$="B" GOTO4430
4360 VDU 5:FOR I=0 TO CN%
4370 MOVE 234+I*XXINT,50:@%=131594:PRINT;MMY+I*YINT :@%=10
4380 NEXT I
4390 FOR I=0 TO MAXY
4400 MOVE 150,80+YYINT*I:PRINT;I
4410 NEXT I:VDU 4
4420 GOTO 4480
4430 VDU 5:FOR I=0 TO CN%
4440 MOVE 234+I*XXINT,50:PRINT;-3+I
4450 NEXT I: FOR I=0 TO MAXY
4460 MOVE 150,80+YYINT*I:PRINT;I
4470 NEXT I:VDU 4
4480 GOSUB 5480
4490 GOTO 5120
4500CLS:PRINT:PRINT"USER SPECIFIED CLASSES":PRINT
4510PRINT"How many classes do you require (minimum is 2, maximum is 10)?"
4520 INPUT CN%:IFCN%<2 OR CN%>10 THEN GOTO 4510
4530FOR I=1TOCN%+1:XCL%(I)=0:YCL%(I)=0:XR(I)=0:YR(I)=0:NEXT I
4540 CLS:PRINT:PRINT
4550FOR I=1TO CN%
4560PRINT"What is the lower class limit of class ";I;
4570PRINT "for variable ";XNAME$:INPUT XR(I):PRINT
4580PRINT"What is the lower class limit of class ";I;
4590PRINT"for variable ";YNAME$:INPUT YR(I):PRINT:NEXT I
4600PRINT"What is the upper class limit of the    last class for variable ";
4610PRINT XNAME$:INPUT XR(CN%+1)
4620PRINT"What is the upper class limit of the    last class for variable ";
```

```
4630PRINT YNAME$:INPUT YR(CN%+1)
4640FORK=1 TOCN%:FOR I=1 TO N
4650IF(X(I)>=XR(K)ANDX(I)<XR(K+1))THENXCL%(K)=XCL%(K)+1
4660IF(Y(I)>=YR(K)ANDY(I)<YR(K+1))THENYCL%(K)=YCL%(K)+1
4670IF(X(I)=XR(CN%+1))THENXCL%(CN%)=XCL%(CN%)+1
4680IF(Y(I)=YR(CN%+1))THENYCL%(CN%)=YCL%(CN%)+1
4690 NEXT I: NEXT K
4700FORI=1TOCN%:XR(20+I)=XR(I+1)-XR(I):YR(20+I)=YR(I+1)-YR(I):NEXTI
4710FORI=1TOCN%:XR(40+I)=XCL%(I)/XR(20+I):YR(40+I)=YCL%(I)/YR(20+I):NEXTI
4720MAXX=XR(41):MAXY=YR(41)
4730FORI=1TOCN%-1:IF(XR(40+I+1)>MAXX)THENMAXX=XR(40+I+1)
4740IF(YR(40+I+1)>MAXY)THENMAXY=YR(40+I+1):NEXTI
4750XXINT=800/(XR(CN%+1)-XR(1)):XYINT=800/(YR(CN%+1)-YR(1))
4760XHINT=700/MAXX:YHINT=700/MAXY
4770 CLS
4780PRINT:PRINT"The variable is ";XNAME$
4790VDU 29,234;80;
4800MOVE0,0:DRAW 800,0:MOVE0,0:DRAW 0,740
4810FORK=1TOCN%:MOVE(XR(K)-XR(1))*XXINT,0:DRAW(XR(K+1)-XR(1))*XXINT,0
4820DRAW(XR(K+1)-XR(1))*XXINT,XHINT*XR(40+K)
4830DRAW(XR(K)-XR(1))*XXINT,XHINT*XR(40+K):DRAW(XR(K)-XR(1))*XXINT,0:NEXTK
4840 VDU 29,0;0;
4850 PRINTTAB(1,8);R$:PRINTTAB(1,9);E$:PRINTTAB(1,10);L$:PRINTTAB(1,11);AA$
4860 PRINTTAB(1,12);T$:PRINTTAB(1,13);II$:PRINTTAB(1,14);V$:PRINTTAB(1,15);E$
4870 PRINTTAB(3,8);F$:PRINTTAB(3,9);R$:PRINTTAB(3,10);E$:PRINTTAB(3,11);Q$
4880 PRINTTAB(3,12);U$:PRINTTAB(3,13);E$:PRINTTAB(3,14);N$:PRINTTAB(3,15);C$
4890PRINTTAB(3,16);Y$
4900VDU 5:FOR I=1TOCN%+1
4910MOVE 234+(XR(I)-XR(1))*XXINT,50:@%=131594:PRINT;XR(I):@%=10:NEXT I
4920 FOR K=0TO4:MOVE190,84+K*(700/4):PRINT;K:NEXTK
4930 VDU 4:PRINTTAB(1,2)
4940GOSUB 5480:CLS
4950PRINT:PRINT"The variable is ";YNAME$
4960 PRINTTAB(1,8);R$:PRINTTAB(1,9);E$:PRINTTAB(1,10);L$:PRINTTAB(1,11);AA$
4970 PRINTTAB(1,12);T$:PRINTTAB(1,13);II$:PRINTTAB(1,14);V$:PRINTTAB(1,15);E$
4980 PRINTTAB(3,8);F$:PRINTTAB(3,9);R$:PRINTTAB(3,10);E$:PRINTTAB(3,11);Q$
4990 PRINTTAB(3,12);U$:PRINTTAB(3,13);E$:PRINTTAB(3,14);N$:PRINTTAB(3,15);C$
5000PRINTTAB(3,16);Y$
5010 VDU 29,234;80;
5020MOVE0,0:DRAW 800,0:MOVE0,0:DRAW 0,740
5030FORK=1TOCN%:MOVE(YR(K)-YR(1))*XYINT,0:DRAW(YR(K+1)-YR(1))*XYINT,0
5040DRAW(YR(K+1)-YR(1))*XYINT,YHINT*YR(40+K)
5050DRAW(YR(K)-YR(1))*XYINT,YHINT*YR(40+K):DRAW(YR(K)-YR(1))*XYINT,0:NEXTK
5060 VDU 29,0;0;
5070VDU 5:FOR I=1TOCN%+1
5080MOVE 234+(YR(I)-YR(1))*XYINT,50:@%=131594:PRINT;YR(I):@%=10:NEXT I
5090 FOR K=0TO4:MOVE190,84+K*(700/4):PRINT;K:NEXTK
5100 VDU 4:PRINTTAB(1,2)
5110 GOSUB 5480
5120 CLS
5130 PRINT
5140 PRINT
5150 PRINT
5160 PRINT
5170 PRINT
5180 PRINT
5190 PRINT
5200 PRINT "Analyses are completed."
5210 END
5220 DEF FNXPLOT(I)=234+((I-MMX)*800/RANGEX)
5230 DEF FNYPLOT(I)=80+(((M*I+C)-MMY)*(740/RANGEY))
5240 REM*MAXIMUM X VALUE
5250 MX=X(1)
5260 FOR I=1 TO N-1
5270 IF X(I+1)>MX THEN MX=X(I+1)
5280 NEXT I
```

```
5290 RETURN
5300 REM*MAXIMUM Y VALUE
5310 MY=Y(1)
5320 FOR I=1 TO N-1
5330 IF Y(I+1)>MY THEN MY=Y(I+1)
5340 NEXT I
5350 RETURN
5360REM*MINIMUM VALUE OF X
5370 MMX=X(1)
5380 FOR I=1 TO N-1
5390 IF MMX>X(I+1) THEN MMX=X(I+1)
5400 NEXT I
5410 RETURN
5420 REM*MINIMUM VALUE OF Y
5430 MMY=Y(1)
5440 FOR I=1 TO N-1
5450 IF MMY>Y(I+1) THEN MMY=Y(I+1)
5460 NEXT I
5470 RETURN
5480 REM*ANY KEY
5490 SOUND 1,-10,450,7
5500 PRINT"PRESS ANY KEY TO CONTINUE":A$=GET$
5510 SOUND 1,-10,450,7
5520 RETURN
5530 REM*SORT X VALUES
5540 FOR I=1 TO N-1
5550 LET CCC=0
5560 FOR J=1 TO N-1
5570 IF X(J)<=X(J+1) THEN 5620
5580 T=X(J)
5590 X(J)=X(J+1)
5600 X(J+1)=T
5610 LET CCC=CCC+1
5620 NEXT J
5630 SSS=SSS+CCC
5640 IF CCC=0 THEN 5660
5650 NEXT I
5660 RETURN
5670 REM*SORT Y VALUES
5680 FOR I=1 TO N-1
5690 LET CCC=0
5700 FOR J=1 TO N-1
5710 IF Y(J)<=Y(J+1) THEN 5760
5720 T=Y(J)
5730 Y(J)=Y(J+1)
5740 Y(J+1)=T
5750 LET CCC=CCC+1
5760 NEXT J
5770 SSS=SSS+CCC
5780 IF CCC=0 THEN 5800
5790 NEXT I
5800 RETURN
```

## V.3 Some additional definitions for the B.B.C. microcomputer

The following definitions are very brief, and the official *User Guide* should be consulted for more details.

*Mode*

Text and graphics can be displayed in four different modes on the model A B.B.C. microcomputer and in eight different modes on the model B B.B.C. microcomputer. For example, Mode 4, which requires 10K of memory permits simultaneous text and graphics displays.

*VDU statements*

This range of statements is part of the 'machine operating system' software. For example,

VDU 2 : turns the printer on
VDU 3 : turns the printer off
VDU 4 : text is written at the text cursor
VDU 5 : text is written at the graphics cursor
VDU 23: re-programs displayed characters
VDU 29: moves the graphics origin

*TAB statement*

For printing in user-defined columns, tabulation.

*@% variable*

This variable allows figures to be printed to user-specified decimal places and field widths. It is noted that @% = 10 sets the print format back to the normal kind.

*MOVE statement*

This statement moves the graphics cursor to a specified location (x,y) on the screen.

*DRAW statement*

This statement draws a straight line from the cursor's current position to the specified location (x,y).

*SOUND statement*

This statement generates sounds from the internal loud-speaker, and it is possible to specify the loudness, pitch and length of the note.

*MOD statement*

This statement gives the remainder after division using integers.

## V.4 Package availability

To avoid having to type in the program, a copy of the package is available on tape from one of the authors (John R. Beaumont) at a small charge.

Appendix VI

THE 1981 CENSUS SCHEDULE

In strict confidence

# 1981 Census England

## H Form for Private Households

***A household comprises either one person living alone or a group of persons (who may or may not be related) living at the same address with common housekeeping. Persons staying temporarily with the household are included.***

**To the Head or Joint Heads or members of the Household**

Please complete this census form and have it ready to be collected by the census enumerator for your area. He or she will call for the form on **Monday 6 April 1981** or soon after. If you are not sure how to complete any of the entries on the form, the enumerator will be glad to help you when he calls. He will also need to check that you have filled in all the entries.

This census is being held in accordance with a decision made by Parliament. The leaflet headed 'Census 1981' describes why it is necessary and how the information will be used. Completion of this form is compulsory under the Census Act 1920. If you refuse to complete it, or if you give false information, you may have to pay a fine of up to £50.

Your replies will be treated in STRICT CONFIDENCE. They will be used to produce statistics but your name and address will NOT be fed into the census computer. After the census, the forms will be locked away for 100 years before they are passed to the Public Record Office.

If any member of the household who is age 16 or over does not wish you or other members of the household to see his or her personal information, then please ask the enumerator for an extra form and an envelope. The enumerator will then explain how to proceed.

When you have completed the form, please sign the declaration in Panel C on the last page.

A R THATCHER
Registrar General

Office of Population Censuses and Surveys
PO Box 200 Portsmouth PO2 8HH
Telephone 0329-42511

**Please answer questions H1 - H5 about your household's accommodation, check the answer in Panel A, answer questions 1-16 overleaf and Panel B on the back page. Where boxes are provided please answer by putting a tick against the answer which applies. For example, if the answer to the marital status question is 'Single', tick box 1 thus:**

1 ☑ Single

**Please use ink or ballpoint pen.**

**To be completed by the Enumerator**

| Census District | Enumeration District | Form Number |
|---|---|---|
| | | |

**Name** ..........................................................

**Address** ..........................................................

..........................................................

.............................. Postcode ☐☐☐☐☐☐☐☐

**Panel A**
**To be completed by the Enumerator and amended, if necessary, by the person(s) signing this form.**

This household's accommodation is:

- In a caravan ☐ 20
- In any other mobile or temporary structure ☐ 30
- In a purpose-built block of flats or maisonettes ☐ 12
- In any other permanent building in which the entrance from **outside** the building is:
  - NOT SHARED with another household ☐ 10
  - SHARED with another household ☐ 11

**H1 Rooms**

Please count the rooms in your household's accommodation.

Do not count:

small kitchens, that is those under 2 metres (6ft 6ins) wide, bathrooms, WCs.

Number of rooms ..........................

**Note**
Rooms divided by curtains or portable screens count as one, those divided by a fixed or sliding partition count as two.

Rooms used solely for business, professional or trade purposes should be excluded.

**H2 Tenure**

How do you and your household occupy your accommodation? Please tick the appropriate box

**As an owner occupier (including purchase by mortgage):**

1 ☐ of freehold property

2 ☐ of leasehold property

**By renting, rent free or by lease:**

3 ☐ from a local authority (council or New Town)

4 ☐ with a job, shop, farm or other business

5 ☐ from a housing association or charitable trust

6 ☐ furnished from a private landlord, company or other organisation

7 ☐ unfurnished from a private landlord, company or other organisation

**In some other way:**

☐ Please give details

..........................................................

**Note**
a If the accommodation is occupied by lease originally granted for or since extended to more than 21 years, tick box 2

b If a share in the property is being bought under an arrangement with a local authority, New Town corporation or housing association (for example, *shared ownership (equity sharing), a co ownership scheme*, tick box 1 or 2 as appropriate.

**H3 Amenities**

Has your household the use of the following amenities on these premises? Please tick the appropriate boxes

- A fixed bath or shower permanently connected to a water supply and a waste pipe

1 ☐ YES – for use only by this household

2 ☐ YES – for use also by another household

3 ☐ NO fixed bath or shower

- A flush toilet (WC) with entrance inside the building

1 ☐ YES – for use only by this household

2 ☐ YES – for use also by another household

3 ☐ NO inside flush toilet (WC)

- A flush toilet (WC) with entrance outside the building

1 ☐ YES – for use only by this household

2 ☐ YES – for use also by another household

3 ☐ NO outside flush toilet (WC)

**H4** Please answer this question if box 11 in Panel A is ticked

Are your rooms (not counting a bathroom or WC) enclosed behind your own front door **inside** the building?

1 ☐ YES    2 ☐ NO

If your household has only one room (not including a bathroom or WC) please answer 'YES'

**H5 Cars and vans**

Please tick the appropriate box to indicate the number of cars and vans normally available for use by you or members of your household (other than visitors).

0 ☐ None
1 ☐ One
2 ☐ Two
3 ☐ Three or more

**Include** any car or van provided by employers if normally available for use by you or members of your household but **exclude** vans used solely for the carriage of goods

**Where boxes are provided please tick the appropriate box** (Please use ink or ballpoint pen)

**1-3 Include on your census form:**

- all the persons who spend Census night 5-6 April 1981 in this household (including anyone visiting overnight and anyone who arrives here on the Monday and who has not been included as present on another census form).
- any persons who usually live with your household but who are absent on census night.
  For example, on holiday, in hospital, at school or college. Include them even if you know they are being put on another census form elsewhere.

Write the names in the top row, starting with the head or a joint head of household (BLOCK CAPITALS please)

Include any newly born baby even if still in hospital. If not yet given a name write 'BABY' and the surname.

| 1st person | 2nd person |
|---|---|
| Name and surname | Name and surname |
| Sex ☐ Male ☐ Female | Sex ☐ Male ☐ Female |
| Date of birth Day Month Year | Date of birth Day Month Year |

**4 Marital status**

Please tick the box showing the present marital status.

If separated but not divorced please tick *'Married (1st marriage)'* or *'Re-married'* as appropriate.

| 1st person | 2nd person |
|---|---|
| Marital status | Marital status |
| 1 ☐ Single | 1 ☐ Single |
| 2 ☐ Married (1st marriage) | 2 ☐ Married (1st marriage) |
| 3 ☐ Re-married | 3 ☐ Re-married |
| 4 ☐ Divorced | 4 ☐ Divorced |
| 5 ☐ Widowed | 5 ☐ Widowed |

**5 Relationship in household**

Please tick the box which indicates the relationship of each person to the person entered in the first column.

Please write in relationship of *'Other relative'* – for example, father, daughter-in-law, brother-in-law, niece, uncle, cousin, grandchild.

Please write in position in household of *'Unrelated person'* – for example, boarder, housekeeper, friend, flatmate, foster child

| 1st person | 2nd person |
|---|---|
| | Relationship to 1st person |
| | 01 ☐ Husband or wife |
| | 02 ☐ Son or daughter |
| | ☐ Other relative, please specify |
| | ☐ Unrelated, please specify |

**6 Whereabouts on night of 5-6 April 1981**

Please tick the appropriate box to indicate where the person was on the night of 5-6 April 1981.

| 1st person | 2nd person |
|---|---|
| 1 ☐ At this address, out on night work or travelling to this address | 1 ☐ At this address, out on night work or travelling to this address |
| 2 ☐ Elsewhere in England, Wales or Scotland | 2 ☐ Elsewhere in England, Wales or Scotland |
| 3 ☐ Outside Great Britain | 3 ☐ Outside Great Britain |

**7 Usual address**

If the person usually lives here please tick *'This address'*. If not, tick *'Elsewhere'* and write in the person's usual address.

The home address should be taken as the usual address for a head of household who lives away from home for part of the week.

For students and children away from home during term time, the home address should be taken as the usual address.

Boarders should be asked what they consider to be their usual address.

| 1st person | 2nd person |
|---|---|
| ☐ This address | ☐ This address |
| ☐ Elsewhere – write the person's usual address and postcode | ☐ Elsewhere – write the person's usual address and postcode |
| Address (BLOCK CAPITALS please) | Address (BLOCK CAPITALS please) |
| including Postcode | including Postcode |

**8 Usual address one year ago**

If the person's usual address one year ago, on 5 April 1980, was the same as that given in answer to question 7 please tick *'Same'*. If not, please tick *'Different'* and write in the usual address.

If everyone on the form has moved from the same address, please write the address in full for the first person and indicate with an arrow that this applies to the other people on the form.

For a child born since 5 April 1980 write 'UNDER ONE'.

| 1st person | 2nd person |
|---|---|
| ☐ Same as at Question 7 | ☐ Same as at Question 7 |
| ☐ Different – write the person's address and postcode on 5 April 1980 | ☐ Different – write the person's address and postcode on 5 April 1980 |
| Address (BLOCK CAPITALS please) | Address (BLOCK CAPITALS please) |
| including Postcode | including Postcode |

**9 Country of birth**

Please tick the appropriate box.

If box 6 is ticked please write in the present name of the country in which the birthplace is now situated.

| 1st person | 2nd person |
|---|---|
| Country of birth | Country of birth |
| 1 ☐ England | 1 ☐ England |
| 2 ☐ Wales | 2 ☐ Wales |
| 3 ☐ Scotland | 3 ☐ Scotland |
| 4 ☐ Northern Ireland | 4 ☐ Northern Ireland |
| 5 ☐ Irish Republic | 5 ☐ Irish Republic |
| 6 ☐ Elsewhere. Please write the present name of the country. | 6 ☐ Elsewhere. Please write the present name of the country |

7th person – please see panel on back page ⟶

| **3rd person** | **4th person** | **5th person** | **6th person** |
|---|---|---|---|
| Name and surname | Name and surname | Name and surname | Name and surname |
| **Sex**<br>☐ Male ☐ Female | **Sex**<br>☐ Male ☐ Female | **Sex**<br>☐ Male ☐ Female | **Sex**<br>☐ Male ☐ Female |
| **Date of birth**<br>Day Month Year | **Date of birth**<br>Day Month Year | **Date of birth**<br>Day Month Year | **Date of birth**<br>Day Month Year |
| **Marital status**<br>1 ☐ Single<br>2 ☐ Married (1st marriage)<br>3 ☐ Re-married<br>4 ☐ Divorced<br>5 ☐ Widowed | **Marital status**<br>1 ☐ Single<br>2 ☐ Married (1st marriage)<br>3 ☐ Re-married<br>4 ☐ Divorced<br>5 ☐ Widowed | **Marital status**<br>1 ☐ Single<br>2 ☐ Married (1st marriage)<br>3 ☐ Re-married<br>4 ☐ Divorced<br>5 ☐ Widowed | **Marital status**<br>1 ☐ Single<br>2 ☐ Married (1st marriage)<br>3 ☐ Re-married<br>4 ☐ Divorced<br>5 ☐ Widowed |
| **Relationship to 1st person**<br>01 ☐ Husband or wife<br>02 ☐ Son or daughter<br>☐ Other relative, please specify<br>☐ Unrelated, please specify | **Relationship to 1st person**<br>01 ☐ Husband or wife<br>02 ☐ Son or daughter<br>☐ Other relative, please specify<br>☐ Unrelated, please specify | **Relationship to 1st person**<br>01 ☐ Husband or wife<br>02 ☐ Son or daughter<br>☐ Other relative, please specify<br>☐ Unrelated, please specify | **Relationship to 1st person**<br>01 ☐ Husband or wife<br>02 ☐ Son or daughter<br>☐ Other relative, please specify<br>☐ Unrelated, please specify |
| 1 ☐ At this address, out on night work or travelling to this address<br>2 ☐ Elsewhere in England, Wales or Scotland<br>3 ☐ Outside Great Britain | 1 ☐ At this address, out on night work or travelling to this address<br>2 ☐ Elsewhere in England, Wales or Scotland<br>3 ☐ Outside Great Britain | 1 ☐ At this address, out on night work or travelling to this address<br>2 ☐ Elsewhere in England, Wales or Scotland<br>3 ☐ Outside Great Britain | 1 ☐ At this address, out on night work or travelling to this address<br>2 ☐ Elsewhere in England, Wales or Scotland<br>3 ☐ Outside Great Britain |
| ☐ This address<br>☐ Elsewhere – write the person's usual address and postcode<br>Address (BLOCK CAPITALS please)<br>including Postcode | ☐ This address<br>☐ Elsewhere – write the person's usual address and postcode<br>Address (BLOCK CAPITALS please)<br>including Postcode | ☐ This address<br>☐ Elsewhere – write the person's usual address and postcode<br>Address (BLOCK CAPITALS please)<br>including Postcode | ☐ This address<br>☐ Elsewhere – write the person's usual address and postcode<br>Address (BLOCK CAPITALS please)<br>including Postcode |
| ☐ Same as at Question 7<br>☐ Different – write the person's address and postcode on 5 April 1980<br>Address (BLOCK CAPITALS please)<br>including Postcode | ☐ Same as at Question 7<br>☐ Different – write the person's address and postcode on 5 April 1980<br>Address (BLOCK CAPITALS please)<br>including Postcode | ☐ Same as at Question 7<br>☐ Different – write the person's address and postcode on 5 April 1980<br>Address (BLOCK CAPITALS please)<br>including Postcode | ☐ Same as at Question 7<br>☐ Different – write the person's address and postcode on 5 April 1980<br>Address (BLOCK CAPITALS please)<br>including Postcode |
| **Country of birth**<br>1 ☐ England<br>2 ☐ Wales<br>3 ☐ Scotland<br>4 ☐ Northern Ireland<br>5 ☐ Irish Republic<br>6 ☐ Elsewhere. Please write the present name of the country | **Country of birth**<br>1 ☐ England<br>2 ☐ Wales<br>3 ☐ Scotland<br>4 ☐ Northern Ireland<br>5 ☐ Irish Republic<br>6 ☐ Elsewhere. Please write the present name of the country | **Country of birth**<br>1 ☐ England<br>2 ☐ Wales<br>3 ☐ Scotland<br>4 ☐ Northern Ireland<br>5 ☐ Irish Republic<br>6 ☐ Elsewhere. Please write the present name of the country | **Country of birth**<br>1 ☐ England<br>2 ☐ Wales<br>3 ☐ Scotland<br>4 ☐ Northern Ireland<br>5 ☐ Irish Republic<br>6 ☐ Elsewhere. Please write the present name of the country |



PLEASE TURN OVER ⟶

**Where boxes are provided please tick the appropriate box** (Please use ink or ballpoint pen)

**1-3 Include on your census form:**

- all the persons who spend Census night 5-6 April 1981 in this household (including anyone visiting overnight and anyone who arrives here on the Monday and who has not been included as present on another census form).
- any persons who usually live with your household but who are absent on census night.
  For example, on holiday, in hospital, at school or college. Include them even if you know they are being put on another census form elsewhere.

Write the names in the top row, starting with the head or a joint head of household (BLOCK CAPITALS please)

Include any newly born baby even if still in hospital. If not yet given a name write 'BABY' and the surname.

| 1st person | 2nd person |
|---|---|
| Name and surname | Name and surname |
| Sex<br>☐ Male ☐ Female | Sex<br>☐ Male ☐ Female |
| Date of birth<br>Day Month Year | Date of birth<br>Day Month Year |

**Answers to remaining questions are not required for persons under 16 years of age (born after 5 April 1965)**

**10 Whether working, retired, housewife, etc last week**

Please tick all boxes appropriate to the person's activity last week.

A **job** (box 1 and box 2) means any type of work for pay or profit but not unpaid work. It includes:

casual or temporary work
work on a person's own account
work in a family business
part-time work even if only for a few hours

A **part-time** job (box 2) is a job in which the hours worked, excluding any overtime, are usually 30 hours or less per week.

Tick box 1 or box 2, as appropriate, if the person had a job but was not at work for all or part of the week because he or she was:

on holiday
temporarily laid off
on strike
sick

For a full-time student tick box 9 as well as any other appropriate boxes.

Do not count as a full-time student a person in a paid occupation in which training is also given, such as a student nurse, an apprentice or a management trainee.

| 1st person | 2nd person |
|---|---|
| 1 ☐ In a full-time job at any time last week | 1 ☐ In a full-time job at any time last week |
| 2 ☐ In a part-time job at any time last week | 2 ☐ In a part-time job at any time last week |
| 3 ☐ Waiting to take up a job already accepted | 3 ☐ Waiting to take up a job already accepted |
| 4 ☐ Seeking work | 4 ☐ Seeking work |
| 5 ☐ Prevented by temporary sickness from seeking work | 5 ☐ Prevented by temporary sickness from seeking work |
| 6 ☐ Permanently sick or disabled | 6 ☐ Permanently sick or disabled |
| 7 ☐ Housewife | 7 ☐ Housewife |
| 8 ☐ Wholly retired from employment | 8 ☐ Wholly retired from employment |
| 9 ☐ At school or a full-time student at an educational establishment not provided by an employer | 9 ☐ At school or a full-time student at an educational establishment not provided by an employer |
| 0 ☐ Other, please specify | 0 ☐ Other, please specify |



**Questions about present or previous employment**

**For persons in a job last week** —**please answer questions 11-15 in respect of the main job during the week**

**For persons wholly retired**
**For persons out of work last week**
**For persons prevented from working because of permanent sickness or disablement** —**please answer questions 11-13 in respect of the most recent full-time job, if any. Leave questions 14 and 15 blank.**

**For other persons including those with no previous job—please write 'Not applicable' at question 11 and leave questions 12-15 blank.**

**11 Name and business of employer (if self-employed the name and nature of the person's business)**

**a** Please give the name of the person's employer. Give the trading name if one is used and avoid using abbreviations or initials.

For members of the Armed Forces, civil servants and local government officers see notes on back page before answering questions 11-15.

**b** Please describe clearly what the employer (or the person if self-employed) makes or does.

For a person employed in private domestic service write 'Domestic Service'.

| 1st person | 2nd person |
|---|---|
| **a** Name of employer | **a** Name of employer |
| **b** Nature of business | **b** Nature of business |

**12 Occupation**

**a** Please give full and precise details of the person's occupation.

If a person's job is known in the trade or industry by a special name, use that name. Precise terms should be used, for example, 'radio-mechanic', 'jig and tool fitter', 'tool room foreman' rather than general terms such as 'mechanic', 'fitter', 'foreman'.

**b** Please describe the actual work done

| 1st person | 2nd person |
|---|---|
| **a** Occupation | **a** Occupation |
| **b** Description of work | **b** Description of work |

**Question 16 should be answered for all persons aged 18 or over**

7th person – please see panel on back page →

| 3rd person | 4th person | 5th person | 6th person |
|---|---|---|---|
| Name and surname | Name and surname | Name and surname | Name and surname |
| Sex ☐ Male ☐ Female | Sex ☐ Male ☐ Female | Sex ☐ Male ☐ Female | Sex ☐ Male ☐ Female |
| Date of birth Day ...... Month ...... Year ...... | Date of birth Day ...... Month ...... Year ...... | Date of birth Day ...... Month ...... Year ...... | Date of birth Day ...... Month ...... Year ...... |

| 3rd person | 4th person | 5th person | 6th person |
|---|---|---|---|
| 1 ☐ In a full-time job at any time last week | 1 ☐ In a full-time job at any time last week | 1 ☐ In a full-time job at any time last week | 1 ☐ In a full-time job at any time last week |
| 2 ☐ In a part-time job at any time last week | 2 ☐ In a part-time job at any time last week | 2 ☐ In a part-time job at any time last week | 2 ☐ In a part-time job at any time last week |
| 3 ☐ Waiting to take up a job already accepted | 3 ☐ Waiting to take up a job already accepted | 3 ☐ Waiting to take up a job already accepted | 3 ☐ Waiting to take up a job already accepted |
| 4 ☐ Seeking work | 4 ☐ Seeking work | 4 ☐ Seeking work | 4 ☐ Seeking work |
| 5 ☐ Prevented by temporary sickness from seeking work | 5 ☐ Prevented by temporary sickness from seeking work | 5 ☐ Prevented by temporary sickness from seeking work | 5 ☐ Prevented by temporary sickness from seeking work |
| 6 ☐ Permanently sick or disabled | 6 ☐ Permanently sick or disabled | 6 ☐ Permanently sick or disabled | 6 ☐ Permanently sick or disabled |
| 7 ☐ Housewife | 7 ☐ Housewife | 7 ☐ Housewife | 7 ☐ Housewife |
| 8 ☐ Wholly retired from employment | 8 ☐ Wholly retired from employment | 8 ☐ Wholly retired from employment | 8 ☐ Wholly retired from employment |
| 9 ☐ At school or a full-time student at an educational establishment not provided by an employer | 9 ☐ At school or a full-time student at an educational establishment not provided by an employer | 9 ☐ At school or a full-time student at an educational establishment not provided by an employer | 9 ☐ At school or a full-time student at an educational establishment not provided by an employer |
| 0 ☐ Other, please specify ...... | 0 ☐ Other, please specify ...... | 0 ☐ Other, please specify ...... | 0 ☐ Other, please specify ...... |



| 3rd person | 4th person | 5th person | 6th person |
|---|---|---|---|
| a Name of employer | a Name of employer | a Name of employer | a Name of employer |
| b Nature of business | b Nature of business | b Nature of business | b Nature of business |
| a Occupation | a Occupation | a Occupation | a Occupation |
| b Description of work | b Description of work | b Description of work | b Description of work |

Question 16 should be answered for all persons aged 18 or over

PLEASE TURN OVER →

**Where boxes are provided please tick the appropriate box** (Please use ink or ballpoint pen)

| | 1st person | 2nd person |
|---|---|---|
| **1-3 Include on your census form:**<br>• all the persons who spend Census night 5-6 April 1981 in this household (including anyone visiting overnight and anyone who arrives here on the Monday and who has not been included as present on another census form).<br>• any persons who usually live with your household but who are absent on census night.<br>For example, on holiday, in hospital, at school or college. Include them even if you know they are being put on another census form elsewhere.<br>Write the names in the top row, starting with the head or a joint head of household (BLOCK CAPITALS please)<br>Include any newly born baby even if still in hospital. If not yet given a name write 'BABY' and the surname. | Name and surname<br>**Sex**<br>☐ Male ☐ Female<br>**Date of birth**<br>Day Month Year | Name and surname<br>**Sex**<br>☐ Male ☐ Female<br>**Date of birth**<br>Day Month Year |
| **13 Employment status**<br>Please tick the appropriate box.<br>Box 3 should be ticked for a person having management or supervisory responsibility for other employees. For a person employed as a quality control inspector and concerned only with the technical quality of a product tick box 2. | 1 ☐ Apprentice or articled trainee<br>2 ☐ Employee not supervising other employees<br>3 ☐ Employee supervising other employees<br>4 ☐ Self-employed not employing others<br>5 ☐ Self-employed employing others | 1 ☐ Apprentice or articled trainee<br>2 ☐ Employee not supervising other employees<br>3 ☐ Employee supervising other employees<br>4 ☐ Self-employed not employing others<br>5 ☐ Self-employed employing others |
| **14 Address of place of work**<br>Please give the full address of the person's place of work.<br>For a person employed on a site for a long period give the address of the site.<br>For a person not working regularly at one place who reports daily to a depot or other fixed address, give that address.<br>For a person not reporting daily to a fixed address tick box 1.<br>For a person working mainly at home tick box 2. | Full address and postcode of workplace<br>Address (BLOCK CAPITALS please)<br>including Postcode<br>1 ☐ No fixed place<br>2 ☐ Mainly at home | Full address and postcode of workplace<br>Address (BLOCK CAPITALS please)<br>including Postcode<br>1 ☐ No fixed place<br>2 ☐ Mainly at home |
| **15 Daily journey to work**<br>Please tick the appropriate box to show how the longest part, by distance, of the person's daily journey to work is normally made.<br>For a person using different means of transport on different days show the means most often used.<br>Car or van includes three-wheeled cars and motor caravans. | 1 ☐ British Rail train<br>2 ☐ Underground, tube, metro, etc<br>3 ☐ Bus, minibus or coach (public or private)<br>4 ☐ Motor cycle, scooter, moped<br>5 ☐ Car or van — pool, sharing driving<br>6 ☐ Car or van — driver<br>7 ☐ Car or van — passenger<br>8 ☐ Pedal cycle<br>9 ☐ On foot<br>0 ☐ Other (please specify)<br>0 ☐ Works mainly at home | 1 ☐ British Rail train<br>2 ☐ Underground, tube, metro, etc<br>3 ☐ Bus, minibus or coach (public or private)<br>4 ☐ Motor cycle, scooter, moped<br>5 ☐ Car or van — pool, sharing driving<br>6 ☐ Car or van — driver<br>7 ☐ Car or van — passenger<br>8 ☐ Pedal cycle<br>9 ☐ On foot<br>0 ☐ Other (please specify)<br>0 ☐ Works mainly at home |
| **16 Degrees, professional and vocational qualifications**<br>Has the person obtained any qualifications after the age of 18 such as:<br>Degrees, Diplomas, HNC, HND,<br>Nursing qualifications, Teaching qualifications,<br>Graduate or corporate membership of professional institutions,<br>Other professional, educational or vocational qualifications?<br>Exclude qualifications normally obtained at school such as GCE, CSE and School Certificates.<br>If box 2 is ticked write in all qualifications even if they are not relevant to the person's present job or if the person is not working.<br>Please list the qualifications in the order in which they were obtained.<br>Write for each qualification:<br>the title<br>the major subject or subjects<br>the year obtained and<br>the awarding institution<br>If more than three, please enter in a spare column and link with an arrow. | 1 ☐ NO — none of these<br>2 ☐ YES — give details<br>Title<br>Subject(s)<br>Year<br>Institution<br>Title<br>Subject(s)<br>Year<br>Institution<br>Title<br>Subject(s)<br>Year<br>Institution | 1 ☐ NO — none of these<br>2 ☐ YES — give details<br>Title<br>Subject(s)<br>Year<br>Institution<br>Title<br>Subject(s)<br>Year<br>Institution<br>Title<br>Subject(s)<br>Year<br>Institution |

**7th person – please see panel on back page →**

| **3rd person** | **4th person** | **5th person** | **6th person** |
|---|---|---|---|
| Name and surname | Name and surname | Name and surname | Name and surname |
| **Sex**<br>☐ Male ☐ Female | **Sex**<br>☐ Male ☐ Female | **Sex**<br>☐ Male ☐ Female | **Sex**<br>☐ Male ☐ Female |
| **Date of birth**<br>Day Month Year | **Date of birth**<br>Day Month Year | **Date of birth**<br>Day Month Year | **Date of birth**<br>Day Month Year |
| 1 ☐ Apprentice or articled trainee<br>2 ☐ Employee not supervising other employees<br>3 ☐ Employee supervising other employees<br>4 ☐ Self-employed not employing others<br>5 ☐ Self-employed employing others | 1 ☐ Apprentice or articled trainee<br>2 ☐ Employee not supervising other employees<br>3 ☐ Employee supervising other employees<br>4 ☐ Self-employed not employing others<br>5 ☐ Self-employed employing others | 1 ☐ Apprentice or articled trainee<br>2 ☐ Employee not supervising other employees<br>3 ☐ Employee supervising other employees<br>4 ☐ Self-employed not employing others<br>5 ☐ Self-employed employing others | 1 ☐ Apprentice or articled trainee<br>2 ☐ Employee not supervising other employees<br>3 ☐ Employee supervising other employees<br>4 ☐ Self-employed not employing others<br>5 ☐ Self-employed employing others |
| Full address and postcode of workplace<br>Address (BLOCK CAPITALS please)<br>including Postcode<br>1 ☐ No fixed place<br>2 ☐ Mainly at home | Full address and postcode of workplace<br>Address (BLOCK CAPITALS please)<br>including Postcode<br>1 ☐ No fixed place<br>2 ☐ Mainly at home | Full address and postcode of workplace<br>Address (BLOCK CAPITALS please)<br>including Postcode<br>1 ☐ No fixed place<br>2 ☐ Mainly at home | Full address and postcode of workplace<br>Address (BLOCK CAPITALS please)<br>including Postcode<br>1 ☐ No fixed place<br>2 ☐ Mainly at home |
| 1 ☐ British Rail train<br>2 ☐ Underground, tube, metro, etc<br>3 ☐ Bus, minibus or coach (public or private)<br>4 ☐ Motor cycle, scooter, moped<br>5 ☐ Car or van — pool, sharing driving<br>6 ☐ Car or van — driver<br>7 ☐ Car or van — passenger<br>8 ☐ Pedal cycle<br>9 ☐ On foot<br>0 ☐ Other (please specify)<br>0 ☐ Works mainly at home | 1 ☐ British Rail train<br>2 ☐ Underground, tube, metro, etc<br>3 ☐ Bus, minibus or coach (public or private)<br>4 ☐ Motor cycle, scooter, moped<br>5 ☐ Car or van — pool, sharing driving<br>6 ☐ Car or van — driver<br>7 ☐ Car or van — passenger<br>8 ☐ Pedal cycle<br>9 ☐ On foot<br>0 ☐ Other (please specify)<br>0 ☐ Works mainly at home | 1 ☐ British Rail train<br>2 ☐ Underground, tube, metro, etc<br>3 ☐ Bus, minibus or coach (public or private)<br>4 ☐ Motor cycle, scooter, moped<br>5 ☐ Car or van — pool, sharing driving<br>6 ☐ Car or van — driver<br>7 ☐ Car or van — passenger<br>8 ☐ Pedal cycle<br>9 ☐ On foot<br>0 ☐ Other (please specify)<br>0 ☐ Works mainly at home | 1 ☐ British Rail train<br>2 ☐ Underground, tube, metro, etc<br>3 ☐ Bus, minibus or coach (public or private)<br>4 ☐ Motor cycle, scooter, moped<br>5 ☐ Car or van — pool, sharing driving<br>6 ☐ Car or van — driver<br>7 ☐ Car or van — passenger<br>8 ☐ Pedal cycle<br>9 ☐ On foot<br>0 ☐ Other (please specify)<br>0 ☐ Works mainly at home |
| 1 ☐ NO — none of these<br>2 ☐ YES — give details<br>Title<br>Subject(s)<br>Year<br>Institution | 1 ☐ NO — none of these<br>2 ☐ YES — give details<br>Title<br>Subject(s)<br>Year<br>Institution | 1 ☐ NO — none of these<br>2 ☐ YES — give details<br>Title<br>Subject(s)<br>Year<br>Institution | 1 ☐ NO — none of these<br>2 ☐ YES — give details<br>Title<br>Subject(s)<br>Year<br>Institution |
| Title<br>Subject(s)<br>Year<br>Institution | Title<br>Subject(s)<br>Year<br>Institution | Title<br>Subject(s)<br>Year<br>Institution | Title<br>Subject(s)<br>Year<br>Institution |
| Title<br>Subject(s)<br>Year<br>Institution | Title<br>Subject(s)<br>Year<br>Institution | Title<br>Subject(s)<br>Year<br>Institution | Title<br>Subject(s)<br>Year<br>Institution |



**PLEASE TURN OVER →**

## Notes

### Armed Forces

For members of the Armed Forces – write 'ARMED FORCES' at 11a; for a member of the Armed Forces of a country other than the UK – add the name of the country.

At 12a give the rank or rating only.

Questions 11b, 12b and 13 should not be answered.

### Civil servants

For civil servants – give the name of their Department at 11a, write 'GOVERNMENT DEPARTMENT' at 11b and 'CIVIL SERVANT' at 12a.

At 12b for a non-industrial civil servant – give the rank or grade only.

At 12b for an industrial civil servant – give the job title only, which should be in precise terms, for example, 'radio mechanic', 'jig and tool fitter', 'tool room foreman' rather than general terms such as 'mechanic', 'fitter', 'foreman'.

### Local government officers

For local government officers and other public officials – give the name of the local authority or public body at 11a and the branch in which they are employed at 11b.

At 12a give rank or grade and complete 12b.

## PLEASE COMPLETE PANELS BELOW

**Panel B**

Is there anyone else you have not included (such as a visitor) because there was no room on the form?

☐ YES ☐ NO

Please ask the Enumerator for another form

Have you left anyone out because you were not sure whether they should be included? If so, please give their name(s) and reason why you were not sure about including them.

Name ____________________

Reason ____________________

Name ____________________

Reason ____________________

Name ____________________

Reason ____________________

Name ____________________

Reason ____________________

May the Enumerator telephone you if we have a query on your form? If so, please write your telephone number here

____________________

**Before you sign the form will you please check:**

- **that all relevant questions have been answered**
- **that you have included everyone who spent the night 5-6 April in your household**
- **that you have included anyone who usually lives here but was away from home on the night of 5-6 April**
- **that no visitors, boarders or children including newly born infants, have been missed.**

**Panel C**

**Declaration**

This form is correctly completed to the best of my knowledge and belief

Signature(s) ____________________

____________________

Date ____________________ April 1981

## REFERENCES

Adderley, K. *et al* (1975) Project Methods in Higher Education, Society for Research into Higher Eduction, SRHE Working Party on Teaching methods: Techniques Group, The University, Guildford

Alder, H. L. and Roessler, E. B. (1972) Introduction to Probability and Statistics, Freeman, San Francisco

Allan, J. A. (1980) 'Remote Sensing in Land and Land-use Studies', Geography, 65, pp. 35-43

Aplin, G. (1983) An Introduction to Nearest Neighbour Analysis, Concepts and Techniques in Modern Geography, Geo Books, Norwich

Auerbach, F. (1913) 'Das Gesetz der Bevolkerungskonzentration', Petermanns Mitteilungen, 59, pp. 74-6

Barrett, E. C. (1975) 'Environmental Survey Satellites and Satellite Data Sources', Geography, 60, pp. 31-39

Barry, R. G. and Chorley, R.J. (1971) Atmosphere, Weather and Climate, Methuen, London

Beaumont, J. R. (1983) Location of Public Facilities: A Geographical Perspective, Croom Helm, London

Beard, R. M. (1969) An Outline of Piaget's Developmental Psychology, Routledge and Kegan Paul, London

Berry, B. J. L. (1964) 'Approaches to Regional Analysis: A Synthesis, Annals of the Association of American Geographers, 54, pp. 2-11

Bibby, J. S. and MacKney, D.(1969) Land Use Capability Classification, Technical Monograpn No. 1, Soil Survey, Harpenden

Blacksell, M. and Gilg, A. (1981) The Countryside: Planning and Change, George Allen and Unwin, London

Bowler, I. R. (ed). (1975) A Register of Research in Rural Geography, Rural Geography Study Group, Institute of British Geographers, London

Bradford M. G. and Kent, W. A. (1977) Human Geography: Theories and Their Applications, Oxford University Press, Oxford

Briggs, D. J. (1977) Sediments, Butterworth, London

References


Campbell, J. B. (1979) 'Spatial Variability of Soils', Annals of Association of American Geographers, 69, pp. 544-556

Carter, H. (1978) 'Towns and urban systems 1730-1900', in R. A. Dodgshon and R. A. Butlin (eds.) An Historical Geography of England and Wales, Academic Press, London

Carter, H. (1981) The Study of Urban Geography, Edward Arnold, London

Chisholm, M. (1968) Rural Settlement and Land Use, Hutchinson, London

Christaller, W. (1966) Central Places in Southern Germany, (translated by C. W. Baskin), Prentice-Hall, Englewood Cliffs

Clark, M. J. (1978) 'Geomorphology in Coastal Zone Environmental Management', Geography, 63(4), pp. 273-82

Clark, M. J. (1982) 'Coastal Geomorphology - Pure and Applied' Geography, 67(3), pp. 235-243

Clarke, J. I. and Fisher, W. B. (eds.) (1972) Populations of the Middle East and North Africa, University of London Press, London

Clayden, B. (1982) 'Soil Classification', in E.M. Bridges and D. A. Davidson (eds.) Principles and Applications of Soil Geography, Longman, London, pp. 58-96

Cloke, P. (1979) Key Settlements in Rural Areas, Methuen, London

Cloke, P. and Griffiths, M.J. (1980) 'Planning Responses to Urban and Rural Problems: A Comparative Study in South-West Wales', Tijdschrift voor Economische en Sociale Geografie, 71, pp. 255-263

Clout, H.D. (1972) Rural Geography: An Introductory Survey, Pergamon, Oxford

Clowes, A. and Comfort, P. (1982) Process and Landform, Conceptual frameworks in Geography, Oliver and Boyd, Edinburgh

Coates, B. E., Johnston, R.J. and Knox, P. L. (1977) Geography and Inequality, Oxford University Press, Oxford

Cooke, R. V. and Doornkamp, J. C. (1978) Geomorphology and Environmental Management, Oxford University Press, Oxford

Coppock, J. T. (1974) 'Geography and Public Policy: Challenges, Opportunities and Implications', Transactions Institute of British Geographers, 63, pp. 1 - 16

Curtis, L. F. and Courtney F.M. and Trudgill, S. (1976) Soils in the British Isles, Longman, London

Daniel, P. and Hopkinson, M. (1979) The Geography of Settlement, Oliver and Boyd, Edinburgh

Darby, H. C. (1973) A New Historical Geography of England, Cambridge University Press, Cambridge

Davies, R.B. and O'Farrell, P.N. (1981) 'A Spatial and Temporal Analysis of Second Home Ownership in West Wales' Geoforum, 12, pp. 161-178

## References


Dewdney, J. C. (1968) A Geography of the Soviet Union, Pergamon, Oxford

Dewdney, J. C. (1981) The British Census, Concepts and Techniques in Modern Geography, 29, Geo Abstracts, Norwich

Doornkamp, J. C., K. J. Gregory, and Burn, A.S. (1980) Atlas of Drought in Britain 1975-76, Institute of British Geographers, London

Dugdale, R. (1981) 'Coastal Processes', in A. Goudie (ed.) Geomorphological Techniques, George, Allen & Unwin, London

Eldridge, J. and Shaldrick, S. (1980) A School Micro: Potential and Limitations, Council for Educational Technology, London, pp. 13-15

Elsom, D.M. (1980) 'Local Air Pollution Studies', Teaching Geography, 6, (2), pp. 57-59

Evans, C. (1979) The Mighty Micro, Gollancz, London

Finlayson, B. and Statham, I. (1981) Hillslope Analysis, Butterworths, London

Fitzpatrick, E. E. (1974) An Introduction to Soil Science, Oliver and Boyd, London

Gilg, A. (1980) Countryside Planning, Methuen, London

Gill, M. (1979) 'Field Research With Sixth-Form Students'. Teaching Geography, 5, pp. 16-18

Gold, J. R. (1980) An Introduction to Behavioural Geography, Oxford University Press, Oxford

Good, C. V. (1973) Dictionary of Education, Mcgraw-Hill, New York

Goudie, A. (ed.) (1981) Geomorphological Techniques, George Allen and Unwin, London

Gould, P. (1981) 'Beginning Geography: A Human and Technical Perspective', Journal of Geography in Higher Education, 5, (1), pp. 45-51

Gould, P. and White, R. (1974) Mental Maps, Penguin, Harmondsworth

Graves, H. (ed.) (1982) New UNESCO Sourcebook for Geography Teaching, Longman/The Unesco Press, Harlow

Green, B. (1981) Countryside Conservation, George Allen and Unwin, London

Green, R. J. (1971) Country Planning, Manchester University Press, Manchester

Gregory, K. J. and Walling, D. E. (1973) Drainage Basin Form and Process, Edward Arnold, London

Griffen, E. (1973) 'Testing the von Thunen Theory in Uruguay', Geographical Review, 63, (4), pp. 500-516

Haggett, P. (1979) Geography: A Modern Synthesis, Harper and Row, New York

Hanwell, J. D. and Newson, M. D. (1973) Techniques in Physical Geography, Macmillan, Basingstoke and London

Haynes R.M. and Bentham, C.G. (1979) Community Hospitals and Rural Accessibility, Saxon House, Farnborough

## References


Hilton, K. (1981) 'Landsat Imagery and Curriculum Considerations in Geography: An Innovation at a Turning Point', in A. Kent (ed.) Recent University Work in Geography and its relation to schools, Department of Geography, University of London Institute of Education, pp. 161-183

Hodge, I. and Whitby, M. (1981) Rural Employment: Trends, Options, Choices, Methuen, London

Hodgson, J.M. (1978) Soil Sampling and Soil Description, Oxford University Press, Oxford

Horton, R. E. (1945) 'Erosional Development of Streams and their Drainage Basins: Hydrophysical Approach to Quantitative Morphology', Bulletin of the Geological Society of America, 56, 275-330

Huckle, J. (1981) 'Geography and Values Education', in R. Walford, (ed.) Signposts for Geography Teaching, Longman, Harlow

Huggett, R. (1980) Systems Analysis in Geography, Oxford University Press, Oxford

Johnston, R. J. (1976) Classification, Concepts and Techniques in Modern Geography, 6, Geo Abstracts, Norwich.

Jones, G. (1979) Vegetation Productivity, Longman, London

Keeble, D. E. 'Employment mobility in Britain', in M. Chisholm and G. Manners (eds.) Spatial Policy Problems of the British Economy, Cambridge University Press, Cambridge, pp. 24-68

Kershaw, K. A. (1973) Quantitative and Dynamic Ecology, Aronld, London

King, C.A.M. (1972) Beaches and Coasts, Arnold, London

Knapp, B. J. (1979) Soil Processes, George Allen and Unwin, London

Kohn, C. F. (1982) 'Real Problem-Solving', in N. J. Graves (ed.) New Unesco Source Book for Geography Teaching, Longman, London, pp. 114-140

Liddle, M. J. (1975) 'A Selective Review of the Ecological Effects of Human Trampling on Natural Ecosystems, Biological Conservation, 7, pp. 17-36

Losch, A. (1954) The Economics of Location, (translated by W. F. Stolper) Yale University Press, Yale

Lynch, K. (1960) Image of the City, Harvard University Press, Cambridge

MacEwan, A. and MacEwan, M. (1982) National Parks: Conservation or Cosmetics? George Allen and Unwin, London

Medawar, P. B. (1961) The Art of the Soluble: Creativity in Science, Methuen, London

Meyer, I. and Huggett, R. (1977) 'Local Micro-Climatology: A School's Heat Island', Teaching Geography, 3, pp. 52-54

Moore, P. D. and Webb, J. A. (1978) Pollen Analysis, Hodder and Stoughton, London

Morisawa, M. E. (1968) Streams: Their Dynamics and Morphology, McGraw Hill, New York

References


Moseley, M. J. (1979) Accessibility: The Rural Challenge, Methuen, London

Moss, G. (1978) 'Rural Settlements', Architects Journal January, pp. 100-139

Moss, M. R. (1975) 'Spatial Patterns of Sulphur Accumulation By Vegetation and Soils Around Industrial Centres', Journal of Biogeography, 2, pp. 205-222

Mottershead, R. (1980) Biogeography, Blackwell, Oxford

Naish, M. (ed). (1983) Geography in Education Now, Bedford Way Papers, London Institute of Education

Pahl, R. E. 'The Rural-Urban Continuum', Sociologia Ruralis, 5, p. 327

Parkes, D. and Thrift, W. (1980) Times, Spaces and Places: A Chorogeographic Perspective, John Wiley and Sons, Chichester

Pears, N. (1977) Basic Biogeography, Longman, London

Pocock, D.C.D. (1975) Durham: Images of a Cathedral City, Occasional Publications (New Series) 6, Durham, Department of Geography, University of Durham

Randall, R. E. (1977) Theories and Techniques in Vegetation Analysis, Oxford University Press, Oxford

Rees, J. and Tivy, J. (1978) 'Recreational Impact on Scottish Lochshore Wetlands', Journal of Biogeography, 3, pp. 365-372

Rhind, D. W. and Hudson, R. (1980) Land Use, Methuen, London

Rich, D. C. (1980) Potential Models in Human Geography, Concepts and Techniques in Modern Geography, 26, Geo Abstracts, Norwich

Robinson, A. H. (1956) 'The Necessity of Weighting Values In Correlation of Areal Data', Annals of the Association of American Geographers, 46, pp. 233-236

Robinson, H. and Bamford, C. G. (1978) Geography of Transport, Macdonald and Evans, Plymouth

Robinson, R., Boardman, D., Fenner, J. and Blackburn, J.D., Data in Geography, Drainage Basins: Seasonal Variation of discharge, Longman, York

Sampson, A. (1981) 'Continuing Change on Blakeney Point: Fieldwork Possibilities', Teaching Geography 6(4), pp. 168-174

Schmid, J. A. (1975) Urban Vegetation: A Review and Chicago Case Study, University of Chicago, Department of Geography Research Paper, 61

Shepherd, I.D.H., Cooper, Z. A. and Walker, D.R.F., Computer Assisted Learning in Geography, Council for Educational Technology with the Geographical Association, London

Short, J. (1980) Urban Data Sources, Butterworths, London

Silk, J. and Bowlby, S., 'The Use of Project Work in Undergraduate Teaching', Journal of Geography in Higher Education, 5, (2), pp. 155-162

## References


Simmonds, I. G. (1979) _Biogeography: Natural and Cultural_ Edward Aanold, London

Smith, D. M. (1977) _Human Geography: A Welfare Approach,_ Edward Awnold, London

Smith, D. M. (1979) _Where the Grass is Greener: Living in an Unequal World_, Penguin, Harmondsworth

Smith, R. M. (1980) 'Landsat Photography as a Resource in Secondary School Geography', _Geography,_ 65 (1), pp. 44-48

Stoddart, D. (1977) 'Biogeography', _Progress in Physical Geography,_ 1, pp. 537-543

Stouffer, S. A. (1940) 'Intervening Opportunities: A Theory Relating Mobility and Distance', _American Sociological Review,_ 5, (6), pp. 845-867

Strahler, A.N. (1952) 'Hypsometric (Area-Altitude) Analysis of Erosional Topography', _Bulletin of the Geological Society of America_, 67, 1117-41

Taylor, P. J. (1975) _Distance Decay in Spatial Interactions_, Concepts and Techniques in Modern Geography, 2, Geo Abstracts, Norwich

Thornes, J. E. (1977) 'The Effect of Weather on Sport', _Weather,_ 32, pp. 258-268

Thrift, N. (1977) _An Introduction to Time-Geography,_ Concepts and Techniques in Modern Geography, Geo Abstracts, Norwich

Tidswell, V. (1978) _Pattern and Process in Human Geography,_ University Tutorial Press, Slouth

Tivy, J. (1982) _Biogeography,_ Longman, London.

Trudgill, S.T. (1977) _Soil and Vegetation Systems,_ Oxford University Press, Oxford

Ullman, E. L. (1956) 'The Role of Transportation and the Bases for Interaction', in W. L. Thomas (ed.) _Man's Role in Changing the Face of the Earth_, University of Chicago Press, Chicago, pp. 862-880

Ullman, E. L. (edited by R. R. Boyce) (1980) _Geography as Spatial Interaction_, University of Washington Press, London

Unwin, D. J. (1978) 'Project Work in Urban Climatology, _Teaching Geography_, 3, pp. 103-105

Unwin, D. J. (1981) _Introductory Spatial Analysis_, Methuen, London

Vale, T. R. and Parker, A. J. (1980) 'Biogeography: Research Opportunities for Geographers, _Professional Geographer,_ 32, pp. 149-157

Walford, R. (1969) _Games in Geography_, Longman, Harlow

Walford, R. (ed.) (1981) _Signposts for Geography Teaching_, Longman, Harlow

Warntz, W. (1965) _Macrogeography and Income Fronts,_ Regional Science Research Institute, Philadelphia

Watts, D. (1978) 'The New Biogeography and its Niche in Physical Geography', _Geography,_ 63, pp. 324-337

References


Webster, M. (1980) 'Buying a Microcomputer', Journal of Geography in Higher Education, 4, (1), pp. 42-50

Weizenbaum, J. (1976) Computer Power and Human Reasoning, Freeman, San Francisco

Whynne-Hammond, C. (1980) Elements of Human Geography, George Allen and Unwin, London

Williams, W. T. and Lambert, J. M. (1959) 'Association Analysis in Plant Communities', Journal of Ecology, 47 pp. 83-101

Williamson, P. (1982) 'Practical Plant Geography For A-Level: Three Projects', Teaching Geography, 8(2), pp. 56-58

Willis, A. J. (1973) Introduction to Plant Ecology, George Allen and Unwin, London

Wilson, A. G. (1971) 'A Family of Spatial Interaction Models and Associated Developments', Environment and Planning, 3, pp. 1-32

Young, A. (1972) Slopes, Longman, London

Young, A. (1974) Slope Profile Survey, British Geomorphological Research Group, Technical Bulletin 11, Geo-abstracts, Norwich

Zipf, G. W. (1941) National Unity and Disunity, University of Illinois, Bloomington

# INDEX